Geosimulation

Geosimulation
Automata-based Modeling of Urban Phenomena

Itzhak Benenson
Tel Aviv University, Israel

and

Paul M. Torrens
University of Utah, USA

JOHN WILEY & SONS, LTD

Other Wiley Editorial Offices

John Wiley & Sons Inc., 111 River Street, Hoboken, NJ 07030, USA

Jossey-Bass, 989 Market Street, San Francisco, CA 94103-1741, USA

Wiley-VCH Verlag GmbH, Boschstr. 12, D-69469 Weinheim, Germany

John Wiley & Sons Australia Ltd, 33 Park Road, Milton, Queensland 4064, Australia

John Wiley & Sons (Asia) Pte Ltd, 2 Clementi Loop #02-01, Jin Xing Distripark, Singapore 129809

John Wiley & Sons Canada Ltd, 22 Worcester Road, Etobicoke, Ontario, Canada M9W 1L1

Wiley also publishes its books in a variety of electronic formats. Some of the content that appears in
print may not be available in electronic books.

Library of Congress Cataloging-in-Publication Data

Benenson, Itzhak.
Geosimulation : automata-based modeling of urban phenomena / Itzhak Benenson, Paul M. Torrens.
      p.   cm.
  Includes bibliographical references (p.).
  ISBN 0-470-84349-7 (cloth : alk. paper)
  1. Urban geography – Simulation methods.   2. Urban geography – Computer simulation.
I. Torrens, Paul M.   II. Title
GF125.B46   2004
307.76′01′13–dc22   2004004938

British Library Cataloguing in Publication Data

A catalogue record for this book is available from the British Library

ISBN 0-470-84349-7

Typeset in 10/12pt Times by Thomson Press (India) Limited, New Delhi
Printed and bound in Great Britain by Antony Rowe Ltd, Chippenham, Wiltshire
This book is printed on acid-free paper responsibly manufactured from sustainable forestry
in which at least two trees are planted for each one used for paper production.

For my parents, Maya and Evsey, with love — Itzhak

Bert and Juicy, this is for you, for all the
times you have rescued me and for making
the good times so much better — Paul

Contents

Preface

Are we witnessing a revolution in urban geography? The answer to that question is almost certainly that, yes, we are.

That is a bold statement to make. But, let's consider the evidence.

During the last four decades, a volume of research on topics of urban geography has been conducted—everything from the geography of urban graveyards to the evolution of world cities and massive Megalopoli. Factual data has not always been available to settle the arguments that discussion has generated either in print or in conversation, but data are, and always will be, in short supply. Regardless of how much data we have, we always thirst for more; it is a hallmark of life in an Information Age. But data aside, the general tone of discussion and views on urban geography appear to be coming full circle. All too often, the general impression is that of discussing the same old issues, although they are often marketed in new forms. This rebranding is undoubtedly important, but we are not so much interested in shifting units, marketing products, as we are in uncovering knowledge. Put briefly, there is a strong sense that all the good theoretical stuff has been said before.

What might free us from ever-wandering around the same Möbius strip; more data, better data?

Not so long ago, the data excuse was a pretty good one. It is not any more. Since the last decade of the twentieth century, an enormous volume of data has become available to us, directly to our desktops and our libraries. These data cover a bewildering array of urban phenomena—information on urban infrastructure and populations at all levels of spatial and temporal resolutions have been generated and accumulated. We have not utilized most of them yet. This is not because they are inaccessible; we simply shy away, for the most part, from getting stuck into these huge reservoirs of remotely sensed data and census databases. Modern statistical and GIS environments enable combination of qualitative and quantitative methods, often freely, and we are no longer critically constrained in terms of computing power.

So, what then; data *analysis?*

The common sense view works something like this: let us take a theory, fit the appropriate data, develop and evaluate a clear and lucid understanding of the phenomenon at hand, and then generate forecasts or what-if scenarios. By these means we might thin out all-embracing descriptions, perhaps even give birth to novel ideas. How has that worked out for us? Has it been successful thus far?

We must admit that most urban models do not work well enough when we deal with real cities. In some cases, the theory turns out to be "too general" to be of use in such exercises; examining closely, we see that phenomenological "regression" between potential factors and observed consequences, but not the theory, is applied. In other, not less frequent, situations, application of theory demands so much "tuning" that the city becomes the product of researcher's intuition—more late-night SimCity than science. We probably have as many models of cities as we have cities; a sure sign of discrepancy between urban theory and urban dynamics. One can pessimistically assert that, with all due respect, we have been kidding ourselves; what we thought of as theory is not, in a natural science sense at least. The theory simply does not explan our observed facts. We have another view.

We say that the lack of success in urban modeling is an interim problem, caused by noncorrespondence between *the theory of urban geography* on the one hand and *modeling tools*, in the form in which they were employed until recently, on the other. This discrepancy has roots in over-simplified representation of elementary components of the urban systems considered.

The city is an artificial creature and urban geography is first and foremost human geography. *Human behavior* and, especially, the *decisions* of humans, drive the city and its dynamics. Cities exhibit properties that resemble chemical reactions, but that does not change the fact that molecules in a chemical reaction do not make decisions. Humans are not dumb particles. Yet, most urban models strip the city of its intelligence.

Under this view, that an individual in the urban collective—and its decision—occupies nothing more than $1/n$-th of the aggregate, one can apply a traditional cybernetic black-box approach and obtain the model, which is by definition convenient for mathematical analysis.

The problem is that the individuality and autonomy of urban objects are lost in this case. This is not a superficial problem—most of these black-box views dictate structure and rules that do not fit at all to a system driven by individual and autonomous decision-makers: the city.

The models, not the geographic theory, are thus backward and must be reworked.

Decision-making, as well as the other forms of individual behaviors of urban objects or entities, has numerous faces. Sometimes we can ignore them altogether, and we may be forgiven for assuming that drivers in a city behave as water in a pipeline. Sometimes the decision-making processes are so complicated that we prefer avoiding formalization. The models should point us where to stop on the way from molecules to decision-makers.

In this book, we treat the city as a creature, the complexity of which is above the complexity of physical and chemical systems, but below the complexity of a human self. We assume therefore, that there is no need to directly account for real complexity of urban inanimate and animate objects when formalizing urban phenomena. Instead, we could succeed with avatars, which exhibit simple human-like or human-driven activities. We believe that an intuitive separation of "simple" and "complex" can be sufficient. Inherent physical constraints, spatial ones, first and foremost, act in the city so strongly that the consequences of really human feelings and reasons, such as love and hate, can be less important, at today's level of understanding, at least for the outer observer.

We, thus, believe that an object-based and "loosely human" approach can explain much in cities. Several advances support this vision of urban system dynamics. And here we turn to the evidence for our bold opening statement, arguing for a revolution in urban geography.

- General system theory, which started 60 years ago with black-box cybernetic models, is developing toward understanding system dynamics as collective phenomena; its modern principles are applicable to multiple interacting decision-making objects, in the same way that those principles apply to collectives of simpler physical, chemical, or ecological particles.
- Formal frameworks are standing by in the sidelines—cellular automata and their extension, multiagent systems, are evidently powerful for modeling and simulating collectives of interacting individual urban autonomous objects.
- High-resolution infrastructure and population GIS and remote sensing databases provide data at the resolution of the urban objects we are interested in, whether they are objects or aggregations of objects.
- Modern programming technology is based on object-oriented paradigms and a number of computer environments for simulating and investigating the dynamics of the collectives of autonomous objects already exist.

All this accompanies a recent boom in urban modeling, with models that tend to act at the resolution of real-world urban objects, and increasingly, describe the behavior of those objects in terms not far-removed from our views. This is an exciting time to be working in this field. All the ingredients for a crystallization of these ideas are there—a burgeoning paradigm shift for urban geography, urban analysis, and urban modeling. We call it urban geosimulation.

Acknowledgements

Many people justly deserve some expression of my appreciation. I wish to thank my friends and colleagues: Juval Portugali, Itzhak Omer, and Erez Hatna. We have worked together for many years, and various parts of the book reflect our common work and discussions. Special thanks go to Erez Hatna who, besides developing the Yaffo model, created several of the figures found throughout the text. I also want to thank Shai Aronovich and Saar Noam, with whom we developed the first version of OBEUS. Lena, Kobi, Pola and Manya Benenson provided crucial support, especially Manya, who built the clay city and located it in the artificial environment of our backyard. The city appears on the book's cover, merged with the image of Tel-Aviv that was so kindly provided by the municipality's GIS Department. Lastly, I want to express my gratitude to my colleagues from the Environmental Simulation Laboratory and the Department of Geography and Human Environment, Tel-Aviv University, for their moral support throughout the writing of this book.

Tel-Aviv, April 2004.   **Itzhak Benenson**

Thanks to Carolina Tobón, Muki Haklay, Martin Dodge, Naru Shiode, and Daryl Lloyd for being such fantastic friends and colleagues and for fueling the creative process with cranberry juice, trips to Bartók and Sak, Refreshers, discount Japanese food, and Caffé Nero americanos. Boards of Canada, Linkin Park, Mogwai, and Dashboard Confessional provided the best soundtrack at crunch-time. Thanks, also and in particular, to the Benenson family.

Salt Lake City, April, 2004.   **Paul M. Torrens**

Permission to reproduce the following illustrations is gratefully acknowledged:

Figure 1.7   CODATA Society.
Figure 2.1   CRC Press LLC.
Figure 2.2   Reprinted from *Computers, Environment and Urban Systems*,
                    Benenson, I., S. Aronovich, and S. Noam, "Let's Talk
                    Objects: Generic Methodology for Urban High-Resolution
                    Simulation". Copyright 2004, with permission from Elsevier.
Figure 2.3   CRC Press LLC.
Figure 2.7   Reprinted from *Computers, Environment and Urban Systems*,
                    24(6), Benenson, I., S. Aronovich, and S. Noam, "Let's Talk
                    Objects: Generic Methodology for Urban High-Resolution
                    Simulation", pp. 559–581. Copyright 2000, with permission
                    from Elsevier.

Figure 2.8   Reprinted from *Computers, Environment and Urban Systems*,
                     24(6), Benenson, I., S. Aronovich, and S. Noam, "Let's Talk
                     Objects: Generic Methodology for Urban High-Resolution
                     Simulation", pp. 559–581. Copyright 2000, with permission
                     from Elsevier.

Figure 2.9   Reprinted from *Computers, Environment and Urban Systems*,
                     24(6), Benenson, I., S. Aronovich, and S. Noam, "Let's Talk
                     Objects: Generic Methodology for Urban High-Resolution
                     Simulation", pp. 559–581. Copyright 2000, with permission
                     from Elsevier.

Figure 3.1   Andow, D.A., P.M. Kareiva, S.A. Levin, and A. Okubo, Spread
                     of invading organisms, *Landscape Ecology*, **4**(2/3), 1990, pp.
                     177–188, reproduced with permission from Kluwer Academic
                     Publishers.

Figure 3.9   Reproduced by permission of The RAND Corporation.
Figure 3.10   Reproduced by permission of The RAND Corporation.
Figure 3.11   Professor J. W. Forrester.
Figure 4.5   Steinitz, C. and P. Rogers, *A System Analysis Model of
                     Urbanization and Change: An Experiment in Interdisciplinary
                     Education*, 1970, reproduced by permission of The MIT Press.

Figure 4.6   Steinitz, C. and P. Rogers, *A System Analysis Model of
                     Urbanization and Change: An Experiment in Interdisciplinary
                     Education*, 1970, reproduced by permission of The MIT Press.

Figure 4.7   University of North Carolina.
Figure 4.8   Steinitz, C. and P. Rogers, *A System Analysis Model of
                     Urbanization and Change: An Experiment in Interdisciplinary
                     Education*, 1970, reproduced by permission of The MIT Press.

Figure 4.9   Tobler, W., Cellular Geography. *Philosophy in Geography*. S.
                     Gale and G. Olison 1979, pp. 379–386, with kind permission
                     of Kluwer Academic Publishers.

Figure 4.10   Clark University.
Figure 4.13   European Commission Joint Research Center.
Figure 4.14   Reproduced by permission of Pion Limited, London.
Figure 4.15   Reproduced by permission of Ferdinando Semboloni.
Figure 4.16   Reproduced by permission of Pion Limited, London.
Figure 4.17   Reproduced by permission of Pion Limited, London.
Figure 4.18   Reproduced by permission of Pion Limited, London.
Figure 4.19   Reproduced by permission of Pion Limited, London.
Figure 4.20   Reprinted from *Computers, Environment and Urban Systems*, **24**,
                     Bell, M., C. Dean, and M. Blake, "Forecasting the pattern of
                     urban growth with PUP: a web-based model interfaced with
                     GIS and 3D animation". Copyright 2004, with permission
                     from Elsevier.

Figure 4.21   Reproduced by permission of Jeannette Candau.
Figure 4.22   Reproduced by permission of Jeannette Candau.
Figure 4.23   Reproduced by permission of Jeannette Candau.
Figure 4.24   Reproduced by permission of Jeannette Candau.

Figure 4.25	Reproduced by permission of Pion Limited, London.
Figure 4.26	Reproduced by permission of Pion Limited, London.
Figure 4.27	Reproduced by permission of Pion Limited, London.
Figure 4.28	European Commission Joint Research Center.
Figure 5.5	Reproduced by permission of Oxford University Press, Inc.
Figure 5.6	OUP.
Figure 5.7a	Reprinted figure 1 with permission from Mandelbrot, B.B., B. Kol and A. Aharony, *Physical Review Letters*, **88**(5) 1–4, 2002. Copyright 2002 by the American Physical Society.
Figure 5.7b	Reproduced by permission of World Scientific Publishing Company.
Figure 5.8	Reprinted with permission from "Modeling urban growth patterns with correlated percolation" by H.A. Makse, et al, *Phys. Rev. E*, 1998, **58**(6):7054–7062.
Figure 5.9	Reprinted with permission from "Modeling urban growth patterns with correlated percolation" by H.A. Makse, et al, *Phys. Rev. E*, 1998, **58**(6):7054–7062.
Figure 5.10	World Scientific Publishing Company.
Figure 5.11	Reproduced by permission of The Royal Society.
Figure 5.12	Reproduced by permission of The Royal Society.
Figure 5.13	Reprinted from *Physica A*, **303**, Schweitzer, F., J. Zimmermann and H. Muhlenbein, "Coordination of decisions in a spatial agent model", pp. 189–216. Copyright 2000, with permission from Elsevier.
Figure 5.14	Reproduced by permission of The Simsoc Consortium.
Figure 5.15	Reproduced by permission of The Simsoc Consortium.
Figure 5.18	Reproduced by permission of Taylor & Francis Ltd. http://www.tandf.co.uk/journals.
Figure 5.19	Reproduced by permission of Taylor & Francis Ltd. http://www.tandf.co.uk/journals.
Figure 5.20	Reprinted from *Computers, Environment and Urban Systems*, **22**, Benenson, I., "Multi-agent simulations of residential dynamics in the city", pp. 25–42. Copyright 2003, with permission from Elsevier.
Figure 5.21	Reprinted from *Computers, Environment and Urban Systems*, **22**, Benenson, I., "Multi-agent simulations of residential dynamics in the city", pp. 25–42. Copyright 2003, with permission from Elsevier.
Figure 5.22	OUP.
Figure 5.23	Reproduced by permission of Pion Limited, London.
Figure 5.24	Reprinted from *Computers, Environment and Urban Systems*, **22**, Benenson, I., "Multi-agent simulations of residential dynamics in the city", pp. 25–42. Copyright 2003, with permission from Elsevier.
Figure 5.25	Reprinted from *Computers, Environment and Urban Systems*, **22**, Benenson, I., "Multi-agent simulations of residential

	dynamics in the city", pp. 25–42. Copyright 2003, with permission from Elsevier.
Figure 5.26	Reprinted from *Computers, Environment and Urban Systems*, **22**, Benenson, I., "Multi-agent simulations of residential dynamics in the city", pp. 25–42. Copyright 2003, with permission from Elsevier.
Figure 5.27	Reprinted from *Computers, Environment and Urban Systems*, **22**, Benenson, I., "Multi-agent simulations of residential dynamics in the city", pp. 25–42. Copyright 2003, with permission from Elsevier.
Figure 5.28	Reprinted from Journal of *Urban Economics*, **45**, Page, S.E., "On the Emergence of Cities", pp. 184–208. Copyright 1999, with permission from Elsevier.
Figure 5.29	Reprinted from *Mathematics and Computers in Simulation*, Vol **27**, Gipps and Marksjo, "A micro-simulation model for pedestrian flows", pp. 95–105, Copyright 1985, with permission from Elsevier.
Figure 5.30	Reprinted from *Mathematics and Computers in Simulation*, Vol **27**, Gipps and Marksjo, "A micro-simulation model for pedestrian flows", pp. 95–105, Copyright 1985, with permission from Elsevier.
Figure 5.31	Reproduced by permission of D. Helbing.
Figure 5.32	Reproduced by permission of D. Helbing.
Figure 5.33	Reproduced by permission of D. Helbing.
Figure 5.34	Reprinted from *Physica A*, **295**, Burstedde, C., K. Klauck, A. Schadschneider, and J. Zittartz, "Simulation of pedestrian dynamics using a two-dimensional cellular automaton", pp. 507–525. Copyright 2000, with permission from Elsevier.
Figure 5.35	Reprinted with permission from *Nature*, **47**(28):487–490, Helbing, D., I. Farkas, and T. Vicsek (2000), "Simulating dynamical features of escape panic". Copyright 2000 Macmillan Magazines Limited.
Figure 5.36	Reprinted with permission from *Nature* **47**(28):487–490, Helbing, D., I. Farkas, and T. Vicsek (2000), "Simulating dynamical features of escape panic". Copyright 2000 Macmillan Magazines Limited.
Figure 5.37	Reprinted with permission from *Nature* **47**(28):487–490, Helbing, D., I. Farkas, and T. Vicsek (2000), "Simulating dynamical features of escape panic". Copyright 2000 Macmillan Magazines Limited.
Figure 5.38	Reprinted from *Parallel Computing*, **27**(5), Wahle, J., L. Neubert, J. Esser, and M. Schreckenberg, "A cellular automaton traffic flow model for online simulation of traffic", pp. 719–735. Copyright 2001, with permission from Elsevier.

Figure 5.39	Reprinted from *Transportation Research Part C*, **10**, Wahle, J., A.L.C. Bazzan, F. Klugl, and M. Schreckenberg, "The impact of real-time information in a two-route scenario using agnet-based simulation", pp. 399–417. Copyright 2002, with permission from Elsevier.
Figure 5.40a	Reprinted from *Parallel Computing*, **27**(5), Wahle, J., L. Neubert, J. Esser, and M. Schreckenberg, "A cellular automaton traffic flow model for online simulation of traffic", pp. 719–735. Copyright 2001, with permission from Elsevier.
Figure 5.40b	Reprinted from *Physica A*, **2857**, Schadscheider, A., "Statistical physics of traffic flow", pp. 101–120. Copyright 2000, with permission from Elsevier.
Figure 5.41	Reprinted from *Parallel Computing*, **27**(5), Wahle, J., L. Neubert, J. Esser, and M. Schreckenberg, "A cellular automaton traffic flow model for online simulation of traffic", pp. 719–735. Copyright 2001, with permission from Elsevier.
Figure 5.42	Reprinted from *Parallel Computing*, **27**(5), Wahle, J., L. Neubert, J. Esser, and M. Schreckenberg, "A cellular automaton traffic flow model for online simulation of traffic", pp. 719–735. Copyright 2001, with permission from Elsevier.
Figure 5.43	Reprinted from *Transportation Research Part C*, **10**, Hidas, P., "Modelling lane changing and merging in microscopic traffic simulation", pp. 351–371. Copyright 2002, with permission from Elsevier.
Figure 5.44	Reprinted from *Transportation Research Part C*, **10**, Hidas, P., "Modelling lane changing and merging in microscopic traffic simulation", pp. 351–371. Copyright 2002, with permission from Elsevier.
Figure 5.45	Reprinted from *Physica A*, **287**, Nagel, K., M. Shubik, M. Paczuski, and P. Bak, "Spatial competition and price formation", pp. 546–562. Copyright 2000, with permission from Elsevier.
Figure 5.46	Reproduced by permission of Pion Limited, London.
Figure 5.47	Reproduced by permission of Pion Limited, London.
Figure 5.48	Reproduced by permission of Pion Limited, London.
Figure 5.49	Reproduced by permission of Pion Limited, London.
Figure 5.50	Reproduced by permission of Pion Limited, London.
Figure 5.51	Reproduced by permission of Pion Limited, London.
Figure 5.52	Reproduced by permission of Pion Limited, London
Figure 5.53	Reprinted from *Parallel Computing*, **27**(5), Wahle, J., L. Neubert, J. Esser, and M. Schreckenberg, "A cellular automaton traffic flow model for online simulation of traffic", pp. 719–735. Copyright 2001, with permission from Elsevier.

Figure 5.54a   Reprinted from *Computer Physics Communications*, **147**, Wang, R. and H.J. Ruskin, "Modeling traffic flow at a single-lane urban roundabout", pp. 570–576. Copyright 2002, with permission from Elsevier.

Figure 5.54b   Reprinted from *Parallel Computing*, **27**(5), Wahle, J., L. Neubert, J. Esser, and M. Schreckenberg, "A cellular automaton traffic flow model for online simulation of traffic", pp. 719–735. Copyright 2001, with permission from Elsevier.

Figure 5.55   Reprinted from *Journal of Transport Geography*, **3**(2), Fox, M. "Transport planning and the human activity approach", pp. 105–116. Copyright 1995, with permission of Elsevier.

Figure 5.56   Reprinted from *Parallel Computing*, **27**(5), Wahle, J., L. Neubert, J. Esser, and M. Schreckenberg, "A cellular automaton traffic flow model for online simulation of traffic", pp. 719–735. Copyright 2001, with permission from Elsevier.

Figure 5.57   Nova Science Publishers.

Figure 5.58   Reprinted from *Landscape and Urban Planning*, **56**(1–2), Ligtenberg, A., A.K. Breft, and R. Van Lammeren, "Multi-actor-based land use modelling: Spatial planning using agents", pp. 21–33. Copyright 2001, with permission from Elsevier.

Figure 5.59   Reprinted from *Landscape and Urban Planning*, **56**(1–2), Ligtenberg, A., A.K. Breft, and R. Van Lammeren, "Multi-actor-based land use modelling: Spatial planning using agents", pp. 21–33. Copyright 2001, with permission from Elsevier.

Figure 5.60   Reprinted from *Landscape and Urban Planning*, **56**(1–2), Ligtenberg, A., A.K. Breft, and R. Van Lammeren, "Multi-actor-based land use modelling: Spatial planning using agents", pp. 21–33. Copyright 2001, with permission from Elsevier.

Figure 5.61   Reprinted from *Landscape and Urban Planning*, **56**(1–2), Ligtenberg, A., A.K. Breft, and R. Van Lammeren, "Multi-actor-based land use modelling: Spatial planning using agents", pp. 21–33. Copyright 2001, with permission from Elsevier.

Figure 5.62   Reproduced by permission of Pion Limited, London.
Figure 5.63   Reproduced by permission of Pion Limited, London.
Figure 5.64   OUP.
Figure 5.65   OUP.
Figure 5.66   OUP.
Figure 5.67   Reproduced by permission of Pion Limited, London.
Figure 5.69   Reprinted from *Computers, Environment and Urban Systems*, **28**(1/2), Semboloni, F., J. Assfalg, S. Armeni, R. Gianassi, and F. Marsoni, "CityDev, an interactive multi-agents urban model

on the web", pp. 45–64. Copyright 2003, with permission from Elsevier.

Figure 5.70   Reprinted from *Agricultural Economics*, **25** (2–3), Berger, T. "Agent-based spatial models applied to agriculture: a simulation tool for technology diffusion, resource use changes and policy analysis", pp. 245–260. Copyright 2001, with permission from Elsevier.

Cover image (map)   Reproduced by permission of Tel-Aviv Municipality.

Cover image (clay city)   Reproduced by permission of Miriam Benenson.

Chapter 1

Introduction to Urban Geosimulation

Geosimulation is a catch-all title that can be used to represent a very recent wave of research in geography. In a broad sense, the field of geosimulation is concerned with the design and construction of object-based high-resolution spatial models, using these models to explore ideas and hypotheses about how spatial systems operate, developing simulation software and tools to support object-based simulation, and applying simulation to solving real problems in geographic contexts.

In this book, we focus on urban applications of geosimulation; our consideration of geosimulation focuses on *urban geosimulation*.

1.1 A New Wave of Urban Geographic Models is Coming

The 1970s were a turbulent time for urban simulation; as a field of research, urban modeling drew heavy criticism in that period and was all but written off as a failure. Publications such as "Requiem for large-scale models" (Lee, 1973), served as harbingers of doom for urban simulations applied in planning contexts. Other critiques were published, bemoaning the inadequacies of simulation as a planning support system and documenting the failures of the field to live up to expectations (Sayer, 1979). To a certain extent, the foundations of those complaints against urban models were weakened by subsequent technological advances, particularly in computer engineering, and developments in computer software, data collection, new simulation techniques, and breakthroughs in understanding of cities (Harris, 1994). However, many of the complaints proffered by Lee and others remained relevant.

In recent years, however, there have been complementary developments that have set the stage for a new generation of urban models. Not only do these models remedy

---

Geosimulation: *Automata-based Modeling of Urban Phenomena*.   I. Benenson and P. Torrens
© 2004 John Wiley & Sons, Ltd   ISBN: 0-470-84349-7

many of the disappointments of previous generations of urban simulation, they also offer unforeseen potential. Models have been developed that enable users to simulate the dynamics of urban systems in the real world with attention to intuitively important details that could not be included before, just because the methodological framework did not support those details. While these "new wave" models have yet to be widely tested in practical contexts, they have definitely passed an initial period of innovation and are beginning to enter into a more mature stage, marked by their application to real-world uses.

This chapter introduces this new generation of urban models, characterized by an innovative paradigm, which we have termed *geosimulation*. The concept of geosimulation is based on geographic, i.e., spatially-related *automata*. Modern GIS and remote sensing databases serve as the information source for geosimulation. Computational implementation of geosimulation models is naturally done within an *Object-Oriented Programming* (OOP) paradigm. Also, modern *system theory* provides the paradigmatic basis and analytical tools for investigating geosimulation models. These are concepts we will explore in more detail in subsequent chapters.

1.2 Defining Urban Geosimulation

1.2.1 Geosimulation Reflects the Object Nature of Urban Systems

Geosimulation differs from conventional urban simulation in its constituent 'elements.' Geosimulation models operate with human individuals and infrastructure entities, represented at spatially nonmodifiable scales such as households, homes, or vehicles. In geosimulation models these objects *behave*. Many of these objects are animated (visually and dynamically), and that animation drives the behavior of inanimate objects in a simulation. That is quite a simple and general overview of the approach, but those features—combined in a unified framework—are what distinguish geosimulation models from other forms of simulation.

Geosimulation models are commonly *generative* in nature; entities at higher levels of geographic representation such as census groupings are mostly derived "from the bottom up," as the product of the interactive dynamics of collectives of animate and inanimate objects observed at elementary scales of spatial representation. In some cases, these larger-scale entities may be considered as emergent (Holland, 1998), in so much as their characteristics may be represented with descriptors that are richer than those that characterize their constituent components at finer scales of geography.

Geosimulation models are often developed to represent phenomena that occur in urban systems—which are usually complex, adaptive, and dynamic—and in a highly realistic manner. When used to model urban systems, geosimulation models focus on representing the elementary units that comprise a system and the interactions that take place between them.

Let us consider the characteristics of geosimulation models in more detail, and explore in some depth what features distinguish them from other modeling methodologies and styles.

1.2.2   Characteristics of the Geosimulation Model

1.2.2.1   Management of Spatial Entities

A basic aspect of geosimulation regards the characterization of *spatial entities* that form the building blocks of a simulation model. Traditionally, urban simulation models have represented units of urban systems—land, real estate, people, etc.—by means of *aggregates* such as geographic zones, tracts, and socioeconomic groups. These aggregate units are *spatially modifiable*, i.e., they can be partitioned geographically in many different ways (Openshaw, 1983). Models designed according to the geosimulation approach, however, are oriented toward *spatially nonmodifiable* objects such as homes, households, and vehicles.

1.2.2.2   Management of Spatial Relationships

The second aspect of geosimulation relates to the portrayal of *spatial relationships* in models. For example, we can consider this in the context of spatial interactions; their representation in traditional urban simulation has been limited to flows between aggregate units. The gravity model (Fotheringham and O'Kelly, 1989) is a typical example; interaction is characterized as flows in a Newton-like manner, as proportional to aggregates' masses and inversely proportional to some power of the distance between them. The gravity formula can be applied to vehicle trips between origins and destinations, migrations of populations between cities or between regions within cities, etc.

Geosimulation models consider interactions as an outcome of the *behavior* of elementary geographic objects. In this way, geosimulation models have the potential to represent spatial interaction of a much wider spectrum of forms, including traditional and far-distance migration. Interaction revealed at higher levels of spatial organization are considered in geosimulation models as the outcomes of collective behavior of urban objects at lower-level geographies. This brings us back to the notion of generative or bottom-up systems; geosimulation models are built from elemental components, elemental spatial components, fused together with specification of their spatial relationships.

1.2.2.3   Management of Time

The third distinguishing characteristic relates to the treatment of *time* in models. Urban systems change over time, and different phenomena occur at different time scales. They dance to different internal clocks. Traffic congestion on individual sections of a highway varies on an hourly basis, while the activity of a shopping center depends on the day of the week. Longer-term cycles of urban decline and gentrification in residential communities may operate over decades, with associated shifts in the fabric of the area and the characteristics of the households that it comprises. Invariably, urban simulations are designed for testing hypotheses relating to the dynamic processes that shape cities.

Geosimulation models treat time through *intuitively justified* units such as housing search cycles. Objects' temporal behavior can be considered as either *synchronous*, when all objects change simultaneously, or *asynchronous*, when they change in turn, with each observing the urban reality as left by the previous one. The spectrum of asynchronous behavior is much wider than that of synchronous behavior. The sequence of objects' changes can be random, follow some predetermined order, or may be event-driven. Geosimulation accommodates and handles all of these schedules; one of the features of the approach is direct representation of object behavior, and geosimulation models can be easily adapted to near-real-time dynamics (Nagel, Beckman, and Barrett, 1999).

1.2.2.4   Direct Modeling

To a certain extent, disappointment with the state of urban simulation as a field of study in the 1970s was a by-product of the expectations of what urban models should actually do (Batty, 1979). Part of the geosimulation approach is a move toward the design of "tools to think with" (Negroponte, 1995). To some degree, the geosimulation approach represents a reevaluation of the goals of urban simulation toward 'artificial experience' (Portugali and Benenson, 1995), as game-playing tools that support the exploration of ideas and testing of hypotheses (which is perhaps the original ethos behind simulation in the first place; it is ironic that we are "rediscovering" this so far along in the development of the field).

Realistic description of objects' behavior in ways that were not previously obtainable, either technologically or intellectually, makes these games worthwhile and, further, allows for direct relation between conceptual and real-world modeling. The idea underlying geosimulation is that the same model can be applied to abstract and real-world phenomena; if modeled phenomena are an abstraction of real-world phenomena, why should modeled objects differ from their counterparts in the real world? As we will explain later, the geosimulation approach is supported by several key developments in the geographical sciences and other fields, particularly mathematics, computer science, and general system studies. The cornerstone of the geosimulation approach, however, is the *automaton*, which has been widely used in simulation and features prominently in geosimulation toolkits.

1.3   Automata as a Basis for Geosimulation

The description of objects' behavior in the geosimulation framework is based on the idea of *automata*. Simply stated, an automaton is a processing mechanism with characteristics that change over time based on its internal characteristics, rules, and external input. Automata are used to process information that is input to them from their surroundings and their characteristics are altered according to rules that govern their reaction to those inputs. "An automaton is a machine that processes information, proceeding logically, inexorably performing its next action after applying data received from outside itself in light of instructions programmed within itself" (Levy, 1992, p. 15). Automata are a useful abstraction of "behaving objects" for many reasons, but principally because they provide an efficient formal mechanism for representing their fundamental properties: attributes, behaviors, relationships, environments, and time.

Formally, a finite automaton A can be represented by means of a finite set of *states* $S = \{S_1, S_2, S_3, \ldots, S_N\}$ and a set of *transition rules* T.

$$A \sim (S, T) \tag{1.1}$$

Transition rules define an automaton's state, S_{t+1}, at time step $t + 1$ depending on its state, $S_t (S_t, S_{t+1} \in \{S\})$, and *input*, I_t, at time step t:

$$T : (S_t, I_t) \rightarrow S_{t+1} \tag{1.2}$$

This basic formulation does not define the nature of the states $S \in S$, or the set of possible inputs $I \in I$. In its simplest case, both states and input sets can be Boolean on/off, true/false. The essence of the basic finite automata approach is in its temporal discreteness and the ability of an automaton to change according to predetermined rules based on internal (S) and external (I) information.

Transition rules T govern how the state of an automaton changes from one moment, unit, or packet of discrete time t to a subsequent period $t + 1$. The result of the transition depends on the state of the automaton at time t, as well as information gleaned from an input I_t derived from neighboring automata at time t.

Automata may be generalized beyond this simple formulation (Salomaa *et al.*, 2001). The most important of these characteristics, for geosimulation, is substitution of the finite set of states of automata by an infinite and continuous set. By doing so, we assume that object's characteristics can vary continuously. At the same time, geosimulation always preserves an assumption that the rule set of the automata is finite, and that automata timing is discrete.

Regarding urban applications, nothing prevents us from considering the entire city as an automaton with myriad states and transition rules. However, to make sense, an individual automaton should be as simple as possible in terms of states, transition rules, and accounting for external information (Torrens and O'Sullivan, 2001).

In the next chapter we introduce a whole class of automata that we term *geographic automata,* which we distinguish based on *spatial* generalizations of this basic description, and we use this as the basis of geosimulation. Two popular automata types that provide the basis for geographic automata are *cellular automata* and *multiagent systems.*

1.3.1   Cellular Automata

Simplicity is a characteristic of the most popular automata tools in urban geography, Cellular automata (CA)—a system of spatially located and interconnected finite automata. Cellular automata are arrangements of individual automata in some form regularly tessellated space, e.g., a rectangular grid (Figure 1.1). They have the same characteristics as the basic automata described above, but the input I is defined in a cellular context. In CA, an individual *cell* represents the discrete spatial confines of an automaton, and an individual automaton is understood as *neighboring* some other automata. *States* are associated with those cells. In CA, input is formed from the state information of neighboring cells. The *neighborhoods* may take on a number of configurations: agglomerations of adjacent cells defined by their distance from an individual automaton, clusters of cells defined by their shape around an automaton, etc. Two standard neighborhood configurations are illustrated in Figure 1.1: the nine-cell "Moore" neighborhood and five-cell "von Neumann" neighborhood.

We can refine general definitions (1.1) and (1.2) to specify an automaton A belonging to a CA lattice as follows:

$$A \sim (S, T, R) \tag{1.3}$$

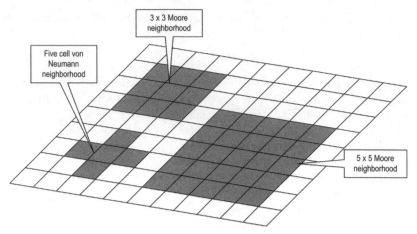

Figure 1.1   *Cellular automata neighborhoods*

where **R** denotes automata neighboring *A*, and defines the boundary for drawing input information **I**, which is necessary for the application of transition rules **T**.

As with basic automata, transition rules **T** determine how automata states adapt over time. These rules can be designed using any combination of conditional statements or mathematical operators; virtually any discrete spatial process may be translated into a transition function for CA.

As will be discussed, an important feature of CA that distinguishes them from other automata-based tools such as multiagent systems is the stationarity of cell location. The position of cells and their neighborhood relations remain fixed over time. Information may be exchanged, or "spread" through neighborhoods, and in this sense CA can support information propagation through space. In the case of one-dimensional 3-cell, or two-dimensional, 3×3-cell Moore neighborhoods, for example, color information may be propagated in space, and reaches cells at a distance *d*, or on the boundary of the $d \times d$-square around an initial cell during *d* time-steps (Figure 1.2).

In urban applications, states are most commonly used to represent possible land-uses of cellular units (White, Engelen, and Uljee, 1997).

1.3.2   Multiagent Systems

Geosimulation has much in common with the individual-oriented *agent-based* modeling approach that is popular in economics (Luna and Stefansson, 2000), political science (Epstein and Axtell, 1996), sociology (Gilbert and Conte, 1995), and ecology (Berec, 2002). The agent-based approach in those fields differs from geosimulation, however, in its treatment of *space*. Agent-based models often represent space in a cursory manner, if at all. The geosimulation paradigm demands explicit representation of space, spatial behavior of objects, and their spatial relationships.

As mentioned, multiagent systems (MAS) are another class of automata systems. They specify automata states and, especially, transition rules, in a way that enables

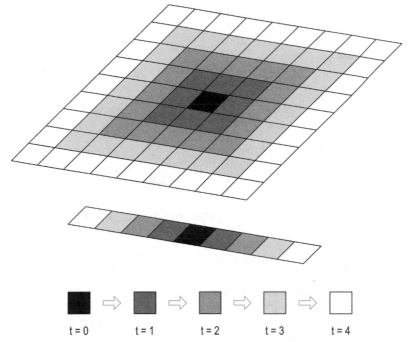

Figure 1.2   *Information exchange in cellular automata*

interpretation of *autonomously behaving agents* (Maes, 1995b). There is a wide debate in the computer science literature, particularly in Artificial Intelligence, as to how human-like agency should be represented in MAS, with agency divided into "weak" and "strong" notions (Ferber, 1999). We will discuss this in detail in Chapter 5 of the book. The issue of agency in MAS for geographic systems has yet to be explored in great depth. One important consideration for geographic applications is that individual agent automata are capable of mobile action; in addition to possessing the ability to change state, a change that may be mediated through their neighborhoods, agent automata may change location in the spaces in which they reside, migrating to other locations at any distance from their current position. Figure 1.3 illustrates this point.

As with CA, the individual automata in MAS have neighbors, but the regularity of CA is usually avoided in MAS, and neighbors can be located arbitrarily regarding the agent. The variety of relationships between automata can be relaxed in MAS, allowing for arbitrary neighborhood connections between automata. It is common for agent automata to be located on nodes of a network, as in Figure 1.4, where neighborhoods are formed by network neighbor relationships.

States of agent-automata are usually designed to convey some form of autonomous behavior and, as we stated above, can be derived from finite or infinite and continuous sets. An agent designed to represent a household, for example, might be characterized by discrete states such as age and number of children and continuous income. Likewise, the transition rules that drive agent automata are commonly designed to resemble human-like behaviors. An agent designed to represent a commuting worker, for example, might

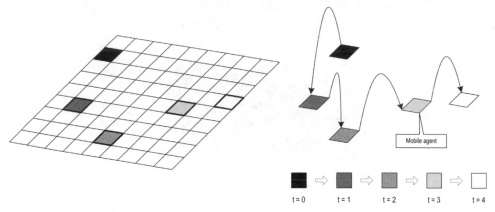

Figure 1.3   *Automata movement in multiagent systems*

be designed with transition rules representing morning and evening trips to and from work, reaction to traffic congestion, etc. Because the individual automata in MAS are free to change position, a range of *spatial* transition rules can be programmed, e.g., navigation behavior, way-finding, and spatial cognition.

1.3.3   Automata Systems as a Basis for Urban Simulation

Automata-based modeling tools hold many advantages for simulation of urban phenomena in space. The decentralized structure of automata systems, their ability to directly

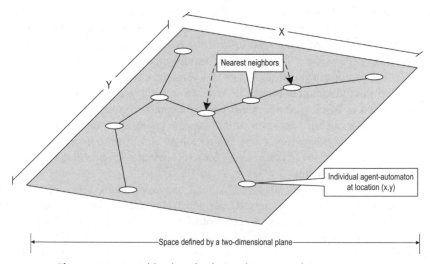

Figure 1.4   *Neighborhood relationships in multiagent systems*

handle individual spatial and nonspatial elements, simplicity of formulation—all of these features offer many benefits to model-builders.

1.3.3.1 Decentralization

Spatial automata systems, such as CA and MAS are, by definition, decentralized. In such systems, individual automata act independently; interactions between automata can be expressed as, and may convey, collaborative actions; these actions are often exhibited without the controlling influence of a centralized executive. The behavior of an individual automaton is governed by the state of the automaton itself and the information encoded in its neighboring automata. Moreover, automata act independently from each other and an arbitrary number of automata may be active at any one point in time. Parallel and sequential automata are extreme cases. The decentralized nature of automata mimics the way in which many systems in the real world are organized.

Of course, many systems *are centralized*, at least partially: the singular experience of self, autocratic political systems, headquarters for commercial companies, etc. However, a great number of systems are decentralized, and are becoming more and more so. Some authors have even argued that we are amid an "era of decentralization" (Resnick, 1994). Cities are fundamentally decentralized, and as they grow and evolve they may continue to decentralize. In many areas, cities are decentralizing at unprecedented rates and in new forms: exurbs, suburban sprawl, and edge cities are some examples. The old monocentric core and its orbital band of related settlement is rapidly giving way to polycentric *city systems* sprawled over a much less centralized network of related elements, and characterized by a spatial structure that is shaped by centrifugal forces rather than centripetal trends. Cities such as Los Angeles can be regarded more as a constellation of subcenters than as a traditional city dominated by a central business district. Automata systems are naturally decentralized and are, thus, an excellent tool for representing decentralized systems of this variety.

1.3.3.2 Specifying Necessary and Only Necessary Details

Automata-based models allow for very detailed representation of the phenomena they are used to simulate—individual land parcels, buildings, businesses, houses, people, vehicles, etc. Not only do automata allow for detailed descriptions of entities' attributes, they also facilitate rich descriptions of automata behavior and linkages that exist between those entities, and are necessary, say, for the representation of movement of pedestrians along streets, the behavior of vehicles on roads, the migration of households within a city, etc. Automata-based models thus offer a mechanism for incorporating detail into simulation. At the same time, the object-basis of automata lends opportunities, in model design, for *encapsulating* what might be regarded as not immediately necessary features in a model. In this way, it is often possible, with automata models, to abstract from characteristics of the phenomenon being simulated, or the model itself, that model-builders or model-users do not want, or do not know how, to employ at a given stage of development. Encapsulation is characteristic of the Object Oriented Programming paradigm that geosimulation is based on, and we discuss this in length in the next chapter.

1.3.3.3   Diversity of Characteristics and Behavior

The autonomy of automata is an important property. Up until recently, it was common practice in urban simulation to model urban systems with a number of aggregate units by assuming that all elements of a unit have the same characteristics and exhibit the same—average—behavior. Characteristics and behavior can, and usually do, vary between individuals of the same group and thus between the agent-automata used to represent them. As will become clear in the forthcoming discussion, unification entails essential bias when the system is nonlinear, and urban systems definitely belong to this class of systems. Examples of nonlinear behavior are numerous: individual households may react sharply to the introduction of a first nightclub to their residential neighborhood, but may be more tolerant to the second and third; young college students renting their accommodation may react to this event favorably, while thirty-something couples with a new-born infant and a case of insomnia may have an entirely different reaction! Any simulation used to model such a scenario with aggregate or typical properties would smooth out these details, forfeiting much of the richness actually at work within the system and distorting the results.

1.3.3.4   Form and Function Come Together

Another advantage that automata offer for simulation is the ability to represent spatial entities and spatial relationships that bind them, and to connect form with function in this way. This aspect of automata-based models is particularly advantageous when modeling geographic phenomena, where actions are often intertwined with the spaces and times in which they take place. Form can be incorporated into automata in a variety of ways; the containers that house individual automata can be designed with an almost limitless range of spatial configurations, e.g., points, lines, polygons, volumes, etc. The functioning of automata depends on their relationships, which, in turn, depend on the form of automata. For example, relationships might be expressed qualitatively, in topological form as intersection, adjacency, or connectivity; alternatively, relationships between automata may be represented as quantitative attributes of these relationships, such as the area of intersection or the length of a common boundary.

1.3.3.5   Simplicity and Intuition

One of the most commonly cited advantages of automata as tools for simulation is their straightforward nature. Compared with most simulation tools, automata-based models can be constructed based on intuitive understanding of the actions of system elements. Traditional urban modeling, as a field of study and particularly as an applied field, has long been criticized for its "black-box" approach (we will return to this idea in much more detail in Chapter 3). From the perspective of users, data are fed into a black-box model and results come out, often with little understanding of what goes on inside the model compartment. Automata-based models offer a remedy to this problem. The automata approach mimics the way in which we, humans, conceive systems, as consisting of objects or elements with associated behaviors or rules. For geographers, the interpretation of automata containers as spaces of various geographic configurations offers an extra degree of intuition to the approach. Also, the "mechanics" of automata-based models may often be easier to comprehend than "flow equations" of traditional urban models. Transition rules can often be expressed using conditional statements of the form "if-then-else" or in any other way that does not have base in continuous analytical dependencies.

The formalism of automata mimics the way in which we study, understand, and describe systems and phenomena in the real world. The properties of system elements are described by their characteristics, e.g., the land-use on a particular parcel of land, the number of floors within a particular building, or the average income of a household's family. Their actions can be represented as transition rules, e.g., rules describing land-use change, building reconstruction, or residential migration. These rules may be designed with consideration of very microscale attributes of these entities: the lifecycle stage of a household, the time-budget of an individual commuter, the occupancy of individual vehicle, etc. The neighborhood function allows for interactions within environments and reflects any alterations in the properties of the environment that those processes may invoke. For example, the decline of particular residential areas in a city could be represented as the diffusion (a process) of urban blight, locally, as adjacent buildings (an environment) fall (a process) into disrepair and abandonment (properties).

This simplicity, however, does not limit the generality of the model. As we will discuss in this book, automata-based models have the power to produce a bewildering range of results. Automata may be simple in their design, but the system that automata represent is most certainly a *complex* system.

1.3.4   Geosimulation versus Microsimulation and Artificial Life

The geosimulation approach is based on, and bears resemblances to, many conventional forms of urban simulation. *Microsimulation* urban models (Clarke, 1996), which form the basis of many popular planning support tools, for example, Paul Waddell's Urbansim (Waddell, 2002), also simulate systems using microscale representations of elementary units. However, microsimulation does not focus on *spatial objects*; the term "micro" is used to refer to disaggregation of data, usually over a regular partition of space at some predetermined scale decided by model needs, but not relative to the objects of analysis. Data about an industrial group, for example, may be disaggregated by type of industrial activity, but data for that group will likely be reported over a very coarse unit of geography; a census tract rather than an individual business, for example. This addresses the ecological fallacy of assigning average group attributes to individual entities, and assuming an ecological correlation between them, but it does not address the Modifiable Areal Unit Problem of decomposing aggregate spatial data, arbitrarily, into smaller geographical units. Objects and entities in geosimulation models, designed in such a way as to be spatially nonmodifiable, each with a unique size and shape, cannot be reduced to a common scale.

Geosimulation models limit themselves to "simple" objects and behaviors. In this respect, the approach differs from Artificial Life models (Maes, 1995a), where there is no conceptual limitation to the complexity of behavioral rules. Artificial Life is largely concerned with simulating long-run evolution, which is usually associated with changes in rules and changes of changes. Its characteristic time-scale is measured in generations. Geosimulation focuses on medium-term dynamics, and its time-units are units of the object activities. For example, regarding humans as householder agents, geosimulation models operate with time-resolution of years, and not decades. On the other hand, geosimulation models of pedestrian movement dynamics may focus on split-second time-resolutions.

1.4   High-resolution GIS as a Driving Force of Geosimulation

The geosimulation approach has been supported, in large part, by enormous advances in geographic dataware—information and data collection resources; tools for querying, manipulating, analyzing, and visualizing data; and methodologies for generating data. These advances have been particularly influential in supporting geosimulation by providing techniques and tools for analyzing and monitoring geographic systems in new ways, and with increased spatial and temporal resolutions. They also allow models of those systems to be fed with new sources of information. Progress in the geographical sciences (GI Science, GIS, and spatial analysis) and Geomatics and Remote Sensing have been particularly influential, as have other, and more recent, developments, such as the derivation of synthetic populations (Bush, 2001).

1.4.1   GI Science, Spatial Analysis, and GIS

GI Science, GIS, and spatial analysis have provided a range of methodologies for handling, interpreting, and producing data for geosimulation. In traditional urban contexts, they have been used to analyze urban data and to provide information—variables—for urban models, or have been used as calibration and verification tools to register simulation runs against real-world conditions. These are loose-couplings, indirect connections. Data may be generated in a GIS, for example, and then fed into an urban simulation. The output that is generated by the simulation could then be fed back into the GIS to be visualized, or to have spatial analysis performed on it.

There also exist opportunities for tight-coupling GI Science, spatial analysis, and GIS within a geosimulation context. We will discuss this further in the next chapter when we introduce Geographic Automata Systems. Models can be designed in such a way that GIS form the database architecture for a simulation. Simulation runs could, for example, be called from within GIS. There is also potential for incorporating GI Science and spatial analysis directly into urban models. GI Science and spatial analysis provide formal methodologies for expressing relationships between geographic objects, as well as representing the behavior of objects and processes in space. All of these functionalities support the exploration of geographic patterns, trends, and relationships in urban systems, as well as providing opportunities for their specification in geosimulation models.

1.4.2   Remote Sensing

Developments in Geomatics and photogrammetric engineering have provided an assortment of tools for collecting data about cities from remote platforms, including airborne and satellite sensors. These advances have provided new information about urban systems, and in many instances remotely-sensed data can be fed directly into urban simulation models.

The resolution and coverage available through remote sensing is growing steadily, offering new insights into urban systems and providing fine-scale datasets for urban models over a wider spatial range. Moreover, these technologies have been in place for a long time, and now offer longitudinal data across long periods of time.

In tandem, innovations in image processing have facilitated the derivation of information from remotely-sensed data. Development of methodologies for automated feature-extraction has been particularly useful. Land cover and land-use classification schemes enable the inference of socioeconomic information regarding urban objects from remotely-sensed images. Likewise, a range of techniques exist for elucidating urban morphology, from the identification of individual structures to the interpretation of digital signatures for various configurations of urban infrastructure (see Webster, 1995; Longley and Mesev, 2001; Torrens, 2004b).

1.4.3   Infrastructure GIS

Remote sensing data provide a powerful source of information for urban high-resolution GIS. The development of urban GIS during the last decade was based on this progress, and is tremendous in scope; many Western municipalities, and almost all of them in some countries, have active geographic databases at a resolution that supports representation of separate buildings and road segments. The precision of these databases varies, but the goal of the municipal GIS as a database for engineering and construction data dictates high standards. Usually, the location of a geographic feature is at a level of meters, sufficient for resolving ownership conflicts, and the attribute information is often updated once a year or even more frequently. The standard basis of urban GIS is built from layers of information pertaining to the road network, land partition at a level of parcels, and building footprints. In many cases layers are included that represent other networks (water, electricity, telephone, sewage), layers of parks, environmental features, and elevation models, terrain models, or other representation of a city's relief. GIS tables contain important attribute data that can inform the delineation of state descriptors in urban geosimulation models, e.g., road width, number of lanes for road segments, number of building floors and year of establishment, etc. As an example, Figure 1.5 illustrates some basic elements of an infrastructure GIS for Tel-Aviv, which is accessible for public use on the Web in a Hebrew version (http://www.tel-aviv.gov.il/hebrew/home.asp); it actually contains a wealth of layers and attribute information.

In many cases, three-dimensional (3D) models of the city are also supplied with links to images of many buildings (Figure 1.6).

In the majority of cases, the enormous volume of information collected remains beyond the reach of modeling research. As compared to the case of urban modeling in the 1970s, amid Lee's critique of data-hungry and continually malnourished simulations, there are ample data today and the current generation of models is not actually hungry enough to consume those data. One of the goals of geosimulation is full use of urban GIS potential.

1.4.4   GIS of Population Census

High-resolution geographic databases of population are less ubiquitous than those that present information relating to the infrastructure of urban environments. The main reason

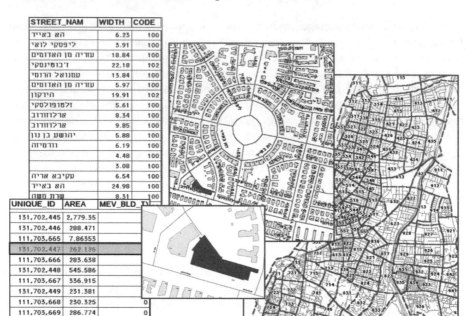

STREET_NAM	WIDTH	CODE
חא באייר	6.23	100
ליפסקי לוצי	3.91	100
עזריה מן האדומים	18.84	100
ד'בוטינסקי	22.18	102
עמנואל הרומי	13.84	100
עזריה מן האדומים	5.97	100
הירקון	19.91	102
זלמופולסקי	5.61	100
ארלוזורוב	8.34	100
ארלוזורוב	9.85	100
יהושע בן נון	5.88	100
ווהמיזה	6.19	100
	4.48	100
	3.08	100
עקיבא אריה	6.54	100
חא באייר	24.98	100
שרת משה	8.31	100

UNIQUE_ID	AREA	MEV_BLD_T)
131,702,445	2,779.35	
131,702,446	288.471	
111,703,665	7.86353	
131,702,447	262.126	
111,703,666	283.638	
131,702,448	545.586	
111,703,667	336.915	
131,702,449	231.381	
111,703,668	230.325	0
111,703,669	286.774	0
131,702,450	231.077	0

Figure 1.5 *Some of the basic elements of Tel-Aviv high-resolution urban GIS: layers of statistical areas, roads, and building footprints are illustrated with some of the attributes presented in a table view*

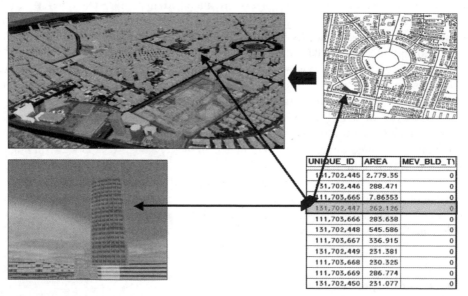

UNIQUE_ID	AREA	MEV_BLD_T)
131,702,445	2,779.35	0
131,702,446	288.471	0
111,703,665	7.86353	0
131,702,447	262.126	0
111,703,666	283.638	0
131,702,448	545.586	0
111,703,667	336.915	0
131,702,449	231.381	0
111,703,668	230.325	0
111,703,669	286.774	0
131,702,450	231.077	0

Figure 1.6 *Typical relationship between layers of 2D and 3D features, representing the same real-world objects of urban infrastructure (Courtesy of the Environment Simulation Laboratory, Porter School of Environmental Studies, University Tel-Aviv)*

for that is evident—people change and move too often; the database investment likely necessary to track households in just one neighborhood on the order of their movements in and out of its housing market would be significant. At the same time, there is not always a necessity for geo-referencing data at the resolution of a person and family, and their independent movements. (Although phone and utility companies may almost certainly have such data.) At the same time, at least in the Western world, personal and family data are always traced or at least reported at a level of country and city; this is best understood in terms of the collection of payments and local taxes. In the UK, this tradition dates back to Domesday. These databases necessarily contain personal address information; it is needed for mailing bills and notices.

The relation between municipal nonspatial databases and GIS becomes, thus, a technical problem of compatibility between systems and of address-matching. Consider some examples of recent advances in the development of GIS-components of popular Database Management Systems (DBMS), such as Oracle Spatial (http://otn.oracle.com/products/spatial/index.html) or use of nonspatial DBMS for spatial data storage, as in the example of ArcStorm (http://www.esri.com/software/arcgis/arcinfo/arcstorm/). Products of this nature render such technical problems relatively easy to negotiate. As a result, it is possible to georeference population data quite precisely. In Israel, for example, the Israeli Central Bureau of Statistics embarked on reconciling its data in this manner, for the entire population of the country, during the population census of 1995 (ICBS, 2000) (you can view an aggregated visualization of ICBS data at http://gis.cbs.gov.il/gis/eng/viewer. htm). From that time on, individual data in the ICBS database have been precisely georeferenced: personal and family records also indicate the house where the person/family live (Figure 1.7).

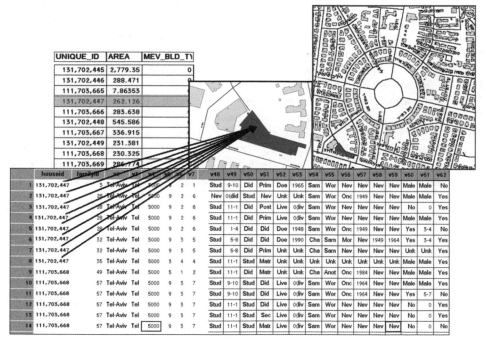

Figure 1.7 *Typical relationship between tables of nonspatial citizens' data and urban GIS*

This individual census record contains a wealth of data items. Records pertaining to the family and individuals also contain data on the residence of those people: number of rooms in the property and home appliances. Data items relating to the family and person are also collected: age, number of children in a family, a settlement the person came from, location of last job, year the person began to work there and traveling time to work, etc. These individual data are available for supervised study in the ICBS offices (Benenson and Omer, 2003) and can be used for statistical analysis of fine properties of urban residential distribution, necessary for initiating geosimulation models (Omer and Benenson, 2002), and calibrating model results (Benenson *et al.*, 2002). Currently this database is static, but, beginning with the next census, the records will be updated on a continuous basis.

1.4.5   Generating Synthetic Data

There are and always will be situations when the resolution of data is insufficient for the goals of a particular research endeavor. Moreover, the use of personal data is usually regulated by laws (data privacy regulations in Europe, for example) and in some cases privacy limitations and the demands of the research are contradictory. One potential solution is the generation of *synthetic* data. Such data imitate *statistical properties* of the investigated populations; the population parameters are usually easily available just because the aggregate data are sufficient to estimate them. Research into the generation of synthetic data sources is also beginning to gather steam. The generation of synthetic population for the application of the TRANSIMS model (Barrett *et al.*, 1999) to Portland (Barrett *et al.*, 2001) is one example. In that case, a range of methodologies was used to generate realistic-looking and statistically fit population data, at a microscale. Synthetic households were generated, with demographic and socioeconomic attributes, georeferenced to microscale geographies. The TRANSIMS Population Synthesizer Module (Barrett *et al.*, 2002) and the Geographic Correspondence Engine (Blodgett, 1998) were used to generate the data. The model took a variety of readily available inputs: Census data, publicly available microsamples (the Public Use Microsample), TIGER polygon boundaries, tax lot data, block-level demographic forecasts, and vehicle records (Bush, 2001). These data are then used to "feed" geosimulation models for simulation purposes. Results generated on the basis of such synthetic data are, like the original data, statistically robust, but without sacrificing the privacy of urban residents.

1.5   The Origins of Support for Geosimulation

Other advances outside geography have provided inspiration for geosimulation-style modeling. Indeed, the idea of using automata as tools for geographic simulation was borrowed from mathematics and what is now called computer science. Developments in these fields also provide significant support for geosimulation.

1.5.1   Developments in Mathematics

Outside geography, much of the inspiration for geosimulation-style modeling has been drawn from mathematics and cybernetics. The idea for the automaton originates in mathematics, where Turing (Turing, 1936) first hypothesized about a simple machine, endowed with a very limited set of simple components that would be capable of representing any algorithm. Given enough time and memory (and a limitless power supply!), such a machine would be capable of supporting any computation.

A machine that can support universal computation is such that it need only be *reprogrammed*, not rebuilt, to perform any calculation that is thrown its way (Wolfram, 1994). Turing's automata—the Turing Machine—harnessed incredible processing power despite simple specifications and a limited range of action. Turing, von Neumann, and others used automata ideas to develop the first digital computers (see Chapter 4) and Turing's ideas for automata are being used to construct molecular computers today (Benenson *et al.*, 2001).

1.5.2   Developments in Computer Science

The Object-Oriented (OO) programming paradigm has also made a crucial contribution to geosimulation, both by providing an intuitive framework for representing real-world objects and by providing object-based simulation software. OO programming techniques focus on objects and the interfaces to objects (Hortsmann and Cornell, 2001). OO programming provides a framework for assembling a set of objects and for using them interactively to accomplish tasks. Objects in the OO paradigm, as in any other software, are inanimate pieces of code, but they can be used to represent properties and activities of real-world entities: cities, buildings, vehicles, people, and their movements, migrations, and use. In OO programming, objects have three fundamental characteristics that can be used to describe them: *identity* (information that distinguishes them from other objects), *properties* (information that associates them with a particular value), and *methods* (statements that govern how operations might be performed on them). Collections of objects that bear similarities are known as a *class*. These characteristics are quite similar to those used to design and build automata, and automata can actually be fashioned quite easily in an OO environment.

Various OO programming languages have been developed: Java, Delphi, C++, and C# being the most popular. All these languages allow for design of *object libraries* that can be integrated into users' programs. Several object libraries have been developed for simulation in the social sciences and have been made freely available for public use. These include the Java libraries RePast and SimBuilder (University of Chicago, 2003), Ascape (Brookings Institution, 2001), and the Objective C library Swarm (Minar, Burkhart, Langton, *et al.*, 1996). However, an equivalent library for geographically based automata simulation has yet to be developed.

Organization of programs built in an OO software environment closely resembles the organization of an urban system as a collection of objects (automata), belonging to several classes, each characterized by variables (states) and associated with methods (transition rules). The OO programming paradigm mimics, thus, the way we consider real-world objects and entities: as distinguishable units, with associated attributes and behaviors, that interact to produce results. This approach directly fits to the needs of geosimulation.

1.6   Geosimulation of Complex Adaptive Systems

As a method of academic inquiry, geosimulation approaches the simulation of systems by modeling *adaptive collectives* of interacting entities. Rather than pursuing a *reductionist* (or top-down) approach, studying systems by dissecting them into logically justified components, geosimulation is characterized by a *generative* (or bottom-up) approach. Phenomena of interest are studied as the product of multiple interactions among simpler elementary units, which correspond to physically existing entities.

Generative systems and phenomena are understood to be, at least partially, *self-organizing* and *emergent* (Holland, 1998). In emergent systems, a small number of rules or laws, applied at a local level and among many entities, are capable of generating complex global phenomena—collective behaviors, extensive spatial patterns, hierarchies—manifested in such a way that the actions of the parts do not simply sum to the activity of the whole. In emergent systems, important attributes and behaviors may not be observable by dissection because the richness of the system lies in the way in which interactions generate adaptations over time. Emergent self-organizing systems are *adaptive* (Holland, 1995), because the global phenomena they exhibit are the result of adaptation of the system to the environment. Urban geography provides many examples of self-organization and emergence, from traffic jams, to urban slums and ethnically homogeneous regions, and we discuss that in length in this book. In all these examples, individual objects adapt and act synergistically, and geosimulation models are often designed to explore, analyze, and, when possible, forecast emergent urban systems.

1.7   Book Layout

The remainder of this book explores the topics raised in this chapter in some more detail. Chapter 2 focuses on the specification of spatially-enabled automata, what we term "Geographic Automata," and explores ways in which these GA may be united as a geographic Automata System of spatially animate and nonanimate urban entities. We devote some attention to elucidating the relationship between GAS and general concepts in Geographic Information Science and Geographic Information Systems. In that chapter we also discuss couplings between automata and GIS more generally, and introduce development resources for coding automata-based simulations.

Chapter 3 focuses on the treatment of urban environments as *systems*. We begin with discussion of the early evolution of ideas about systems dynamics and their relationship to social science and geography, before focusing attention on ideas about complex adaptive systems. Systems, and ideas about the evolution of systems, provide much of the background for contemporary development of automata principles, as well as understanding about the dynamics of urban environments. Indeed, exploration into cities as complex adaptive systems is in its relative infancy as a field of inquiry and there is much potential for automata-based tools to contribute to developments in that area. Geographers arrived somewhat late to the party and there are abundant opportunities for geosimulation to influence the debate.

Chapter 4 focuses squarely on cellular automata. The development of cellular automata is traced from its early pioneering days in mathematics through to its introduction to social science, geography, and urban studies.

In much the same vein, Chapter 5 turns the reader's attention to multiagent systems: their origins, development, and recent popularity, as well as their introduction to urban simulation.

The book concludes with Chapter 6, where we consider geosimulation, somewhat synoptically, in the context of a potential paradigm shift—a revolution even—in geography and urban studies. We argue that the use of geosimulation is more than simple tool-smithing; rather, it represents a fundamental shift in the way we conceptualize, models, and think about urban environments. These developments, we believe, are just beginning to mature and the potential for significant advances in the study of urban environments is promising.

Chapter 2

Formalizing Geosimulation with Geographic Automata Systems (GAS)

2.1 Cellular Automata and Multiagent Systems—Unite!

Geosimulation requires a genuinely *geographic* framework for modeling urban systems, one as formulated on the basis of *objects located in space*. Ideally, such an approach would allow for simulated geographic entities to be specified as *automata*; moreover, Cellular Automata (CA) and Multiagent Systems (MAS) concepts could ideally be combined by considering collections of interacting geographic automata. In this chapter, we introduce such a framework, which considers geographic objects, interacting to form Geographic Automata Systems (GAS) and urban phenomena as a whole are considered as the outcomes of collective dynamics among multiple animate and inanimate geographic automata.

2.1.1 The Limitations of CA and MAS for Urban Applications

Cellular Automata (CA) or Multiagent Systems (MAS) formalisms are already tantalizingly close to fulfilling the requirements of geosimulation. Applied in isolation, CA and MAS approaches have been used to simulate a wide variety of urban phenomena (O'Sullivan and Torrens, 2000; Torrens, 2002; Benenson and Torrens, 2004). Nevertheless, the amalgamation of CA and MAS tools for urban simulation necessitates certain awkward methodological compromises and most combined CA–MAS computer environments and applications exploit a strict CA view of the geographic systems that they model. Either CA cells are granted some degree of agency in their state descriptions and are

Geosimulation: *Automata-based Modeling of Urban Phenomena*.   I. Benenson and P. Torrens
© 2004 John Wiley & Sons, Ltd   ISBN: 0-470-84349-7

simply reinterpreted as artificial agents (Box, 2001) and/or MAS are imposed on top of CA and simulated agents are interpreted as responding to averaged cell conditions (Portugali, 2000; Polhill *et al.*, 2001). While these frameworks are certainly useful, they are based on pragmatic rather than theoretical considerations: the combination of CA and MAS in this manner is a function of the limitations of the available tools rather than being informed by knowledge or theory regarding how real urban systems function in space.

From the geosimulation point of view, a reliance on CA cells as the basis for defining space and spatial interaction is superfluous, and a reliance on regular partitions of space (Figure 2.1a) can be regarded as a weakness for urban applications of CA; not all urban spaces are regular in their delineation. The assumption of regularity is largely superficial, however, and CA have been implemented on a variety of nonregular tessellations: arbitrary networks (Figure 2.1b), irregular partitions or Voronoi tessellations (Figure 2.1c) (Semboloni, 2000b; Shi and Pang, 2000; O'Sullivan, 2001; Benenson *et al.*, 2002). In this case, the form of the neighborhood and the number of neighbors varies between automata in the CA. An assortment of definitions of neighborhoods, based on connectivity, adjacency, or distance can be applied to these generalized CA.

Another more important weakness of the CA approach is the inability of automata cells to move within the lattice in which they reside. Despite repeated attempts to interpret units' mobility (Portugali *et al.*, 1994; Schonfisch and Hadeler, 1996; Wahle *et al.*, 2001), the genuine inability to allow for automata movement in the CA framework catalyzed geographers' recent interest in MAS. This tendency is especially strong in urban geography, where the CA framework is regarded as insufficient in dealing with mobile objects such as pedestrians, migrating households, or relocating firms.

The main geographic advantage of *agent automata* lies in their ability to transmit information by themselves, moving to another location, which can be at any distance from an agent's current position. Agents' spatial behavior can manifest more complex forms than simple relocation. For example, landlord agents might perform spatially mediated sale and purchasing of real estate; the spatial behavior of agents designed to represent car drivers could include the choice of links and turning opportunities at junctions, etc.

Generally, agent automata employed in social science research (Epstein, 1999; Kohler, 2000) are used to represent individual decision-makers (or, sometimes, groups of decision-makers). Consequently, the states that are attributed to agent automata are usually designed to represent socioeconomic characteristics, and agent transition rules commonly correspond to human-like *behaviors*. For the most part, however, work in agent-based simulation in the social sciences outside geography is *non-spatial* in nature, as are the tools that are used. There is a compelling justification for developing MAS that are flexible enough to describe the sorts of spatial behaviors and interactions mentioned in the previous paragraph. Many of the decisions and behaviors of geographic agents are spatial in nature, and this distinguishes agent tools used in geographic applications.

Intuitively, we might identify several geographic mechanisms that are essential to simulating an urban system:

- A typology of entities;
- The space in which they are situated;

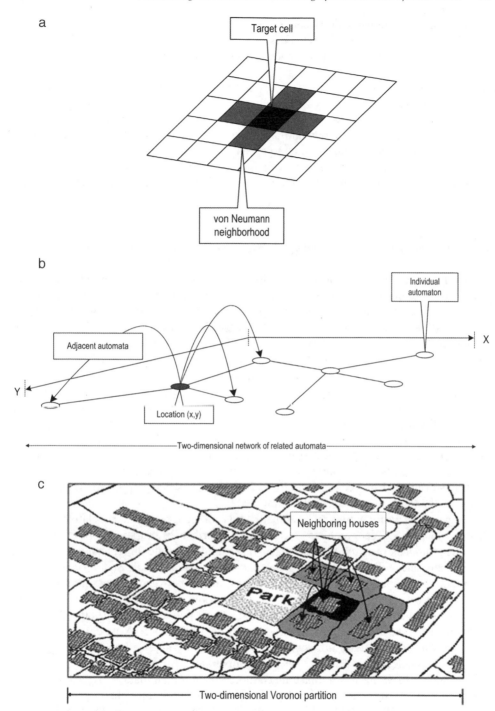

Figure 2.1   *Different partitions of space for Cellular Automata definition: (a) two-dimensional regular lattice, with the von Neumann neighborhood represented; (b) automata defined on a two-dimensional network; (c) automata defined on a Voronoi partition of two-dimensional urban space, constructed based on GIS coverage of building footprints*

- The spatial relationships between entities;
- The processes governing the changes of entities' characteristics; and
- The processes governing the changes of their location in space.

Simulating such a system, then, involves explicit formulation of all these five components.

Neither CA nor MAS can fully provide these requirements in isolation. The geography of the CA framework is problematic for urban simulation because CA are incapable of representing autonomously mobile entities. MAS are weak as a single tool because of the generality of the concept and the broad problem that existing MAS tools and methodologies still underestimate the importance of space and relocation behavior. In many cases, the simulated entities represented by CA and MAS models do not behave as we understand they should, largely because the modeling framework will not permit them to.

2.1.2   The Need for Truly Geographic Representations in Automata Models

Despite the widely acknowledged suitability of automata tools for geographic modeling (Gimblett, 2002), there has been relatively little exploration into the development of patently *spatial* automata tools for urban simulation. This is partly due to the history of automata development. The idea of abstract automata dates to Alan Turing's work on Turing Machines in the 1930s (Levy, 1992). The CA framework was popularized by John von Neumann and Stanislaw Ulam in the 1950s during the development of the first digital computers (Levy, 1992), and the idea of connected and interacting spatial units was developed by Norbert Wiener's work on cybernetics (Wiener, 1948/1961). Originally, CA were pioneered for the description of networks of units influencing each other by means of signals transferred along links. These networks were used as an abstraction of several phenomena: universal computational devices, neural networks, the human brain, cellular tissue, ecological webs, etc. Interest in CA remained obscure for two decades after these initial developments, until interest was revived by the popularity of John Horton Conway's Game of Life (Gardner, 1970, 1971), as well as many applications in physics, chemistry, biology, and ecology (Wolfram, 2002).

However, the introduction of automata tools in geography is a relatively recent phenomenon. It took geography, as a discipline, a further 20 years to adopt the concept, long after automata research had permeated other fields. Despite direct analogies between land parcels and cells on the one hand and land-uses and cell states on the other, geographical applications of CA models were few and far between. A few sound examples published in the 1970s (Chapin and Weiss, 1968; Albin, 1975; Nakajima, 1977; Tobler, 1979) were nonetheless ignored, before interest was revived in the 1980s (Couclelis, 1985; Phipps, 1989). However, it was not until the 1990s that CA modeling became a widespread research activity in geography, popularized by applications in urban geography (Batty, Couclelis, and Eichen, 1997; O'Sullivan and Torrens, 2000).

The study of MAS has taken place much more recently than that of CA. Human-based interpretations of MAS have their foundation in the work of Schelling and Sakoda (Schelling, 1969, 1971, 1974, 1978; Sakoda, 1971). Just as with CA, the tool began to feature prominently in the geographical literature only in the mid-1990s (Portugali *et al.*, 1997; Sanders *et al.*, 1997; Benenson, 1999; Dijkstra *et al.*, 2000), following its introduction in ecology and economics at the end of the 1980s (De Angelis and

Gross, 1992; Tesfatsion, 1997). Until recently, the mainstream of MAS research in geography involved populating regular CA with agents of one or several kinds, which could migrate between CA cells, or simply reinterpreting CA as agent-based models, by attributing anthropomorphic state variables to cells. Often, it is assumed that agents' migration behavior depends on the properties of neighboring cells and neighbors (Epstein and Axtell, 1996; Benenson, 1998; Portugali, 2000). Very recent explicit agent-based models locate agents in relation to real-world geographic features, such as houses or roads, the latter stored as GIS layers (Benenson *et al.*, 2002) or landscape units—pathways and view points (Gimblett, 2002). These models clearly demonstrate the potential of MAS for modeling the intricacies of human spatial behavior.

As a spatial science, geography concerns itself with the behavior and distribution of objects in space. In urban geography, these are necessarily urban objects—households, pedestrians, vehicles—and urban features—land parcels, shops, roads, sidewalks, etc. In dynamic spatial systems, all these objects change their properties and/or location; the goal of a geographic model is to mimic these activities and their consequences, often at multiple scales. In the discussion that follows, we present a framework that aims to infuse spatial properties into automata tools. The framework adopts an *object-based* and explicitly *spatial* view of urban systems. It assumes that urban objects—agents and features—are all individual automata and, as characteristic of automata, their rules of behavior can be defined *a priori*. Urban objects are also conceptualized as *geographic automata*, with focus on their spatial properties and behaviors. Under this framework, a city system can be modeled as a collection of geographic automata, as a *Geographic Automata System*.

2.2   Geographic Automata Systems (GAS)

The Geographic Automata System (GAS) framework unites CA and MAS formalisms in such a way as to directly reflect a geographic and object-based—more specifically, automata-based—view of urban systems. However, this necessitates a re-working of the automata idea to incorporate the five essential components mentioned in Section 1 of this chapter.

2.2.1   Definitions of Geographic Automata Systems

We consider a distinct class of automata, Geographic Automata Systems (GAS), as consisting of geographic automata of various types. In general, automata are characterized by states and transition rules. In the case of geographic automata, we also introduce functionality to enable the explicit consideration of *space* and *spatial behavior*. In CA, pre-defined partitions are often used as a *proxy* for geography. The approach that we adopt with GAS differs; instead of pre-defined partitions, we introduce an independent set of *geo-referencing rules* for situating geographic automata in space. Likewise, we define *neighborhood rules*, rather than relying on fixed neighborhood patterns that are incapable of being varied in space or time once delineated. Considering the mobile functionality introduced by the agent-based paradigm, we also consider a set of independent *movement rules* that allow for the independent navigation of geographic automata in their simulated environments.

Formally, a Geographic Automata System (GAS), G, may be defined as consisting of seven components:

$$G \sim (K; S, T_S; L, M_L; N, R_N) \tag{2.1}$$

Here, K denotes a set of *types* or *ontologies* of automata featured in the GAS and three pairs of symbols denote the rest of the components noted in Section 1 of this chapter, each representing a specific spatial or non-spatial characteristic and the rules that determine its dynamics.

The first pair denotes a set of *states* S, associated with the GAS, G consisting of subsets of states S^k of automata of each type $k \in K$, and a set of state transition rules T_S, used to determine how automata states should change over time. The second pair represents location information. L denotes the geo-referencing conventions that dictate the location of automata in the system and M_L denotes the movement rules for automata, governing changes in their location. According to general definition (1.1) and (1.2) of Chapter 1, state transitions and changes in location for geographic automata depend on automata themselves and on input (I), given by the states of neighbors. The third pair in (2.1) specifies this condition. N represents the neighbors of the automata and R_N represents the neighborhood rules that govern how automata relate to the other automata in their vicinity.

General automata can be easily specified in terms of the framework laid out above. If the state of a geographic automaton, G, at time t is S_t, and the automaton is located at L_t, then external input, I_t, is defined by its neighbors N_t. The state transition, movement, and neighborhood rules—T_S, M_L, and R_N—define G's state, location, and neighbors at time $t+1$. To animate, or spatially enable the GAS, the following rules are applied to each of its automata:

$$T_S: (S_t, L_t, N_t) \rightarrow S_{t+1}$$

$$M_L: (S_t, L_t, N_t) \rightarrow L_{t+1} \tag{2.2}$$

$$R_N: (S_t, L_t, N_t) \rightarrow N_{t+1}$$

Exploration with GAS then becomes an issue of qualitative and quantitative investigation of the spatial and temporal behavior of G, given all of the components defined above. In this way, GAS models offer a framework for considering the *spatially enabled* interactive behavior of elementary geographic objects in a system.

2.2.1.1 Geographic Automata Types

As mentioned, GAS may be composed of automata of different types or ontologies. At an abstract level, we can distinguish between *fixed* and *non-fixed* geographic automata. Fixed geographic automata represent objects that do not change their location over time and thus have close analogies with CA cells. For example, in the context of urban systems, a variety of urban objects may be specified as fixed geographic automata: road links, building footprints, parks, etc. Fixed geographic automata may be subject to any of the transition rules outlined already, except rules of motion, M_L. Non-fixed geographic automata symbolize entities that change their location over time. The full range of rules for GAS can be applied to non-fixed geographic automata, including movement rules. Typical examples of non-fixed urban automata include pedestrians, vehicles, and householders (Figure 2.2).

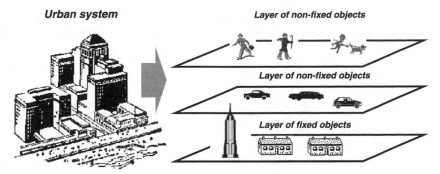

Urban system   **Layer of non-fixed objects**

Layer of non-fixed objects

Layer of fixed objects

Figure 2.2   *An urban system represented as a set of layers of objects of different types*

Any piece of an urban system usually contains objects of both fixed and non-fixed type. For example, in a GAS model of housing dynamics, apartments and houses might be represented by fixed geographic automata, with state variables describing several of their characteristics that are important for residential choice: number of rooms, floor level, value, architectural style, the year of establishment, presence of elevators, etc. Non-fixed geographic automata in the GAS model might represent households, with state variables including economic status of a family, mean age of parents, and number of children.

The characteristics of fixed and non-fixed automata often depend on each other. For example, the value of an apartment depends on the real estate in the property and property's neighborhood and on the population of the neighborhood.

2.2.1.2   *Geographic Automata States and State Transition Rules*

As with the general automata discussed in earlier sections and the previous chapter, a variety of state variables S may be assigned to the individual geographic automata that comprise a GAS, and these states describe the characteristics of the automata in that system. Any variable can be used to derive state values, including variables of geographic significance: height, accessibility, visibility, etc. In the case of non-fixed automata, state variables of relevance to the movement rules of the system may be introduced, for example, heading, speed, progress toward destination.

State transition rules T_S are based on geographic automata of *all* types from K. It is worth mentioning that, in the context of the GAS framework, CA are artificially closed, simply because cell state transition rules are driven only by cells. In contrast, the states of urban infrastructure objects represented by means of geographic automata depend on the neighboring objects of the infrastructure, but are also driven by non-fixed geographic automata—agents—that may be responsible for governing object states such as land-use, land value, etc. In this way, urban objects do not simply mutate (O'Sullivan and Torrens, 2000); rather, state transition is governed by all relevant objects. This is a crucial concept for simulating *human-driven* urban systems, in which people interact and are affected by their environments.

For example, a transition rule, $T_{value\_update} \in T_S$, that describes the change in value of real estate in the model should depend on the states of the real estate objects and, more importantly, on various attributes of the households that occupy them. In terms of (non-fixed) householder geographic automata, a transition rule $T_{economic\_update} \in T_S$, describing changes in the economic status of a householder could be defined. Similarly, we could specify a rule, $M_{householder\_locate} \in M_L$, describing the way households change their residence, and $M_{householder\_neighbor} \in R_N$, describing how householders' neighbors are determined.

2.2.1.3   Geographic Automata Spatial Referencing and Migration Rules

Geo-referencing conventions L govern how geographic automata should be registered in space. For fixed geographic automata, geo-referencing is a straightforward process in most instances; these automata can be geo-referenced by recording their position coordinates. However, for non-fixed geographic automata, geo-referencing is dynamic; automata may move and this demands specific conventions. Also, their location in relation to other automata, represented in simulated goals, destinations, opportunities, etc., may be dynamic in space and time. It is also worth noting that there are instances in which geo-referencing is dynamic for fixed geographic automata also, for example, when land parcel objects are sub-divided during simulation.

Formally, we say that automata in a GAS can be geo-referenced to simulated spaces *directly* and *indirectly* (Figure 2.3).

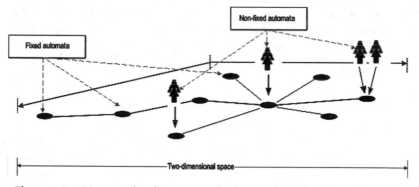

Figure 2.3   *Direct and indirect geo-referencing of fixed and non-fixed GA*

Fixed geographic automata are usually located by means of direct geo-referencing. Direct methods of geo-referencing follow a vector GIS approach, using coordinate lists. Such a list indicates all spatial details necessary to represent an object—automata boundaries, centroids, nodes' location, etc. The details of the particular rules depend on the automata employed in a modeling exercise. For typical urban objects such as buildings or street segments, 2D footprint polygons or 3D prisms may be used to register objects in space. Varying resolutions may be employed, depending on the model application. For example, when modeling housing dynamics at a "microscopic" scale, data such as building perimeter, outer borders, and road segments may be required (Figure 2.4a). However, in other cases, this amount of detail may not be needed and the centroid of a building and centerlines of road segments may be enough to register automata in the model (Figure 2.4b). In abstract models, cell-based approximations may suffice (Figure 2.4c).

The second method by which geographic automata might be geo-referenced, indirectly, is by *pointing* to other automata. For example, in the instance of a model of property

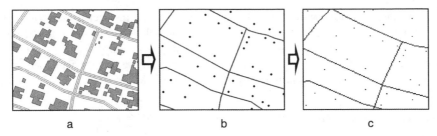

Figure 2.4 *Direct geo-referencing. (a) Buildings are represented by means of foundation contours; road segments by means of road boundaries. (b) Buildings are represented by means of foundation centroids; road segments by means of a road segment centerline. (c) Building centroids and roads are represented by cells*

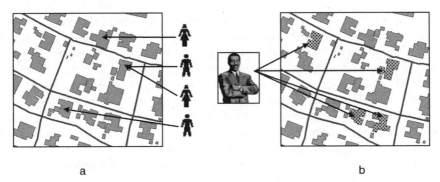

Figure 2.5 *Indirect geo-referencing by pointing. (a) Locating householder agents by pointing to the houses they occupy. (b) Locating a landowner agent by pointing to its properties*

dynamics, we can geo-reference householders by address (Figure 2.5a). Landlords provide a more complex example: they can be geo-referenced by their home addresses, but also by pointing to the properties that they own (Figure 2.5b).

Indirect referencing, by pointing, is mostly relevant for non-fixed geographic automata, but can also be used for fixed geographic automata, for example, for apartments in a house (Torrens, 2001). Indirect geo-referencing is convenient for dynamic modeling, with references varying as a simulation evolves. There is a growing volume of research into geo-referencing formalisms, particularly associated with location-based services (Goodchild, 2001).

The movement rule-set, M_L, is important for the specification of non-fixed automata and GAS-based research into different formulations of M_L offers great potential for geographic inquiry. Particularly in terms of urban simulation, automata-based models have most often been formulated as CA. Even when cells are used to encode movement-like behaviors, as is often the case in traffic models, transition rules are used to mimic movement (Portugali *et al.*, 1994; Schonfisch and Hadeler, 1996), rather than representing the motion of traveling objects as vehicles, pedestrians, householders, institutions, etc. in real urban systems. In other areas of research, realistic rules, M_L, for encoding automata movement, based on repel–attract–synchronize interactions between close neighbors, are being developed, for example, in Animat research (Meyer, 2000) and in the gaming industry (Reynolds, 1999). There is much opportunity for geographers to contribute to this line of research. Traditionally, attention has focused on the generation of realistic choreographies for automata, particularly in traffic models, through the specification of rules for collision avoidance, obstacle negotiation, lane-changing, flocking, behavior at junctions, etc. (Torrens, 2004). However, there remain many relatively neglected areas of inquiry: spatial cognition, migration, way-finding, navigation, etc.

2.2.1.4   Geographic Automata Neighbors and Neighborhood Rules

Another component of GAS that needs explicit definition is the set of neighbors of automata, N, and the rule set for determining the change in neighborhood relationships between automata, R_N. The set of neighbors of different types is necessary for the application of transition rules T_S (state transition) and M_L (movement), which depend on properties of geographic automata and their neighbors. In contrast to the static and symmetrical neighborhoods employed in traditional CA models (Figure 2.1a), spatial relationships between geographic automata vary in *space* (Figure 2.1b,c) *and time*, and, thus, R_N rules should be formulated in such a way as to account for geographic automata locations neighbors at each time step in a model's evolution.

Neighborhood rules for fixed geographic objects are relatively easy to define (simply because the objects are static in space). There are a variety of geographical ways in which neighborhood rules N can be expressed for them—via adjacency of the units in regular or irregular tessellations, connectivity of network nodes, proximity, etc. Similarly, other spatial notions of neighborhood, related to the incorporation of human-like automata into GAS, such as accessibility, visibility, and mental maps can be formally encoded as R_N rules.

Non-fixed geographic automata pose more of a challenge when specifying neighborhood rules, because the objects—and hence their neighborhood relations—are dynamic in space and time. It can be done straightforwardly, via distance and nearest-neighbor relations, as used in Boids models (Reynolds, 1987), but can become very heavy computationally when more complex definitions of visibility or accessibility are involved. In this case, it is more appropriate to base neighborhood rules on indirect location, as defined in the sections above, and consider two indirectly located automata as neighbors when *the automata they point to are neighbors*.

For example, in Figure 2.6, two householder agents are established as neighbors by assessing the neighborhood relationship between the houses in which they reside.

Figure 2.6 *Definition of neighboring relations for indirectly located geographic automata. Two householders are neighbors if they are located in the same property or in neighboring properties*

2.2.2 GAS as an Extension of Geographic Information Systems

GAS have strong affiliations with Geographic Information Systems (GIS). Indeed, re-formulation of automata tools as GAS opens up some exciting opportunities for tightly-coupling automata, Geographic Information Science (GISci.), and GIS. On a basic level, GIS is the natural environment for preparing and visualizing GAS models. Advancing further, GAS are based on the ability of GIS to register data spatially, and to use spatial analysis to shape data as layers of entity-level objects and to estimate relationships between them (Torrens and Benenson, 2005).

2.2.2.1 GAS as an Extension of the Vector Model

Geographic automata models are tightly bound to vector GIS (Figure 2.1).

First, geographic automata of many types correspond to GIS features, which can be used to derive automata location. For fixed geographic automata, it is done directly, by using the coordinate representation of a corresponding GIS feature.

Second, the majority of relationships between geographic automata can be naturally evaluated within vector GIS: standard overlay operators such as point-in-polygon, buffering, intersection, etc., make it possible to determine how automata are situated in relation to other automata. More specifically, neighborhood rules are readily available for evaluating adjacency, contiguity, continuity, distance, accessibility, visibility, and so on.

Third, and as mentioned, GIS are an excellent tool for visualizing and querying the outcomes of GAS simulations.

Finally, recent advances in GIS technology guarantee a functional GIS background for potential GAS computational environments. A number of GIS libraries can be interfaced with other software through the Component Object Model (COM) (Microsoft Corporation and Digital Equipment Corporation, 1995; Ungerer and Goodchild, 2002) or technologies such as JavaBeans[TM] (Sun Microsystems, 2002).

However, in many ways, GAS move far beyond vector GIS, just as CA models go far beyond raster GIS. First, this relates to GAS object types—GIS features do not cover all the variety of geographic automata types. Landlords, as part of housing GAS, are a typical example of this kind: their own location might not be important for the model, while the location of the real estate that they possess really is. Consequently, the location of landlords in such a housing model should be implemented by pointing to real estate holdings. Second, GAS are dynamic, while GIS are not. The essence of GAS is in the rules of state, location, and neighborhood transitions: T_S, M_L, R_N, which do not have analogies in GIS; GIS would benefit from the introduction of automata-like functionality.

Indeed there are opportunities to extend regular vector GIS, especially open source GIS (Baylor University, 2002; Centre for Computational Geography, 2003) toward GAS; recent raster GIS extensions to CA provide a proof-of-concept for such development (Clark Labs, 2002).

2.2.2.2 GAS and Raster Models

While GAS have obvious affiliations with vector GIS; they are also functionally connected to raster GIS in several ways. Each pixel of a raster layer can be regarded, at least morphologically, as an automata cell, geo-referenced by column and row positions within a GIS scene. Based on these coordinates, one can easily consider a pixel-cell as a point or square feature of vector GIS, and the latter fully enable vector GIS functionality, including estimation of relations between objects. Nonetheless, the conceptual difference between features originating from cells of raster and features of vector GIS layers still remains; the latter are normally chosen to represent real-world objects, while the former are not. The choice of a raster or vector view is beyond the GAS scheme and evidently depends on the goal of a model. Irregular tessellations based on land partition fit naturally into real-world simulations, while raster representations essentially simplify neighborhood definitions and other relationships.

2.3   GAS as a Tool for Modeling Complex Adaptive Systems

Urban systems are widely regarded as complex adaptive systems (Portugali, 2000); GAS offer much potential for simulating their various properties, their spatial attributes in particular. GAS models of urban systems are built as collectives of interacting geographic automata and these interactions may be specified in such a way that they facilitate the emergence of higher-level entities, phenomena, and events, from the bottom-up. Complex systems exhibit self-organization phenomena—emergence, bifurcations, catastrophes—and GAS models can be used to reflect phases of continuous *quantitative* development in urban systems as well as recognizing and modeling possible abrupt and *qualitative* changes in their dynamics. The relevance of GAS to applications of complex system theory is geographic in nature; the focus in GAS models is, primarily, on investigating *self-organization in space*.

Spatial systems can self-organize in two ways, and this reflects the aforementioned dichotomy of fixed and non-fixed objects encapsulated in the GAS framework. First, fixed elements can change their properties in a way that entails emergence of assembled spatial objects or the dissolution of existing ones; models of voting are an obvious example (Stauffer, 2001). Second, the same can happen when non-fixed elements change not only their properties, but also their locations, as in residential dynamics models initiated by Schelling and Sakoda (Schelling, 1969, 1971, 1974, 1978; Sakoda, 1971). The alteration of states and locations is an especially important consideration in geographic automata that represent human individuals. The behavior of the latter is essentially based on the ability to recognize emergence or, in the opposite instance, disintegration among ensembles of objects (as, for example, concentration of householders in particular groups). These abilities can be formally considered as positive feedbacks, which accelerate or decelerate the self-organization processes in urban systems.

2.4   From GAS to Software Environments for Urban Modeling

2.4.1   Object-Oriented Programming as a Computational Paradigm for GAS

Operationally, geosimulation modeling is "Simulation of GAS." Any geosimulation environment should therefore provide functionality for the representation of all three GAS components defined in Section 2 of this chapter: to implement and locate urban objects; to determine neighborhood and other spatial relationships between them; and to formulate state transition, migration, and neighborhood rules. The *Object-Oriented Programming* (OOP) paradigm provides an excellent framework for facilitating these sorts of representations, and, furthermore, allows for the depiction of single automata and collectives of automata in a relatively seamless manner. In what follows, we concentrate on the logic of the GAS implementation as a OOP simulation environment framework, discussing abstract software classes that form the logical core of such a software environment; for the most part, we ignore issues such as implementing specific models, user interfaces, performance, data storage, visualization, etc.

2.4.2 From an Object-Based Paradigm for Geosimulation Software

Any simulation software environment must come equipped with some clear determination of its own position between the extremes of abstract concepts and rigid formalizations. Ideally, an environment for geosimulation should be "open," as open as the concepts of urban dynamics it is used to simulate. However, the greater the freedom afforded to the user, the lower the potential benefit to be gleaned from specialized simulation software. General-purpose programming languages implemented within user-friendly OOP development environments—such as Java, VB. Net, Delphi, C++ or C#—are standard tools, always at hand and directly "under user control." Simulation environments based on predetermined formalization of laws of system dynamics promise significant reduction of development time; however, as a tool, those environments are invariably bound to the laws of system dynamics, as well as specific kinds of data. The software cooperates only if you ask it to, in the right way, and feed it the right data snacks. The assumptions of the underlying concept also constrain the tool, functionally. Invariably, there is significant overhead in terms of learning time and more often than not this is accompanied by a level of anxiety, on the part of the user, related to discovering that something "does not fit," too late in the process.

Simulation practitioners are aware of this, and there is a relatively recent trend or shift toward favoring development environments that support specific *concepts* of the system in question. Where software is based on automata concepts, the resulting environments are generally quite extensible in functional terms. We follow this line of reasoning, and assert that two fundamental features of a software environment for geosimulation should be as follows:

(a) Openness for different formalizations of objects' behavior, provided that the user accepts a GAS-based view of the system;
(b) Users' full responsibility for the formalizations' correctness.

This sort of view has already been realized in some popular agent-based simulation environments. The first such environment to enjoy popular use was SWARM, followed by SWARM-based systems such as MAML (Gulyas *et al.*, 1999) and EVO (Krumpus, 2001). The new kid to the block is Repast (RePast, 2003), which was developed from scratch, around a SWARM-like conceptual framework.

Geosimulation is a modeling paradigm and its realization is bound to concept-oriented software. We argue that a reference to *urban* systems is sufficient to specify a universal object-oriented paradigm in a way that facilitates "object-conversant" students of urban studies in formalizing and studying the dynamics of urban phenomena through simulation-based experimnentation and exploration. Let us demonstrate what we mean.

2.4.3   GAS Simulation Environments as Temporally Enabled OODBMS

Geosimulation considers urban infrastructure and social objects as *spatially located* discrete entities, characterized by several properties, which can be directly interpreted as software objects. Computer avatars of real-world urban entities comprise only one of the three basic components of GAS; to reflect the full range of components, the environment should interpret relationships between entities, and self-organizing ensembles of entities, as software objects.

Vector GIS, to which the GAS approach can be considered as tightly-coupled, points in the direction of avenues for implementing GAS as software. GIS is an extension of a relational database, and data storage, updating, querying, and computation within GIS environments is organized in the framework of the *entity-relationship data model* (ERM). The main concept of ERM lies in consideration of real-world objects—entities— separately from relationships between them. To preserve GIS structure, GAS formalism should also have a basis in this delineation. OOP and ERM are the cornerstones of Object-Oriented Database Management Systems (OODBMS) (Booch, 1994); it makes sense that a "straightforward" geosimulation environment could conveniently rely on OODBMS logic.

Consideration of OODBMS theory immediately raises a basic problem likely to be encountered in the development of any geosimulation software environment: generally, ERM and OOP ideas contradict each other (Booch, 1994)! The source of contradiction is the absence of a general solution for managing relationships between objects. A range of partial solutions—*software patterns*—have been proposed, all of which depend on the nature of objects and the semantics of their relationships, and one can refer to computer science publications for complex examples demanding complex solutions (Peckham *et al.*, 1995). Below we discuss the simplest of patterns; luckily for us, it is nonetheless sufficient for all geosimulation models we are aware of.

2.4.4   Temporal Dimension of GAS

A basis for fixed and non-fixed objects, and direct and by-pointing locating, implies specification of Geographic Automata Systems within Object-Oriented GIS. However, the dynamic nature of GAS also implicates *temporal* dimensions of GIS databases, and, thus entails its own limitations. Given these considerations, transition rules T_S, M_L, and R_N, should be defined in a way that avoids conflicts when states, locations, or neighborhood relations are created, updated, or destroyed.

A triplet of transition rules determines the states S, locations L, and neighbors N of automata at time $t + 1$ based on their values at time t. It is very well known that different interpretations of the "hidden"—time—variable in a discrete system can critically influence model formulation and resulting dynamics (Liu and Andersson, 2004). There are several ways to implement time in a dynamic system. On the one hand, we consider time as governed by an *external* clock, which commands simultaneous application of rules (2.2) to each automaton and at each "tick" of some artificial, simulated clock. On the other hand, each automaton can have its own *internal* clock and, thus, the units of time in (2.2) can have different meaning for different automata. Formally, these approaches are expressed as *synchronous* or *asynchronous* modes of updating of automata states. System dynamics strongly depend on the details of the mode employed (Berec, 2002).

2.5 Object-Based Environment for Urban Simulation (OBEUS)—a Minimal Implementation of GAS

Under GAS, an abstract concept can be formalized in different ways. We assert that GAS simulation software should be "down to earth," that is, implementation of an abstract idea should be appropriate for users who are not programming experts. In what follows, we describe the Object-Based Environment for Urban Simulation (OBEUS), which has been developed as a simplest implementation of the GAS concept in a software environment (Benenson *et al.*, 2004). OBEUS is sufficient for all urban geosimulation models we are aware of.

There are three categories of classes in the OBEUS scheme: *Universal, Model,* and *User-defined.* Classes that belong to the Universal category are considered as those that are necessary for simulating *any urban process.* The Model category of classes inherits abstract Universal classes and is necessary for specifying *any model of a specific class.* In what follows, a model of housing dynamics is considered as an example of such a class. User-defined classes reflect specificity in users' models and are constructed anew, if necessary, in each application. In the discussion that follows, we only focus on abstract classes.

2.5.1 Abstract Classes of OBEUS

OBEUS is a software environment based on a GAS conceptual core. As such, the basic components of GAS are defined in OBEUS with respect to automata types $k \in K$, its states S^k, location L, and neighborhood relations N to other objects. These are implemented by means of three abstract root classes:

- *Population*, which contains information regarding the population of objects of given type k as a whole;
- *GeoAutomata*, acting as a container for geographic automata of given type k;
- *GeoRelationship*, which facilitates specification of (spatial, but not necessarily) relationships between geographic automata of the same or different types.

Figure 2.7 illustrates the hierarchy of OBEUS abstract classes and their main methods by means of a UML diagram (Booch *et al.*, 1999).

The *location* information for geographic automata essentially depends on whether the object we consider is fixed or non-fixed. This dichotomy is handled using abstract classes, *Estate* and *Agent.* The *Estate* class is used to represent fixed geographic automata (land parcels and properties in a residential context). The *Agent* class represents non-fixed geographic automata (householders and landlords in a residential context).

Fixed objects are located *directly,* in a manner similar to vector GIS, that is, by means of a coordinate list. Features of planar GIS layers or 3D CAD models can represent urban objects in spatially explicit models. For theoretical models, the points of a regular grid usually suffice.

Non-fixed urban objects are located by pointing to one or several fixed objects. For example, householders are located by pointing to their habitats, as it was shown in Figure 2.5a. Non-fixed urban objects, the location of which is relatively unimportant but

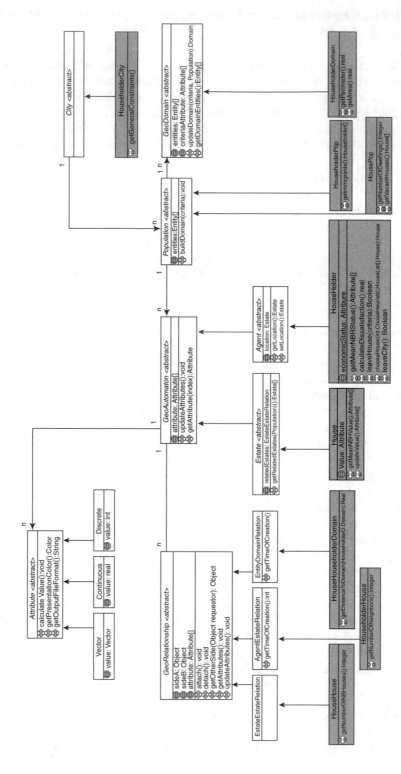

Figure 2.7 Hierarchy of OBEUS abstract classes (transparent blocks) and extension of OBEUS for residential dynamics modeling (gray blocks)

which retains important properties, are related to possessions—fixed objects—by pointing; for example, landlords point to their properties, as in Figure 2.5b.

Following from this, three abstract relationship classes can be specified: *EstateEstate*, *AgentEstate*, and *AgentAgent*. The latter is not implemented; the only way of locating *Agents* modeled in OBEUS is by pointing to *Estates*; consequently, direct relationships between non-fixed objects are not allowed (Figure 2.8). The reasons are nothing but utilitarian.

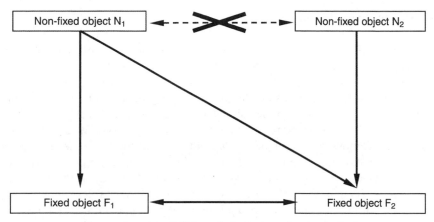

Figure 2.8   *Possibilities of object location in OBEUS*

First, spatial relations between fixed objects are also fixed after these objects are established; thus, they can be estimated and updated only when new fixed objects are created. Second, relationships between non-fixed objects can be easily obtained as a superposition of relationships of non-fixed–fixed and fixed–fixed type. Third, re-estimation of non-fixed–non-fixed relationships can become unwieldy if many non-fixed objects change their location simultaneously. All urban models we are aware of can be interpreted in terms of the fixed–fixed and non-fixed–fixed relationships and a majority of them accounts for only one, two, or three relationship classes at most.

2.5.2   Management of Time

The OBEUS architecture utilizes both *Synchronous* and *Asynchronous* modes of updating time.

In *Synchronous* mode, all objects are assumed to change simultaneously. The calling order of the objects has no influence in this mode, but conflicts arise when agents compete over limited resources, as in the case of two householders trying to occupy the same apartment. Resolution, if ever, of these conflicts depends on the model's context, a decision OBEUS leaves to the modeler. It is worth noting that the logic of synchronous updating often passes conflicts further in time. If two mutually avoiding agents occupy adjacent locations and simultaneously leave them at a given time-step, then in synchronous mode nothing can prevent occupation of these locations by another pair of avoiding agents.

In *Asynchronous* mode, objects change in turn, with each observing an urban reality as left by the previous object. Conflicts between objects are thereby resolved; instead, the order of updating (often, but not necessarily, random) is critical as it may influence results. OBEUS demands that the modeler define an order of object actions according to several templates; random sequence, pre-defined sequence, sequence in order of some characteristic are currently being implemented.

2.5.3 Management of Relationships

Relationships in GAS models can change in time and this might cause conflicts, when, in housing applications, for example, a landlord wants to sell his property, while the tenant does not want to leave the apartment. This example represents the general problem of consistency in managing relationships. It has no single general solution; there are plenty of complex examples discussed in the computer science literature (Peckham *et al.*, 1995). In OBEUS, we follow the development pattern proposed by Noble (2000). To maintain a consistency in relationships, an object on one side, termed the *leader,* is responsible for managing the relationship. The other side, the *follower,* is comprised of passive objects. The leader provides an interface for managing the relationship, and invokes the followers when necessary. There is no need to establish leader or follower "roles" in a relationship between fixed objects once the relationship is established, while, in relationships between a non-fixed and a fixed object, the non-fixed object is always the leader and is responsible for creating and updating the relationship. For instance, in a relationship between a landlord and her property—Figure 2.5b—when ownership cannot be shared, the landlord initiates the relationship and is able to change it.

Relationships between fixed objects should be initialized when fixed objects are initiated, and only then retrieved. Relationships between non-fixed and fixed objects are initiated or eliminated according to requests from the former because they are always leaders within the relationships.

The distinction between objects and relationships, and employment of the leader–follower pattern, offers immediate advantages for software applications. For example, it provides an identical framework for CA and explicit GIS-based land-use models, both of which are based on neighborhood relationships. For a square CA grid and von Neumann or Moore neighborhoods, the degree of neighborhood relationship is either 1:4 or 1:8, while for White and Engelen's (1993) constrained CA, the degree of neighborhood relationship is 1:113, based on neighborhoods of radius 6 around the cell. The only difference between CA models and GIS-based representation of land units based on Voronoi or other partitions of a plane (Flache and Hegselmann, 2001; Benenson *et al.*, 2002) is the variation in degree of neighborhood relationship from parcel to parcel, which is identical for cells in the CA. The distance between the cell and its neighbor, the length of the common boundary, and other factors, can be considered attributes of a *neighborhood relationship* object.

The leader and follower should be defined separately for each relationship. In addition to that consistency, the latter ensures that the algorithm implemented for relationship updating will result from *intentional* thought. There is no proof that the majority of real-world situations can be imitated by the *leader–follower* pattern, although we are not aware of any natural instance, in urban models, where this pattern is insufficient.

2.5.4 Implementing System Theory Demands

Systems theory suggests another challenge for automata modeling in which the usefulness of the GAS–OBEUS approach appears to offer advantages.

The idea of self-organization is external to GAS and it is not necessary to incorporate it into software implementation. Nonetheless, self-organization is too often important for studying urban systems; it cannot be ignored, even at the first step of GAS software implementation. Emerging *spatial ensembles* of geographic automata are supported in OBEUS by means of an abstract class *GeoDomain*. The simplest approach to emergence, determined by the set of *a priori* given predicates defined on geographic automata is implemented; domains are thus limited to capturing "foreseeable" self-organization of specific types.

Domains intuitively represent industrial or commercial areas, rich or poor neighborhoods, or areas displaying a specific architectural style. We assume that domains can be determined by fixed predicates, defined on unitary objects. Let us give an example; the set of unitary objects D_C form a domain, satisfying criterion C if

1. For each $d \in D_C$, a sufficient number of d's neighbors (but not necessarily d itself) satisfy criterion C;
2. D_C contains a sufficient number of unitary objects.

Figure 2.9 presents a simple illustration of "dark gray" domain D_C, with criterion C demanding that "50% or more of the neighbors within a 3×3 Moore neighborhood are dark gray."

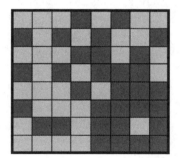 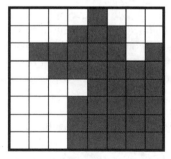

Figure 2.9 *Illustration of the domain criterion. A cell belongs to "dark gray" domain D_C if 50% or more of its neighbors within a 3×3 Moore neighborhood are dark gray*

The importance of domains for representing urban phenomena is evident. A suburb containing many expensive apartments is considered an expensive neighborhood; the value of other real estate in that market consequently increases. One can investigate the consequences, for the city, of the positive feedback generated by the unsatisfied demand of wealthy agents who enter the area, buy properties, and subsequently reinforce these areas as "expensive" domains. Alternatively, an expensive domain can disappear when, for example, wealthy agents become aware of a nearby polluting factory and migrate out of the area. Another domain criterion may be based on the age of urban objects (e.g., the age of buildings), an idea utilized in deltatron CA (Clarke, 1997; Candau *et al.*, 2000).

As with non-fixed objects, domains in OBEUS can be related to fixed objects only. These relationships can capture properties such as distance between unitary fixed objects and the domain; several definitions of distance based on objects' and domains' centroids, boundaries, etc., can be evidently applied. Analysis of available urban models demands consideration of domains of fixed as well as non-fixed objects (low-cost dwellings and poor households, for example).

Domains can always change, which automatically makes them leaders in the relationship with unitary objects. As in the case of non-fixed objects, OBEUS ignores direct relationships between domains and non-fixed objects or domain-domain relationships.

2.5.5   Miscellaneous, but Important, Details

Some of the properties of an urban system relate naturally to the city as a whole. These properties concern relationships between populations' attributes, as, say, the mean number of apartments in a building. The software class *City* (Figure 2.7) represents the city as a whole. The software manages a single instance of the City class, a *singleton* according to Gamma *et al.* (1995).

An abstract class Attribute (Figure 2.7) is not directly related to the idea motivating this chapter but it is useful and convenient, so we will mention it anyway. In OBEUS, attributes are typed, and common data types are supported, as are arrays. Attribute arrays are of the same scalar type so as to support internal comparisons between the array's different elements (e.g., different components of householder income).

OBEUS is only one example; many other software systems exist, and several have rather large and active development and use bases. For the most part, these environments are mostly aimed at MAS simulations, and some offer GAS-like or close-to-GAS abilities. Their description is somewhat of a sideline to the discussion relevant to this book. No doubt any description we could document here will be out-of-date quite quickly; such is the nature of software development in a quickly growing area.

2.6   Verifying GAS Models

GAS models, as any other models, should be verified, adjusted to a particular system, place, and time to compare the performance of the model to the properties of the *real* system being simulated. Specified by direct application of the object-based view of geographic reality, geosimulation models are applied in the following sections to a broad spectrum of urban phenomena, which differ greatly in the level of their abstraction. The models vary from very abstract description of collectives of objects randomly wandering over rectangular grids, to detailed quantitative varieties that imitate residential dynamics in the Yaffo area of Tel-Aviv during the period 1951–1995, at the resolution of separate buildings.

Geosimulation deals with objects in space and time; consequently, employing it in full, one could, for example, relate a geosimulation model to time-series of urban 2D maps, which might represent objects involved in the description of the system or a variety of systems that the model talks about. As in the case of any spatial model, to do that, it is necessary to establish a few things:

(1) *Initial conditions*—the states of model objects, the coordinate location of fixed objects, neighborhood relationships and relationships necessary for indirect location that define the values of S, L, at a zero moment t_0 in which the simulation is begun;
(2) *Boundary conditions*, or, in geographic terms, *spatial constraints*—the conditions that should be held during the entire simulation run and;
(3) The values of model *parameters*, which should be employed to run the simulation ahead in time—the parameters that specify the rules of state transitions T_S, migrations M_L, and neighborhood rules R_N.

One of the goals of scientific research is to develop theory that explains the whole spectrum of phenomena being considered; that is why we are usually interested in as broad a spectrum of initial conditions, constraints, and parameters as possible—characteristic of all possible instances of the phenomena we try to explain with the help of the model.

Dataware resources are evidently crucial for this purpose. As discussed in the introductory chapter of this book, recent progress in development of high-resolution geographic data supply is significant, and remotely-sensed imagery and GIS databases are ready for feeding geosimulation models with necessary information (Gimblett, 2002; Benenson and Omer, 2003; Torrens, 2004b). Quite often, the situation looks close to ideal—the simulated area is covered with a series of remotely sensed images and/or high-resolution GIS layers, say, layers of houses and land parcels and images of their use at different moments in time. Despite progress in this arena, however, a closer view of these kinds of data invariably reveals some incompleteness, and a number of compromises are often necessary if the model is to work with the data.

This state of dataware is relatively new, and brings standard technical problems of verification into the focus of geographic research. A huge general literature on the topic based mostly on non-geographic examples is at hand (Albrecht *et al.*, 1993; Brown, 1993; Smith and Kandel, 1993; Roache, 1998), and geography, as a field, should still adopt those developments. In this section, we will briefly examine examples of validation and verification. The issue as a whole will, undoubtedly, be one of the hot topics in future geosimulation research. We will discuss model-specific issues when they occur in later chapters.

2.6.1 Establishing Initial and Boundary Conditions

For explicit models, we can be certain of data demands, while at the same time the problem of incompleteness in those resources may often need to be addressed. Consider an example. In a model of Arab-Jewish ethnic residential dynamics in the Israeli city of Yaffo, developed for simulation of a period spanning from 1951 to 1995 (Benenson *et al.*, 2002), two high-resolution GIS layers of buildings (with attributes *capacity* and *architectural style*) and road networks (with an attribute *width*) are used as constraints. In the model, these layers are constructed in 1995, while applied from the beginning of 1951; the belief being that the infrastructure did not change very much in that time. The initial condition demanded in the model is the distribution of Arab and Jewish families at resolution of separate buildings in 1951, which is unavailable. Those conditions are constructed in a synthetic way; the population fractions of Arabs and Jews are randomly distributed over the areas they were assumed to populate in 1951, proportional to buildings' capacity for inhabitants. The uncertainty of the population distribution in 1951 should thus be investigated (see Chapter 5).

That was an explicit model; what about abstract models? Abstract models concentrate on features that are characteristic of many instances of the phenomena being considered. Let us consider models aimed at explaining the emergence of urban structure as a typical flavor (Batty and Xie, 1997, 1999; Wu, 1998c; Andersson *et al.*, 2002). Generally, space is represented in such models as a rectangular grid of points; initially, that space is either "empty" or randomly filled with the objects being considered. The constraints employed are very general and regard the size, resolution, or form of the grid; investigation through simulation generally concentrates on self-organization of spatial patterns as a consequence of model rules.

The models we consider in this book cover a spectrum between explicit and abstract formulation. Discussion in Chapters 4 and 5, devoted to Cellular Automata and Multiagent Systems, begins with abstract models and proceeds from there in a more and more specific fashion, toward explicit models; in parallel, initial and boundary conditions become more and more detailed and more and more real-world data are employed.

2.6.2   Establishing the Parameters of a Geosimulation Model

The objects of GAS systems change their state, location, and relations with other objects according to state transition, location, and neighborhood rules. To establish the model the analytical form of the rules should be provided; when defined, the parameters of the rules might be estimated based on comparison of model outcomes and real-world data or their abstraction. As we mentioned above, series of 2D maps or coverages are common output from geosimulation models, and comparison can be performed according to every characteristic of these maps.

Of course, there is no substitute for visual comparison, really, in terms of expertise. The model developer is her own best expert system, especially at initial stages of model investigation. "The eye of the human model developer is an amazingly powerful map comparison tool, which detects easily the similarities and dissimilarities that matter, irrespective of the scale at which they show up" (Straatman *et al.*, 2004). Quantitative methods of comparison of this nature characterize model maps and real maps and provide a measure of similarity between them. Visually, we do that simultaneously and implicitly. Statistical methods do that explicitly. As typical examples let us consider two approaches, pixel-to-pixel comparison of maps on the basis of χ^2 criteria and χ^2-based statistics such as the kappa index of agreement (KIA) κ, and fractal dimension.

Criterion κ is very popular in remote sensing (Congalton and Mead, 1983), and is applied to raster maps. The pixels-values of two maps, one representing model output characteristics and one representing pixels of a real-world map of the same characteristics should be classified into the same categories $i = 1, 2, \ldots K$. Then a K-by-K cross-table that gauges the number of pixels that belong to category i on one map and j on the other, should be performed, and the criteria χ^2 should be used to reveal whether the correspondence between the two maps is non-random.

Significance of κ is the same as that of χ^2 and the value of κ is a convenient measure of correspondence between model output and a map of real-world conditions:

$$\kappa = (P_O - P_C)/(1 - P_C) \qquad (2.3)$$

where P_O is an *observed correspondence*—a number of pixels that belong to identical categories on both maps—and P_C is a *chance agreement*; it is calculated as follows:

$$\kappa = \left(N \sum_i x_{ii} - \sum_i x_{i+} x_{+i} \right) \Big/ \left(N^2 - \sum x_{i+} x_{+i} \right) \qquad (2.4)$$

where $i = 1, 2, \ldots K, x_{ii}$ is the number of pixels in cell i along the diagonal (that is, belonging to the same category i on both maps), x is the total number of observations in row i, and x_{+i} in column i of the cross-table, and N is the total number of pixels.

The value of κ varies within the interval $[-1, 1]$ and the closer κ is to unity, the better the simulation is judged as representing reality. It is worth noting that, to measure correspondence for one category only, κ can be applied to each of the K categories separately (Rosenfield and Fitzpatrick-Lins, 1986) and, for completeness, that negative κ means that two maps non-randomly "oppose" each other; this is a non-typical case in modeling research.

Analysis or measurement of fractal dimension may be applied to one map. In order to compare model output and the real-world map, it is necessary to calculate the fractal dimension of each of them and to compare the resulting two numbers (Batty and Longley, 1994).

Fractal dimension as a measure of comparison has been used in many cases (White and Engelen, 1993, 1994, 1997; Batty and Longley, 1994; Benguigui *et al.*, 2001), as has the kappa-statistic (White and Engelen, 1997; Wu, 1998b).

When measures of correspondence between the maps are defined, we can try to establish the parameters of the rules employed. In the example of the Yaffo model mentioned previously and discussed in detail in Chapters 5, one of the rules dictates that a simulated Jewish householder leave a neighborhood when the fraction of Arab neighbors there is too high. Analytically, this rule can be expressed as a curve, and that curve specifies the probability of departure from a neighborhood, depending on the fraction of the strangers there. The simplest analytical form of this curve—a linear function—has two parameters, and to verify the model it is necessary to specify their numeric values, or, more generally, the intervals these parameters belong to. The same verification procedure may be followed in the case of an abstract model; intervals of variation, rather than exact numeric values, are usually of interest in that case.

How would one go about finding, or settling on, the values of parameters or intervals of their variation, such that they yield a "fit" model? If we are to apply strict mathematical methods, the rules have to be formulated in quite a strict way; for instance, as probabilities of changes that depend linearly on modeled factors. In many cases, we want to employ a conditional, if-then-else, approach, non-smooth dependencies, etc. That is one reason why approaches to validation or verification that involve as little mathematics as possible are most popular in geosimulation contexts.

The typical, and sometimes dangerous; procedure may be politely called "intuitive tuning." Under this approach, the researcher "plays" with the model until values of parameters are uncovered that provide reasonable fit to "reality." It is clear that nothing can be guaranteed with these trials, and better parameter sets may be missed. Nonetheless, in many model examples, the experience of the researcher, and her expert knowledge of the system and conditions of the system, is sufficient to reach quite a good fit. Some of the most popular urban geosimulation models have been tuned in this way, for example Roger White, Guy Engelen, and co-authors have done this quite successfully for their many-parametric land-use model, under application to many real-world situations (White and Engelen, 1993, 1994; Engelen *et al.*, 1995, 2002; White and Engelen, 1997, 2000; White *et al.*, 1997; White, 1998).

More rigorously, one can rerun the model with all possible sets of parameter values to get the set that maximizes some criteria of correspondence. The latter, for example, can be a mean value of κ over all the time moments for which the model and reality can be compared. It is clear that the number of model parameters should be low in this instance, to make this approach realistic, and the research completely depends on the hardware power available for doing so. Keith Clarke and co-authors have developed these sorts of approaches in their SLEUTH model of land-use change (Clarke, 1997; Clarke *et al.*, 1997; Clarke and Gaydos, 1998). They limit themselves to ten or so parameters.

In a case of simple analytical formulation of the rules, mathematical methods of dynamic programming can be performed and recent early attempts to apply these methods in geosimulation modeling have been developed (Arai and Akiyama, 2004; Straatman *et al.*, 2004).

2.6.3   Testing the Sensitivity of Geosimulation Models

None of the components we determine when verifying a geosimulation model are exact. That is why questions of what may happen if the initial conditions, constraints, or model rules are changed slightly or essentially are of principal value in geosimulation. Investigation of model *sensitivity* to variation in each of these components is necessarily part of a verification process and this is a traditional topic of elaboration in the general literature on simulation. Until now, the focus, in geosimulation modeling, has been on investigating one aspect of sensitivity—to stochastic perturbation of model components.

This is usually approached by adding stochastic terms to initial and boundary conditions, as well as to transition rules, and running a simulation several times to obtain the distribution of the results. The stochastic terms also serve as a proxy for unexplained factors in a model, equivalent to the error term in a regression equation (White and Engelen, 1993, 1994, 1997, 2000; Engelen *et al.*, 1997, 2002; White, 1998; Ward *et al.*, 2000; Yeh and Li, 2002).

2.7   Universality of GAS

The Geographic Automata Systems framework is a unified scheme for representing discrete, object-based geographic systems. The framework is designed as a skeleton of the geosimulation paradigm and we dare to assert that it is sufficient for formalization and analysis of *any* urban system. The logical chain between geographic systems and GAS representation is as follows:

Geographic system → Priority of location information and spatial relations between elements → Collective dynamics of geographic automata in space

Indeed, geographic automata are located and behave in space—urban space—in most of the examples discussed throughout this book. The depiction of space necessitates *location conventions*, which differentiate between fixed (houses, land parcels, road segments) and moving urban objects (householders, pedestrians, cars). A minimal realization of GAS can borrow location conventions from vector GIS regarding fixed objects, and utilize the latter as anchors for moving, non-fixed, objects. This has close analogies with the ways in which we understand real urban entities to move within fixed urban infrastructure, such as he example of a pedestrian shopper moving from Calvin Klein by foot, on toward Diesel by taxi cab, and on to Versace by limousine. The minimal location conventions presented above are also sufficient for representing neighborhood and other spatial relationships. Thinking empirically, a tautological statement such as, "householders living in nearby houses are my neighbors" could easily be represented in a GAS context, involving little more than the expression of neighborhood relations between non-fixed householders via fixed real estate units.

Formally, we use the notions of direct and indirect location to encapsulate these sorts of expressions in a simulation framework; informally, we claim that, for the majority of situations, this is just the way *humans describe space and spatial behavior.* It is not surprising, therefore, that a minimal GAS environment is sufficient for interpreting most, if not all, urban CA and MAS simulations we know. The priority of geography in the framework reinforces the strengths of the simulation environment for spatial purposes; as geographers, of course, we would argue its strengths for all purposes. Once we think of the main elements of a system in non-spatial terms, for example, broker agents in a stock exchange model (Luna and Stefansson, 2000), or Internet users on an Information Autobahn (Leonard, 1997), the problem of selecting the relationships that are important for the model arises immediately. The common-sense geographic background for studying complex urban spatial phenomena fades away in non-spatial systems.

A further advantage of the GAS framework is that it is reusable; the minimal GAS skeleton allows for a degree of standardization between automata models and other systems, not least of which are GIS. It also provides a mechanism for *transferability.* Until now, the majority of—if not all—urban simulations, could be investigated only by their developers. The development of a software environment for GAS would breach this barrier, potentially offering opportunities to turn urban modeling from art into science. As mentioned, it has the additional benefit of allowing for exploration into the complex adaptive properties of systems.

Two additional steps are necessary for full implementation of the GAS framework; none, we think, demand decades of development. The first requirement is that the GAS framework should be transformed into a software environment. We have begun to build such an environment with OBEUS (Benenson *et al.*, 2004); similar approaches, also based on object-based views of environmental processes, have recently been developed in ecology (Ginot *et al.*, 2002). The second requirement is that a high-level, preferably geography-specific, *simulation language* based on the GAS approach should be developed. The goal is to enable the formulation of simulation rules in terms of objects' spatial behavior. We believe that the continued development of simulation languages—for description rather than modeling (Schumacher, 2001)—that has gathered steam in the last decade, coupled with advances in GISci. and spatial ontologies (Kuhn, 2001), could answer this requirement in the near future.

We have formulated the general framework of geosimulation and described its formalization in Geographic Automata Systems (GAS). The latter combine CA and MAS into a united framework. Before proceeding to disscussion of urban geosimulation models (Chapters 4 and 5), let us review general systems theory, which is a necessary foundation what we might refer to as pre-geosimulation models, those mostly formulated in the seventies and eighties, were based on systems theory and we discuss such models in the next chapter.

Chapter 3

System Theory, Geography, and Urban Modeling

This chapter serves two main purposes. First, it provides a general overview of complex system theory, focusing on aspects of relevance to urban modeling. Second, it offers an investigation of the achievements and failures of a first wave of urban modeling, that dating to the period between the 1970s and 1980s. The chapter is not intended as an in-depth review of complex system theory, nor does it exhaustively examine urban modeling and simulation over that period. Rather, the intention, in this chapter, is to look at how the methodology of complex system theory was used to specify a first generation of urban models, and to understand what came out of this work. Our book is about models that belong to a second generation with origins in the 1990s and 2000s; the review in this chapter is limited to establishing a background for purposes of comparison—between a "traditional" approach and a new geosimulation approach.

3.1 The Basic Notions of System Theory

The 1960s bore witness to somewhat of a paradigmatic change in the scientific mindset. Two decades of broad discussion involved scientists, from various disciplines, in shaping "System Theory;" this theory offered a common view of natural systems across all levels of organization and observation, from social and psychological to atomic and molecular.

The roots of the system paradigm can be traced back to the beginning of the twentieth century and even earlier, while the idea gained momentum only after World War II. Some

Geosimulation: *Automata-based Modeling of Urban Phenomena*.   I. Benenson and P. Torrens
© 2004 John Wiley & Sons, Ltd   ISBN: 0-470-84349-7

important understanding came to the fore in that time: very different natural systems often have much in common, and knowledge of their functioning can be shared across systems and is nontrivial.

The general system theory was shaped, as a result of attention focused on the *dynamics of systems*. Some very clever minds of the twentieth century devoted a large proportion of their intellectual resources to thinking about transitioning from a static to a dynamic vision of natural phenomena. Much of the work originated in physics and chemistry, where the dynamics of nonlinear molecular, mechanical, and engineering systems were studied. One of the major breakthroughs took place when the range of systems under scrutiny was broadened, and Norbert Wiener's idea for Cybernetics came to the fore (Wiener, 1948/1961). Many other theories that came into the light during the first part of the twentieth century revealed basic common phenomena—across systems—again and again. For example, Vernadsky's views of lithosphere evolution (Vernadsky, 1926/1997); ecological theory and the models of Lotka and Volterra (Lotka, 1925/1956; Volterra, 1926); the automata theory offered by Turing and von Neumann (Turing, 1936; 1950; von Neumann, 1951; 1961; 1966); Shannon's information theory (Shannon and Weaver, 1949); as well as Lewin's work on human behavior (Lewin, 1951).

Ludwig von Bertalanffy, the founder of system theory, cited the objectives of system theory as "the tendency to study systems as an entity, rather than as a conglomeration of parts" (von Bertalanffy, 1968, p. 9). In what follows, we examine the basic concepts of system theory, and introduce several mathematical notions that will appear in the remainder of the book. (Like having your wisdom teeth removed, this might sting a little bit at first, but will feel much better in the long run.)

3.1.1   The Basics of System Dynamics

3.1.1.1   Differential and Difference Equations as Standard Tools for Presenting System Dynamics

The concept of "system" can be considered philosophically and as a commonsense notion. Until the 1960s, an ordinary understanding prevailed: a system was regarded as an ensemble of interacting components. The mathematics of the nineteenth and twentieth centuries involved much work with dynamics of multicomponent systems, which means "dynamics" by any change—changes to the characteristics of components, change in components' location, or both.

The tradition of describing the dynamics of multicomponent systems by means of differential or difference equations is associated with the very beginning of the seventeenth century (see Gleick, 2003), and the names of Newton, Leibnitz, and Laplace are most commonly associated with work in that period. Following that tradition, we can consider system dynamics in the following context. If we regard K as the number of system components and, for the sake of simplicity, the *state* of the ith $(i = 1, 2, \ldots, K)$ component is determined only by one variable $X_i(t)$, then the standard description of system dynamics is given by differential equations in the case of "continuous" time

$$dX_1(t)/dt = F_1(X_1(t), X_2(t), \ldots, X_K(t))$$
$$dX_2(t)/dt = F_2(X_1(t), X_2(t), \ldots, X_K(t)) \tag{3.1}$$
$$\cdots$$
$$dX_K(t)/dt = F_K(X_1(t), X_2(t), \ldots, X_K(t))$$

or difference equations in the case of "discrete" time,

$$\Delta X_1(t) = F_1(X_1(t), X_2(t), \ldots, X_K(t))$$
$$\Delta X_2(t) = F_2(X_1(t), X_2(t), \ldots, X_K(t)) \tag{3.2}$$
$$\cdots$$
$$\Delta X_K(t) = F_K(X_1(t), X_2(t), \ldots, X_K(t))$$

where $dX_i(t)/dt$ denotes the time derivative of $X_i(t)$ and $\Delta X_i(t) = X_i(t+1) - X_i(t)$. In a discrete case, equations for dynamics are also often written as

$$X_i(t+1) = F_i(X_1(t), X_2(t), \ldots, X_K(t)), \tag{3.3}$$

the letter obtained from Eq. (3.2) by transferring $X_i(t)$ to the right-side term. In the discussion that follows, we will note explicitly if this second form is used.

Equations (3.1) represent the changes that occur in the system during an infinitesimally small time-interval, while Eq. (3.2) considers a finite time-interval.

Both the differential and difference systems of Eqs. (3.1) and (3.2) describe an *autonomous* or *closed* system, because functions F_i represent $X_i(t)$ only as a function of component states; there is no exchange between the system and its environment.

It is easily noted that Eqs. (3.1) and (3.2) provide implicit description of the components' state; $X_i(t)$ are unknowns here. To obtain numeric values for $X_i(t)$, as the explicit time-dependent functions $X_i(t) = f_i(t)$, one must *solve* these equations. To do that, the *initial conditions* $X_i(0)$, i.e., the value of each of the components X_i at the beginning of the process, $t = 0$, should be given. Values of $X_i(t)$ are calculated by iteration in the case of Eq. (3.2) or integration in the case of Eq. (3.1). Iteration of Eq. (3.2) is straightforward; explicit solution of Eq. (3.1), however, demands use of methods of analytic or numerical integration (Davis and Rabinowitz, 1975).

3.1.1.2 General Solutions of Linear Differential or Difference Equations

The *general solution* of Eq. (3.1) or (3.2) results in a K-dimensional vector $X(t) = (X_1(t), X_2(t), \ldots, X_K(t))$. That vector can be determined analytically when the functions $F_i(X_1, X_2, \ldots, X_K)$ depend linearly on all X_i, i.e.,

$$F_i(X_1, X_2, \ldots, X_K) = a_{i1}X_1 + a_{i2}X_2 + \cdots + a_{iK}X_K, i = 1, 2, \ldots, K \tag{3.4}$$

Linear systems serve as an anchor for system theory. Let us describe their general analytical solutions in a discrete case. In this case, each $X_i(t)$ is a linear combination of exponential functions $e_i(t) = \lambda_i^t$, where λ_i are 'eigenvalues' of the matrix $B = A + I$. The matrix B is the sum of matrix $A = \|a_{ij}\|$ of the coefficients of Eq. (3.4) and the unit matrix

1, with units on the diagonal and zeros in the other places; that is the diagonal elements $b_{ii} = a_{ii} + 1$, while for $i \neq j, b_{ij} = a_{ij}$.

Eigenvalues are calculated according to a standard procedure; they are complex numbers, and their number always equals the number of equations (Rabenstein, 1975).

Given K eigenvalues $\lambda_i, i = 1, 2, \ldots, K$, the general solution of Eq. (3.2)–(3.4) is as follows:

$$X_i(t) = \beta_{i1}\lambda_1^t + \beta_{i2}\lambda_2^t + \cdots + \beta_{iK}\lambda_K^t, \tag{3.5}$$

where β_{ij} are determined by initial conditions $X(0) = (X_1(0), X_2(0), \ldots, X_K(0))$. Coefficients β_{ij} are numerical constants in a common case or polynomial functions in degenerate situation of equal eigenvalues (see Rabenstein (1975) for further details).

The analytical form of (3.5) offers full understanding of the solution of Eq. (3.2). No matter what the initial conditions may be, after an initial period of relaxation, the solution depends less and less on all values of λ_i, besides one $(\lambda_{\max})$, the module of which is maximal among $|\lambda_i|$. The case in which all $b_{ij} = a_{ij} + 1$ are positive (positively defined $\|b_{ij}\|$), that is of a positive input of each component of the system into the others, is of particular importance. In this case, $\lambda_{\max}$ is necessarily real and positive, and according to Eq. (3.5), it is almost evident that the ratio of any other term of the sum to the term determined by the $\lambda_{\max}$ tends to zero when $t \to \infty$. Consequently, possible types of behavior of the solution of the system of linear-difference equations, when $t \to \infty$, are few and far between: the solution either geometrically converges to a zero vector $0 = (0, 0, 0, \ldots, 0)$ when $\lambda_{\max} < 1$, or grows geometrically when $\lambda_{\max} > 1$. In both cases, *each* component $X_i(t)(i = 1, 2, 3, \ldots, K)$ decays or grows at the rate of $\lambda_{\max}$, and the ratio $X_i(t)/X_j(t)$ tends to a constant, that is the *structure of the system stabilizes*. In the case $\lambda_{\max} = 1$, the module of solution remains constant, but this is a degenerate case that demands exact numeric relations between elements of the matrix $\|a_{ij}\|$ (Rabenstein, 1975). The following point is worth noting; depending on the other eigenvalues, convergence to a zero vector or unlimited growth can be either monotonous or oscillating.

When $\|b_{ij}\|$ is not positively defined, i.e., the overall influence of the component j on component i is not necessarily positive, in addition to geometric growth or decay the possibility of fluctuations of geometrically increasing amplitude exists; again, details can be found in the literature (Rabenstein, 1975).

In the continuous case (3.1), the components λ_i^t in the formulae of solution should be substituted by $\exp(\lambda_i t)$, where $\exp(z)$ denotes, as usual, e^z:

$$X_i(t) = \beta_{i1}\exp(\lambda_1 t) + \beta_{i2}\exp(\lambda_2 t) + \cdots + \beta_{iK}\exp(\lambda_K t), \tag{3.6}$$

and $\lambda_i, i = 1, 2, \ldots, K$ are the eigenvalues of a matrix $\|a_{ij}\|$. Just as in the discrete case, the solution either converges to a zero vector or grows without limit; or, most generally, fluctuates with exponentially growing amplitude. As identical to the discrete case, the components $X_i(t)$ of the solution either all decay, grow, or fluctuate unlimitedly, in proportion to $\exp(\lambda_{\max}t)$, and the ratio $X_i(t)/X_j(t)$ tends to constant.

Malthusian growth (after the eighteenth/nineteenth century English mathematician and sociologist, Thomas Robert Malthus) is the simplest example of the above theory. Malthus published a book, *On the Growth of Population*, in 1798. In that book, he investigated potential results of expansion of population—increase in population numbers $N(t)$ at a constant rate. In terms of differential equations, that means that

$$\mathrm{d}N(t)/\mathrm{d}t = rN(t) \tag{3.7}$$

or in discrete time that

$$\Delta N(t) = N(t+1) - N(t) = rN(t) \quad \text{or} \quad N(t+1) = (r+1)N(t) \tag{3.8}$$

Equations (3.7) and (3.8) are so simple that we do not need any theory to solve them. Nonetheless, to illustrate the general theory, the matrices $\|a_{ij}\|$ and $\|b_{ij}\|$ have one element only—r in a continuous case and $R = r + 1$ in a discrete case—and, correspondingly, the only eigenvalue ($\lambda_{\max}$) equals r for Eq. (3.7) and $r + 1$ for Eq. (3.8). According to Eqs. (3.5) and (3.6), if the initial population numbers are $N(0) = N_0$, then the solution of Eq. (3.7) is

$$N(t) = N_0 e^{rt}, \tag{3.9}$$

while for Eq. (3.8) it is

$$N(t) = N_0 R^t. \tag{3.10}$$

The solutions grow exponentially when r is positive; this actually forms the cornerstone for the well-known paradigm proposed by Malthus: the juxtaposition of population growth versus linear growth of food resources.

Note that correspondence between continuous and discrete time, and between the rates of population growth r and R, is established when we consider that $e^{r\tau} = R$, where τ refers to the duration of a discrete unit of time, say, a year in continuous time units (measured in days). In the latter case, the relation between r and R is given by $e^{365r} = R$, and 5% annual growth, i.e., $R = 1.05$, corresponds to $r = 0.00013367$, i.e., 1.3 hundredth of percentage daily increase.

The model proposed by Malthus is a minimal example of the general linear model of age-structured population dynamics (Royama, 1992) that is applied extensively in demography and ecology. To make the model realistic, age-specific birth and death rates are assumed to depend on conditions of living, cultural traditions, and many other factors. In any case, if coefficients a_{ij} of Eq. (3.4) remain constant, then its solution is given by Eqs. (3.5) and (3.6).

Let us proceed now to the nonlinear case, which is particularly important for system theory. Regarding (3.7), for example, nonlinearity means that population growth rate r in Eq. (3.7) depends on the population numbers themselves, i.e. $\mathrm{d}N(t)/\mathrm{d}t = r(N(t))N(t)$, where $r(N(t))$ changes with $N(t)$, say, decreases, when the population numbers are high.

3.1.1.3 *Equilibrium Solutions of Nonlinear Systems, and Their Stability*

The existence of a general analytical solution in a linear case is somewhat of a luxury, a luxury that does not exist in the context of *nonlinear* cases. Mathematical theory relating to nonlinear cases, then, focuses on *equilibrium* or *steady-state* solutions.

By definition, the equilibrium solution does not change in time, and formally that means that it satisfies either the set of conditions $X_i(t+1) = X_i(t)$, or $dX_i/dt = 0$ for each i. If we denote an equilibrium as $X^* = (X_1^*, X_2^*, \ldots, X_K^*)$, then for (3.1) − (3.2) it can be determined from *static* equations,

$$0 = F_1(X_1^*, X_2^*, \ldots, X_K^*)$$

$$0 = F_2(X_1^*, X_2^*, \ldots, X_K^*) \tag{3.11}$$

$$\ldots$$

$$0 = F_K(X_1^*, X_1^*, \ldots, X_K^*)$$

It is worth noting (and easy to test) that the zero vector $X^* = (0, 0, \ldots, 0)$ is the only equilibrium solution of the linear differential or difference equation (3.4). It is not so in nonlinear cases. Equation (3.11) can have none, one, or several solutions, all being steady states, and system theory is based on notions of their *stability*. The notion of stability is applicable for any solution of Eq. (3.1) and (3.2), while definition is simplest for equilibrium solution (3.11). Namely, equilibrium X^* is *globally stable* if any other solution $X(t)$ tends to X^*, that is, $|X(t) - X^*|$ becomes infinitesimally small when $t \rightarrow \infty$, or

$$\text{for any } X(0), |X(t) - X^*| \rightarrow 0 \text{ when } t \rightarrow \infty \tag{3.12}$$

The equilibrium solution X^* is *locally stable* if any solution $X(t)$ with initial conditions $X(0)$ that are sufficiently close to X^*, tends to X^*, that is,

$$\text{there exists } \varepsilon > 0 \text{ that } |X(t) - X^*| \rightarrow 0, \text{ when } t \rightarrow \infty, \text{ given } |X(0) - X^*| < \varepsilon \tag{3.13}$$

How can one decide, then, whether an equilibrium or other solution is locally or globally stable? Numerical recipes are of little use here, because convergence should be proven for each $X(t)$. Regarding local stability, a general solution to this question is as follows: reduce investigation of the steady state of the general system of Eqs. (3.1) and (3.2) to investigation of a linear system, derived from the general one, for which solutions can be determined explicitly. This linear system is obtained through expansion of the functions $F_i(X_1, X_2, \ldots, X_K)$ into a Taylor series in the vicinity of an equilibrium point X^*, and omitting derivatives of orders higher than one (La Salle and Lefschetz, 1961).

Linear approximation is valid in the vicinity of the equilibrium solution, but becomes invalid when initial conditions $X(0)$ are *far from* X^*, and we cannot ignore derivatives of higher order. Consequently, we cannot base global stability of solutions of Eqs. (3.1) and (3.2) on Taylor linear expansion. Lyapunov theory provides a method for investigating global stability of solutions as well, but they are not so general as the methods for investigating local stability above, and can only be applied with specific functions $F_i(X_1, X_2, \ldots, X_K)$ (La Salle and Lefschetz, 1961). We do not consider the situation of unstable equilibrium yet.

3.1.1.4   Fast and Slow Processes and Variables

Natural systems usually involve processes that are evolving on different time scales. Intuitively, car traffic processes are "faster" than residential dynamics, and the latter are faster than urban infrastructure evolution. Fast and slow processes can be found in any natural system, considered as a whole. This naturally introduces problems of formal investigation: what should be the time resolution of the model? Consider the problem of car traffic. If we choose, as a base case, a time-unit that is characteristic of a fast process, say minutes, during which the density of cars on roads can change essentially, then infrastructure changes simply do not occur during that interval, and vice-versa. At the same time, car traffic definitely influences infrastructure development, and we might want to consider traffic in a model of urban development.

The formal representation of the problem (Arnold, 1973) is beyond the level of discussion in this book, while the approach of mathematical science to the solution of the problem is intuitively clear. Namely, general system theory proposes to deal with the processes that have different internal, or *characteristic*, time *separately*. Moreover, if the pace of the modeled phenomena is established, the theory also proposes to account for the average outcome of the faster processes, while setting the parameters that come from "slower" process as constant. Consider the car traffic example once again; for a model of urban traffic, the infrastructure (slow process) should be set constant, while the parameters of traffic—faster processes—should be set as average over the time unit of the infrastructure model, say a year.

The above scheme, although intuitive, works well if we think about the process times in advance, before formulating the model. By contrast, when we try to combine all the variables that might be important in the models, we can obtain a model with disproportional influence of parameters—some of them do influence the results; others do not at all.

3.1.1.5 The Logistic Equation—The Simplest Nonlinear Dynamic System

One of the most important, while simple, examples of nonlinear models is the *logistic equation*—the basic ingredient of mathematical ecology and demography. The Belgian mathematician Pierre François Verhulst proposed this model more than 150 years ago, when the Malthusian paradigm of exponential growth, formally expressed by Eqs. (3.7) and (3.8), dominated the view of dynamics in populations of humans and animals. Malthus took for granted that birth and death rates are properties of humans as biological species; Verhulst was one of the first demographers that digressed from that assumption. He proposed a mechanism, *density-dependent* population regulation, whereby the rate of population growth decreases when population numbers are high. Analytically, Verhulst used the simplest expression of this dependence, namely, in Eq. (3.7) he substituted r by $r(1 - N(t)/K)$, where K is the *capacity* of an area occupied by the population:

$$dN(t)/dt = r(1 - N(t)/K)N(t) \tag{3.14}$$

The equilibrium solution of Eq. (3.23) is easily obtained, solving

$$r(1 - N/K)N = 0 \tag{3.15}$$

This yields two equilibrium solutions: $N^*(t) \equiv 0$ or $N^*(t) \equiv K$. Investigation of the local stability demonstrates that the zero solution $N^* \equiv 0$ is locally and globally stable, when

$r < 0$ (just as in the case of the zero solution of the Malthusian equation), while nonzero equilibrium solution $N^* \equiv K$, which does not exist in the Malthusian case, is globally stable when $r > 0$.

The paradigm of density-dependent population growth proved to be valuable in both a theoretical and practical sense, because estimates of parameters r and K can be obtained very easily. In formulating a population forecast for Belgium, for example, Verhulst used data for Belgium's population that was available before 1849, and juxtaposed values of relative annual increase $Y(t) = (N(t+1) - N(t))/N(t)$ against values of $N(t)$. According to Eq. (3.15), this dependence is linear, i.e., $Y(t) = r - (r/K)N(t)$, and if the relation is given by linear regression $Y(t) = a + bN(t)$, then $r = a$, and $K = -a/b$. Estimates for the case of Belgium were obtained in this way, yielding a value of $K = 9.4$ million. This prediction, performed in 1846, works even today; the present population of Belgium is close to 10.3 million!

Testing for local stability was considered as being 'sufficient' for understanding dynamics of solutions of Eqs. (3.1) and (3.2) until the middle of the twentieth century; models were introduced, thereafter, with striking examples of equations with *stable, but nonequilibrium* solutions of the non-linear systems. These examples were discovered in seemingly "simple" cases. The discussion in this chapter will turn to these shifts shortly, as they are somewhat revolutionary in this context; first, let us present a standard mathematical approach to formalizing dynamics of systems *in space*.

3.1.1.6   Spatial Processes and Diffusion Equations

The formalizations (3.1) and (3.2) do not deal with consideration of whether the processes exist in space or not. For example, K components X_i could represent concentrations of chemical compounds, numbers of plant and animal populations, revenues in a country's economic sectors, for *both* spatial and nonspatial cases. To make it "spatial," one should only assume that components $i = 1, 2, \ldots, K$ denote *spatial units*. In this interpretation, it is evident that space is considered implicitly; there is no way to distinguish where units are located, say nothing about the unit's boundaries or how the "entire amount" of X_i is distributed within the i-th unit. The formalizations are, thus, weak in their treatment of space; nonetheless, as we will see below, the mainstream of urban modeling in the period from the 1970s to 1980s was based on this implementation.

Representation of spatial processes is based on another analytical description, diffusion equations, and this constitutes an explicit approach. Models of diffusion have origins in pseudorandom motion of particles, "Brownian motion," following famous work in 1727 by the British botanist Robert Brown, observing and describing zigzag motion of pollen grains suspended in water. Albert Einstein provided the first explanation of the observed phenomena in 1905, considering pollen as big particles, being randomly bombarded by many small ones (and he won the Nobel Prize for that work). In the formal one-dimensional, discrete, model of Brownian motion, each particle is located at a discrete point of numbered axe and can move one point aside, either left or right; in a discrete chunk of time (we will call it a tick), all particles move simultaneously. The move to the left or to the right is chosen randomly, with equal probability ½.

If we use the notation $N(x, t)$ for the number of particles at coordinate x at moment t, then the equation that describes the dynamics of particles' distribution in space $N(x, t)$ can be written as follows:

$$(N(x, t + \Delta t) - N(x, t))/\Delta t = (\Delta x^2/\Delta t)(N(x + \Delta x, t) - 2N(x, t) + N(x - \Delta x, t))/\Delta x^2 \tag{3.16}$$

Equation (3.16) is also discrete. It can be transformed into a continuous equation if we assume that the length of spatial unit and the time interval between particle movements both tend to zero, i.e., $\Delta x \to 0$ and $\Delta t \to 0$. In this case, the left term of Eq. (3.16) tends to the partial time derivative of $N(x, t) - \partial N(x, t)/\partial t$, while the right term tends to the partial second derivative $\partial^2 N(x, t)/\partial x^2$ of $N(x, t)$ by the spatial coordinate x. The dynamics of spatial distribution of particles, given by the $N(x, t)$ is described by a partial differential equation—a *diffusion equation*,

$$\partial N(x, t)/\partial t = D\partial^2 N(x, t)/\partial x^2 \tag{3.17}$$

where $D = \lim_{\Delta x \to 0, \Delta t \to 0}(\Delta x)^2/\Delta t$ is called a *diffusion coefficient*.

To solve Eq. (3.17), initial distribution of particles—$N(x, 0)$—should be given, and in the simplest case one can consider all of them as concentrated at one point x_0:

$$N(x_0, 0) = N_0, N(x, 0) = 0 \quad \text{for } x \neq x_0 \tag{3.18}$$

In Brownian motion, it is usually assumed that the initial number of particles N_0 is large, say, of an order of Avogadro number 10^{23}.

As with the linear case (3.4) of the equations (3.1) and (3.2), the diffusion equation in its simplest form (3.17) and (3.18) can be solved analytically. The solution $N(x, t)$ can be formulated as a function of the distance $x - x_0$ between an arbitrary point and the point of initial concentration x_0:

$$N(x, t) = N_0 \exp(-(x - x_0)^2/4Dt)/\sqrt{4\pi Dt} \tag{3.19}$$

If the space is considered as unlimited, it is easy to note that, in the course of time, the particles disperse further and further from the point of initial concentration (x_0). In this case, it does not make sense to talk about equilibrium solution of Eqs. (3.17) and (3.19), just because $N(x, t) \to 0$ at each point x when $t \to \infty$. If, in an opposite case, diffusion is considered on a limited interval, then the distribution tends to uniformity over the interval, the latter evidently being a case of equilibrium.

The diffusion model can be combined straightforwardly with any nonspatial model. For example, if we assume that the (diffusing) particles can give birth to others, i.e., in the nonspatial situation, the population growth at each point x of space is described by an equation

$$dN(x, t)/dt = f(N(x, t)) \tag{3.20}$$

Then, in one-dimensional case the spatiotemporal population dynamics is given by

$$\partial N(x, t)/\partial t = f(N(x, t)) + D\partial^2 N(x, t)/\partial x^2 \tag{3.21}$$

Equation (3.21) can be generalized for a two-dimensional case. For example, if we assume that diffusion does not depend on direction, but only on distance from the origin it becomes:

$$\partial N/\partial t = f(N) + D/\rho(\partial N/\partial \rho + \rho \partial^2 N/\partial \rho^2) \qquad (3.22)$$

where ρ is the distance between the given point and the point of initial concentration and $N = N(\rho, t)$.

The solution of Eqs. (3.21) and (3.22), as well as its behavior, depends on the law of population growth (3.20). The spatial nature of the system raises specific questions regarding the rate of spread of the population over the area, and over time. In an interesting, and basic, case for demography and ecology—the case of Malthusian growth at every point, i.e., $f(N(\rho, t)) = rN(\rho, t)$—the solution of the two-dimensional Eq. (3.22) is as follows:

$$N(\rho, t) = N_0 e^{(rt - \rho^2/4Dt)}/(4\pi Dt) \qquad (3.23)$$

The spread of the muskrat in Europe (Figure 3.1) was one of the first applications of Eq. (3.22) beyond physics and chemistry (Skellam, 1951; Andow, *et al.*, 1990). The spread of the muskrat in this context followed an occasional escape of five animals of this Australian species from a Bohemian farm in 1905 and is described by Eq. (3.23) fairly well.

Namely, according to the model (3.22), the front of the solution (3.23) is expanding at a constant rate $v \sim \sqrt{4rD}$ (Skellam, 1951). Careful comparison provides, depending on direction of expansion, estimates of v as 10–25 km/year, r as 0.2–1.1 per year, and D as 50–250 km²/year, and these estimates do not vary in time (Andow *et al.*, 1990).

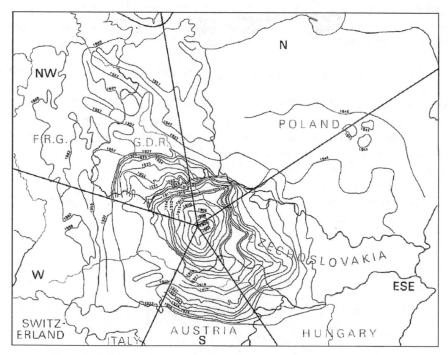

Figure 3.1   *Spread of the muskrat in Europe after occasional release; source Andow et al. (1990)*

While the diffusion equation demands more careful treatment than models (3.1) and (3.2), the notion of stability can be applied to its solutions in the same way as it is applied to Eqs. (3.1) and (3.2). It is also worth noting that diffusion can be considered not only in one and two, but also in three dimensions, as in chemistry, where diffusion-reaction equations can describe the kinetics of compound concentration $N(x, y, z, t)$ at point (x, y, z) of a chemostat at a moment t. In ecology and demography a two-dimensional equation is usually applied, with $N(x, y, t)$ representing the number of individuals at a point (x, y) at a moment t.

As we mentioned above, the nineteenth century view of natural systems, regardless of whether it was spatial or nonspatial, was based on dichotomy between exponential growth and convergence to equilibrium discovered with the linear and nonlinear differential and difference equations. System science came into being in the middle of the twentieth century, when striking examples of seemingly simple equations were discovered, the solutions of which "behave" in a qualitatively more complex manner.

3.1.2   When a System Becomes a "Complex" System

Put very generally, to become complex, a system should exhibit properties that are beyond convergence to a globally stable equilibrium. In this milieu, simple systems, i.e., systems that always tend to steady state, are very "convenient" regarding their reaction to environmental changes, because the effects of external factors and internal properties of the system become separable. If we understand how external conditions influence equilibrium, then the system will simply follow these changes when the environment changes. Of course, frequent or irregular changes of environment might cause problems, because the system requires time to converge to a new equilibrium, but the tendency to conform remains inherent. One might study, for example, limitations on the rate and frequency necessary for external conditions to change in such a way that safe proximity of the system to equilibrium is maintained, and based on this knowledge may direct the system. Put plainly, behavior of simple systems is always "controllable."

Toward the end of the 1960s, the "steady state" paradigm exhibited signs of cracking. A number of important examples accumulated, examples in which the system's behavior appeared to be more complex than convergence to a steady state. Von Bertalanffy relates these complex situations to qualitative changes in the system's structure: "Concepts and models of equilibrium, homeostasis, adjustment, etc., are suitable for the maintenance of the system, but (turned to be) inadequate for phenomena of change, differentiation, evolution, neg-entropy, production of improbable state, creativity, emergence, etc." (von Bertalanffy, 1968, p. 28).

There is a path that we can follow, in terms of system theory, from simplicity to complexity. Along the way, the borders of linearity and autonomy are crossed, before entering the universe of nonlinear and open systems. These steps are vital for study of urban systems, because the object of study—the city—is full of nonlinearly interacting components and is open in many respects. The interactions are nonlinear because the capacity of geographic space for urban uses is limited, and the city is usually close to these limits; moreover, the very existence of cities is based on exchange with the environment. As we will see, further, formal investigation of open non-linear systems does reveal creativity, emergence, and other non-equilibrium phenomena. In this sense, then, urban dynamics cannot be limited to simple systems.

What methodology is available, then, for describing systems beyond steady-state logic? Let us begin with some examples.

3.1.2.1   How Nonlinearity Works

3.1.2.1.1   Cycles in Predator-Prey Communities. The dynamics of predator-prey fauna communities can be more complex than stabilization. Numbers of species in these communities can fluctuate, and the reason for the fluctuations relates to the nonlinear character of interactions between species. This explanation was arrived at in the mid-1930s, when Lotka (1925/1956) and Volterra (1926) formalized predator-prey relationships in terms of two differential equations:

$$dX/dt = f(X, Y)$$
$$dY/dt = g(X, Y)$$

(3.24)

where $X(t)$ and $Y(t)$ denote numbers or densities of species over an area.

Let $X(t)$ denote the number of prey and $Y(t)$ the number of predators in a community. Prey can survive by themselves in the system; predation results in a decrease in their numbers. Predators, on the other hand, are completely dependent upon prey. If the number of prey drops to a level that is not sufficient to support a predator, predators will decline in number; if the drop continues, they may vanish from that community altogether. The more prey in the system, the better the conditions are for the predator.

The work by Lotka and Volterra catalyzed modern mathematical ecology, with the simplest formalization of these arguments by means of the following equations:

$$dX/dt = r_1 X - cXY$$

$(3.25a)$

$$dY/dt = -r_2 Y + dXY$$

$(3.25b)$

Let us assume for a moment that the predator Y is an arctic fox and its prey X are lemmings. Examining Eqs. (3.25), if arctic foxes and lemmings are considered separately, i.e., $c = d = 0$, then lemming numbers grow exponentially as $X(t) = X(0) \exp(r_1 t)$, while the number of foxes exponentially decays as $Y(t) = Y(0) \exp(-r_2 t)$.

The dynamics change, qualitatively, when species interact, i.e., $c \neq 0$ and $d \neq 0$. First of all, a nonzero equilibrium solution of Eq. (3.25) exists in this case:

$$Y(t) = r_1/c, \quad X(t) = r_2/d$$

(3.26)

The striking result offered by Lotka and Volterra was that this equilibrium solution is unstable—none of the other solutions tends to it. Instead, the numbers of foxes and lemmings oscillate (Figure 3.2), that is, the basic predator-prey interactions result in nonequilibrium, but cyclic dynamics of the species! The dynamic of the predator-prey community is presented in Figure 3.2 in two forms. One is based on $(X(t), t)$ and $(Y(t), t)$ coordinate systems, while the second uses *phase space*, given by the pair of coordinates $(X(t), Y(t))$.

The results gleaned from the work of Lotka-Volterra relate to a system that is quite far from the ways in which real lemming–arctic fox communities function. Under the

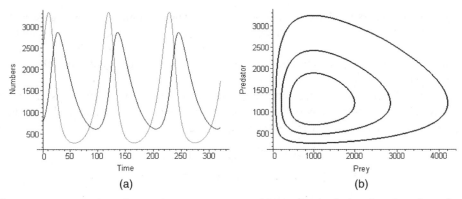

Figure 3.2 *Dynamics of a predator-prey system. (a) Numbers of prey (gray) and predators (black) in time; (b) numbers of predators and prey presented in a phase space for different values of initial conditions*

Lotka-Volterra specification, lemmings in Eq. (3.25a) have unlimited food supply; if there were no foxes, the lemming population $X(t)$ would grow at a constant rate r_1; similarly, if there were no lemmings, decay in the number of foxes would take place at a constant rate r_2; the number of lemmings eaten is proportional to the numbers of foxes, etc. Numerous generalizations of the Lotka-Volterra predator-prey model were proposed after the 1930s and, depending on anlytical form of $f(x,y)$ and $g(x,y)$ in Eq. (3.24) and parameters, these models provide both persistent oscillations, or converge to equilibrium (Gilman, Hails, 1997). All these regimes were also revealed in the field and laboratory experiments with two- or three-species systems; oscillatory dynamics are particularly interesting, yielding an agreement with experimental data in several cases (Huffaker, 1958).

For us, the significant point of the predator-prey system is in the very possibility of endogenous persistent oscillations in predator-prey systems. This led to a crucial shift in understanding the dynamics of natural systems in the 1930s and 1940s. It was not enough, however. The real revolution began in the 1960s, when other decisive examples of seemingly simple, but mathematically very complex (formal) systems came to the fore.

3.1.2.1.2 Fluctuations in Chemical Reaction. The example of the predator-prey system remained somewhat isolated in the literature until the beginning of the 1960s; at this stage, several other exciting examples elevated complex system theory to a position whereby it was considered as the hot topic in scientific discussion. One of these catalyzing examples came from chemistry, where the system of Eqs. (3.1) had been used from the middle of the nineteenth century, as a formal expression of the law of acting masses. Following common views, it was taken for granted among chemists that concentrations of compounds of chemical reaction converge to equilibrium until what is now called a Belousov-Zhabotinsky reaction was revealed (see the introduction at www.online.redwoods.cc.ca.us/instruct/darnold/DEProj/Sp98/Gabe/intro.html). Differing from the example of the (mathematical) predator-prey system, the Belousov-Zhabotinsky reaction was discovered, not on paper, but in chemical experiments. This experiment can be performed with different compounds; for example, if malonic acid ($CH_2(COOH)_2$), potassium bromate ($KBrO_3$), and a catalytic amount of cerium sulfate ($Ce_2(SO_4)_2$), are all dissolved in sulfuric acid (H_2SO_4), the color of the solution begins to oscillate with a period of a few seconds, and it takes several minutes for those fluctuations to decay. The phenomena were so unusual that an experimental paper by Boris Belousov, forwarded for consideration to a scientific journal in 1951, was rejected on the basis that "it simply cannot be." Zhabotinsky, who rediscovered this effect in ten years, also had problems with publishing the results (Zhabotinsky, 1964).

Formally, the chemical system studied by Belousov and Zhabotinsky can be described by three differential equations (which, let us recall, describe a real chemical system, as opposed to the theoretical predator-prey system (3.25)):

$$dX/dt = (qY - XY + X(1 - X))/\varepsilon$$
$$dY/dt = (-qY - XY - fZ)/\varepsilon' \qquad (3.27)$$
$$dZ/dt = Z - X$$

where ε, ε', f and q are parameters.

Just as in the case of the predator-prey system, the XY terms, included in the first two equations (3.27), cause steady oscillation in chemical solution. There is an important difference between the predator-prey and chemistry examples here: the predator-prey system (3.25) oscillates for *all* values of the parameters; that is not the case for the Belousov-Zhabotinsky reaction. The parameter space of Eq. (3.27) can be divided into domains: where the solution oscillates, and where it converges to equilibrium. A nonlinear system, thus, can behave in a *qualitatively* different fashion, depending on parameters' values.

So (seemingly analytically simple) chemical and ecological systems do not necessarily converge to equilibrium—they can also oscillate. That is remarkable, but there is more to the story; the nonlinear systems can exhibit dynamics that are even more complex. Examples of externally "simple" systems, for which solutions neither converge to a steady state nor to cycles, were discovered in the beginning of the 1960s.

3.1.2.1.3 Lorenz Attractors.
In 1961, meteorologist Edward Lorenz explored possibilities of weather changes, using mathematical models. He considered atmosphere in a physics context, as a gas system in a three-dimensional (3D) rectangular box, heated from the bottom. Temperature, pressure, and other physical parameters of the gas within the box change with heating, and vary in 3D space; Lorenz made use of the Navier-Stokes equation to explain those changes. The Navier-Stokes equation was introduced in the middle of the nineteenth century and describes nonturbulent fluid dynamics. It considers space explicitly, as the diffusion equation does, and is very difficult to solve numerically, say nothing about investigating it analytically.

Making use of what we would, today, consider to be a weak computer, Lorenz approached a numerical solution of the equation in the following way. He employed several loops of simplification, ending up with a set of three differential equations that describe dynamics of the three integral characteristics $X(t)$, $Y(t)$, $Z(t)$ of a gas over the entire box (the meaning of the variables is not important at this point in the discussion):

$$dX/dt = -c(X - Y)$$
$$dY/dt = aX - Y - XZ \qquad (3.28)$$
$$dZ/dt = b(XY - Z)$$

Equations (3.28) look quite similar to the model of the Belousov-Zhabotinsky reaction, and one might expect either steady or oscillatory behavior of its solutions. That is not what happens. The scientific community was surprised; Lorenz had discovered that for some values of parameters, a, b, and c, solutions of Eq. (3.28) behave in a very complex fashion—they do not converge to a steady state, and do not converge to a cycle (Alligood *et al.*, 1996).

According to mathematical folklore, this discovery was made by chance. Lorenz wanted to reexamine a sequence of results from his model with $b = 8/3$, $c = 10$, and $a \sim 20$. To save computer time, instead of restarting the entire run, he began with the values of $X(t)$, $Y(t)$, $Z(t)$ recorded in the middle of the previous run. His subsequent observations seemed like hardware failure: the new run matched the previous one during some number of iterations, but then diverged more and more until coherency was lost (Figure 3.3)!

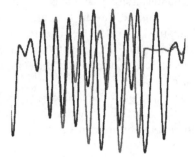

Figure 3.3 *Diverging trajectories of the Lorenz system*

The hardware was tested and found to be reliable; the source of divergence was, thus, in the model itself. Mathematical investigation very soon revealed that the divergence was actually the result of a new phenomenon. For $b = 8/3$, $c = 10$, and a value of a close to 20, the solutions of Eq. (3.28) are *unstable regarding initial conditions* (Alligood *et al.*, 1996). What happens can be illustrated if one plots these solutions in two and three dimensions. (Figure 3.4, built with the help of Java applet from http://www.sekine-lab.ei.tuat.ac.jp/~kanamaru/Chaos/e/Lorenz/)

As Figure 3.4 demonstrates, the solution of Eq. (3.28) enters a specific 3D domain and remains there, never repeating itself. This domain is called a "strange" attractor, which is neither a stable equilibrium point, nor a cycle—they are both "normal" attractors.

(a) (b)

Figure 3.4 *Trajectories of the Lorenz equation in (a) x-z, and (b) 3D, phase spaces*

For different values of a, b, and c, equations (3.28) demonstrate qualitatively different attractors, see http://www.ci.ee.act.ac.za/postgrads/ryorke/gallery/lorentz.html.

Lorenz referred to the sensitivity of solutions of Eq. (3.28) to initial conditions as the "butterfly" effect; for lovers of science fiction, it might remind them of Ray Bradbury's "The Sound of Thunder," a time-safari from the future into the prehistoric past, in order to hunt dinosaurs. In the story the safari team are told not to disturb anything, but one of the party accidentally steps on a butterfly. Cascading through time, this event ends up altering the whole shape of the future. When the hunters return from the expedition, they discover that a recent presidential election, in which a Hitler-like fascist was narrowly defeated, has produced a different winner, and their own native language has been significantly transformed, just like English after the Norman invasion—reasonable deviation, actually, with the entire system still remaining within the strange attractor of the trajectory of world development.

Strange attractors were counterintuitive to ideas circulating at the time in which Lorenz's conclusions were published, even to researchers already familiar with predator-prey, Belousov-Zhabotinsky, and other oscillating systems. If an outwardly simple, autonomous, and "minimally" nonlinear portrayal of nature can exhibit very complex dynamic behavior, what does that say about the complexity of the real world, and how complex *is* the real world?

There is one more step required to establish a link between formal systems and nature, and it involves invoking artificial isolation of models from the "external" world. Natural systems are never fully autonomous; by inference, neither should models be.

3.1.2.2   How Openness Works

3.1.2.2.1   Control Parameters. Intuitively, it is clear that most natural systems *are* open. What do we mean when we refer to a formal system as 'open?' The usual interpretation of an open system is something like this: a system in which some of the *parameters can change under the influence of external factors*. These parameters are referred to as "control parameters" (Haken, 1983), and each parameter can become a control one. For example, we can interpret lemmings' net growth rate r_1 in a predator-prey system (3.24) as a control parameter if we consider food supply for lemmings as being dependent on external conditions (weather, for example) and assume that the greater the abundance of food, the higher the lemmings' population growth rate will be.

The physical meaning of the control parameter is vitally important. Without knowing the meaning of parameters a, b, or c in the Lorenz system, investigation of the influence of external factors remains a formal exercise. The situation changes completely when we are told that a and c are important physical constants—c is a so-called "Prandtl number," characterizing the regime of convection, and a is a "Rayleigh number," which determines turbulence; both of these numbers vary within intervals of several orders (Alligood *et al.*, 1996).

When we know the meaning of parameters and the range of their variation, we can study the implications of changing environmental conditions for the system. Actually, we know part of the answer—with a change in parameters might come qualitative change in the dynamics of complex systems, from convergence to steady state, to convergence to stable cycles, or to strange attractor, for example. The rest of the answer is this: these changes—*bifurcations*—are always abrupt.

3.1.2.2.2   Bifurcations in Open Systems. Described roughly, a bifurcation is a *qualitative* change in an attractor's structure when model parameters are varied *smoothly*. To elaborate, we can think of a limited example, the three types of attractor structures we have already discussed: stable equilibrium; stable cycles of predator-prey or Belousov-Zhabotinsky systems; and strange attractors, characteristic of the Lorenz system. One can easily imagine, for a certain set of a, b, and c, for example, that the solution of a Lorenz equation does tend to equilibrium; the weather in a Lorenz box stabilizes and no longer changes. If changes in the environment (for example, increase in the temperature of heating) cause parameters to change toward $b = 8/3$, $c = 10$, and $a = 20$, then steady weather will somehow break from that equilibrium and fluctuations of strange attractor type should appear.

The Lorenz equation is too complex a mathematical object for demonstrating how bifurcations occur. To make it easier, we are going to consider another famous example, which came from ecology and was popularized by Robert May in 1967. This example is based on the discrete logistic equation (DLE), the continuous version (3.14) of which (see http://fisher.forestry.uga.edu/popdyn/DensityDependence.html) we introduced already. The DLE remained a powerful, but regular, ecological model until the beginning of the 1970s, at which stage the extraordinary dependence of its solutions on the growth rate had been discovered (May, 1976; May and Oster, 1976). Different dynamic regimes of DLE cover the entire qualitative spectrum we introduced above—steady state, oscillations, and strange attractors—and in what follows we demonstrate how they substitute each other with change of parameters.

Introducing density-dependent regulation into discrete Malthusian equation (3.8) we obtain:

$$N(t + 1) = R(1 - N(t)/K)N(t) \qquad (3.29)$$

where K is the carrying capacity of the environment. If we consider population numbers $N(t)$ in units of capacity K, i.e., substitute $X(t) = N(t)/K$, then the DLE model becomes one-parametric:

$$X(t + 1) = RX(t)(1 - X(t)) \qquad (3.30)$$

Let us note that Eq. (3.30) is only valid for values of $R \le 4$; for $R > 4$, its solutions (i.e., population numbers per unit of capacity) tend to $-\infty$.

In Eq. (3.30), population is closed; let us open it, i.e., make its only parameter r *externally* altered. This is quite natural, actually—low net rate of population growth might be characteristic of bad environmental conditions; similarly, a high rate may be indicative of good conditions. By opening the system, we can study, for instance, changes in dynamics of population when the environmental conditions improve. Let us do that when the only parameter of the system—R—varies within $[0, 4]$.

Two equilibrium solutions of Eq. (3.31) are easily obtained by solving an equation

$$X^* = RX^*(1 - X^*) \qquad (3.31)$$

Those solutions are as follows:

$$X(t) \equiv X^*_{low} = 0, \quad X(t) \equiv X^*_{high} = 1 - 1/R \qquad (3.32)$$

For given initial conditions $X(0)$, solutions of Eq. (3.30) are obtained by simple iteration and Figure 3.5 illustrates them for different values of R, beginning at $R = 2$ and growing toward 4. For $R = 2$ the system behaves in a simple way, namely, each solution converges to the equilibrium solution, X^*_{high}, which in this case equals

$$X(t) \equiv X^*_{high} = 1 - 1/R = 1 - 1/2 = 0.5 \tag{3.33}$$

Let us note that the other—zero—equilibrium solution X^*_{low} is unstable for $R = 2$, and this can be tested "experimentally," beginning with close to zero initial conditions of, say, $X(0) = 0.01$, as demonstrated in Figure 3.5(a). We can say that for $R = 2$, solution $X(t) \equiv X_{low} \equiv 0$ "repels" the other solutions of Eq. (3.30), while $X(t) \equiv X^*_{high}$ "attracts" them.

Let us consider what happens when the conditions improve and R grows. For $R = 2.9$ (Figure 3.5(b)), nonzero equilibrium X^*_{high} equals $X^*_{high} = 1 - 1/2.9 \approx 0.655$, and solution with $X(0) = 0.01$ converges to equilibrium nonmonotonously.

The next value of R we will consider is $R = 3.2$. Nonzero equilibrium solution in this case is $X^*_{high} = 1 - 1/3.2 = 0.6875$.

Let us begin, as usual, with $X_0 = 0.01$. This yields the series $X(1) = 0.03168$, $X(2) = 0.09816\ldots$, $X(3) = 0.2833\ldots$, $X(4) = 0.6497\ldots$, $X(5) = 0.7283\ldots$, $X(6) = 0.6333\ldots$, $X(7) = 0,7432\ldots$. It does not converge to equilibrium! $X(6)$ is smaller than $X(4)$ and $X(7)$ is bigger than $X(5)$! Indeed, as opposed to the case for $R = 2$, the solution does converge to a regular regime, which switchs between two numbers, and we can see in Figure 3.5(c) that one is close to 0.513, and the other is close to 0.799. Mathematical investigation confirms that, for $R = 3.2$, solutions converge to a *cycle of period* 2. This cycle attracts all solutions of Eq. (3.30), i.e., it is a *globally stable cycle*.

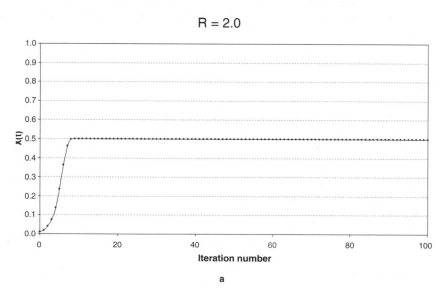

R = 2.0

a

Figure 3.5   *Solutions of DLE for different values of R: (a) R = 2.0; (b) R = 2.9; (c) R = 3.2; (d) R = 3.5; (e) R = 3.8*

R = 2.9

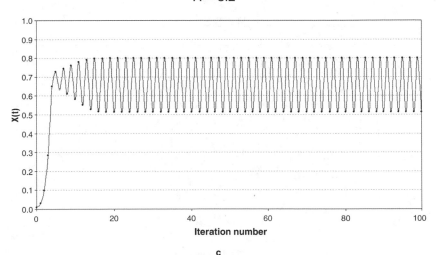

Iteration number

b

R = 3.2

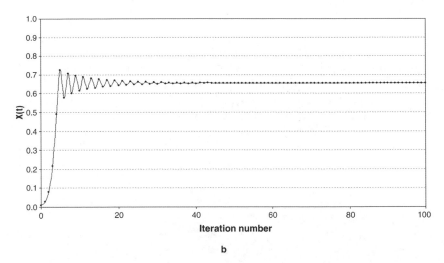

Iteration number

c

Figure 3.5 *(Continued)*

The values of $R = 2, R = 2.9$ and $R = 3.2$ provide, thus, qualitatively different, but globally stable regimes; for $R = 2$ and for $R = 2.9$ it is an equilibrium and for $R = 3.2$ it is a cycle of period 2. What happens when R changes smoothly between 2 and 3.2?

Mathematical investigation shows that the critical value of R is 3: when R grows from $R = 2$ to $R = 3$, the equilibrium solution $X(t) = X^*_{high} = 1 - 1/R$ also grows from 0.5 to 0.6666..., but *remains globally stable*. At $R_0 = 3$ this smooth behavior is broken; the equilibrium solution $X(t) = X^*_{high}$ becomes unstable. At the same time, a

R = 3.5

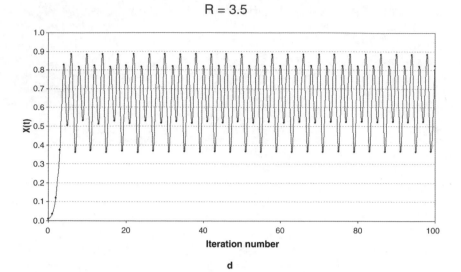

d

R = 3.8

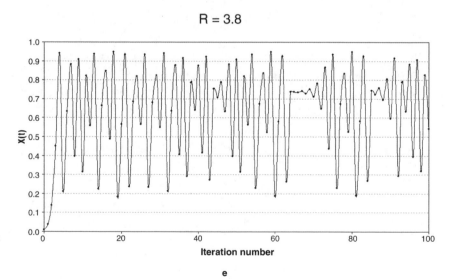

e

Figure 3.5   *(Continued)*

qualitatively new solution is yielded and becomes globally attractive. This solution is a cycle of period 2, the amplitude of which grows with growth of *R*, reaching a value of about 1.6 for *R* = 3.2 (Figure 3.5(c)).

The change in solution behavior at *R* = 3 is called *bifurcation*. The value of *R* = 3 is not the only *point of bifurcation* of Eq. (3.30). Figure 3.5(d) illustrates solution of Eq. (3.30) for the other value of *R*, *R* = 3.5. One can see that, for this rate of growth, population numbers also cycle, but the *period of oscillation is 4.*

Mathematical analysis demonstrates that this cycling solution is also globally stable. How does the cycle of length 2 turn into a cycle of length 4? Just as there is bifurcation at $R_0 = 3$, there is another bifurcation point between $R = 3.2$ and $R = 3.5$, where the cycle of period 2 becomes unstable, while the stable cycle of period 4 comes up. Calculation reveals this point to be at $R_1 = 3.449\,490\ldots$, while the next point of bifurcation, is at $R_2 = 3.544090\ldots$. At R_2, 4-cycle becomes unstable, while a stable 8-cycle emerges, and, in turn, 8-cycle becomes unstable at $R_3 = 3.564\,407\ldots$ where the stable cycle of period 16 emerges. What happens if we go further?

The events that occur when R passes bifurcation points $R_0, R_1, R_2, \ldots, R_k$ are *period doubling;* the cycle of period 2^k becomes unstable, while the stable cycle of period 2^{k+1} emerges. Further mathematical inquiry demonstrates that the series of period doubling bifurcations does not continue to the upper limit of $R = 4$, but has its limiting point at $R_c = 3.569\,945\ldots$. When R passes R_c, the solutions of Eq. (3.30) become even more complex. The graph for $R = 3.8$ illustrates the complications coming up—Figure 3.5(e)—the solution of DLE fluctuates irregularly!

Probing further and mathematically (May and Oster, 1976), it has been proven that this irregular behavior is typical when R is greater than R_c. For R between R_c and 4, the structure of attractor essentially depends on the exact value of R and can be very complex; we are in trouble with graphs of the type illustrated in Figure 3.5, they are too many and uninformative. The solution comes in the form of a *bifurcation diagram,* which presents the change in attractor structure with growth in R as shown in Figure 3.6 built with the help of http://www.cs.laurentian.ca/badams/LogBif/LogBifApplet.html. In this diagram, the phenomena we discuss are visible: from the accumulation of period-doubling bifurcation points for $R < R_c$, to complex structures for $R > R_c$.

We are now able to roughly characterize bifurcations of the solution of DLE with growth in R from 0 to 4. When R grows from 0 to 1, any solution of Eq. (3.30) tends to zero equilibrium point X^*_{low}. The value $R = 1$ is the first point of bifurcation; when R

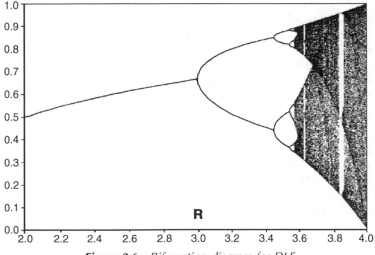

Figure 3.6 *Bifurcation diagram for DLE*

passes this value, X^*_{low} becomes a repelling equilibrium and X^*_{high} becomes an attracting equilibrium. The first bifurcation at $R = 1$, then, is the switch between stability and instability of two existing equilibrium points. The next bifurcation point is $R = 3$, where *the equilibrium solution is not an attractor* any more, and with further growth in R the system enters an interval of period-doubling bifurcations. After R passes R_c the solution behaves in a more complex fashion; we discuss that issue further in Section 3.1.2.2.3 of this chapter.

While it is beyond our introductory discussion, the period-doubling cascade is a *universal property*. For any discrete model of the form

$$X(t + 1) = RF(X(t)) \qquad (3.34)$$

we can formally check whether it has a series of period-doubling bifurcations with change in R. If yes, the length of the interval between consecutive points of period-doubling bifurcation diminishes geometrically, at a constant rate of 4.669 201... no matter what the function $F(X)$ is! In 1975, Mitchell Feigenbaum discovered that this is a new constant of nature, just as π or e (Feigenbaum, 1980). It is also important to note that the notion of bifurcation is much broader than the example we presented above (see Alligood *et al.* (1996) for an introduction).

Back to DLE; the period-doubling bifurcation points $R_0, R_1, R_2, \ldots$ accumulate toward $R_c = 3.569\,945\,6\ldots$. With values of $R > R_c$, we enter the domain of *chaotic* dynamics, which demands special consideration.

3.1.2.2.3 Deterministic Chaos. In everyday language, 'chaos' implies lack of order. The formal meaning of chaos, however, is narrower and concentrates on lack of *simple regularity* in dynamic behavior of a system. System theory employs the notion of *deterministic chaos*, talking about irregular and *unpredictable* behavior of deterministic models, as DLE with $R = 3.8$, for example. The Lorenz system for $b = 8/3$, $c = 10$, and $a = 20$ is another example of deterministic, but chaotic, dynamic behavior.

The very idea of deterministic chaos is a paradigmatic change; this was the "disruptive technology" that challenged the deterministic paradigm of natural science in the seventeenth and eighteenth centuries, referred to as "Laplace's Demon."

Marquis Pierre Simon de Laplace formulated his principle of determinism on the eve of the nineteenth century. In modern language, "An intelligence knowing all the forces acting in nature at a given instant, as well as the momentary positions of all things in the universe, would be able to comprehend in one single formula the motions of the largest bodies as well as the lightest atoms in the world, provided that its intellect were sufficiently powerful to subject all data to analysis; to it nothing would be uncertain, the future as well as the past would be present to its eyes..." (Laplace, 1820, as quoted in Pitowsky, 1996).

Deterministic chaos chased away the Laplace demon. Long-term behavior of chaotic systems (e.g., of Lorenz equations or DLE) is impossible to predict, because deterministic chaotic systems are "infinitely sensitive" to initial conditions. For chaotic systems, no matter how close, two sets of initial conditions diverge more and more with time.

Let us assume for a moment that DLE does describe the population dynamics of real-world species, say the tropic fly, which reproduces quickly with a seasonal reproduction rate R close to 4, and for which the reproduction rate depends on density in a DLE fashion. Slightly erroneous data on population numbers will result in rapid divergence between the expected and actual numbers for this species. There is always limited accuracy of measurement; this makes chaotic systems essentially unpredictable.

Nonlinear deterministic models of a form (3.34) and exhibiting chaotic regimes successfully describe many experimental and field data, which range from physiology epidemics and ecology (Ellner *et al.*, 2001; Alligood *et al.*, 1996), to economy and chemical reactions (Johnson, 2002) (see http://hypertextbook.com/chaos/ and http://www.sekine-lab.ei.tuat.ac.jp/~kanamaru/Chaos/e/ for more examples).

Chaos is not the only explanation of deterministic, but still completely or partially unpredictable, behavior in natural systems. Several other mechanisms make models' behavior unstable and complex. The most popular are *positive feedbacks* and *delays*.

3.1.2.2.4 Positive Feedbacks and Delays. Positive feedbacks, as with each of the system theory notions we have introduced thus far, have common sense meaning—positive feedback simply refers to "more of the same," say, the growth of a system's characteristics leading to even more growth (Milsum, 1968). Positive feedbacks oppose *negative feedbacks*—a standard mechanism of control and stability widely used in engineering and natural systems (Figure 3.7) (Jones, 1973).

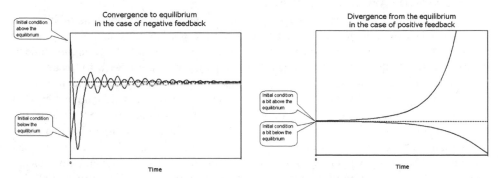

Figure 3.7   *Schematic representation of the action of positive and negative feedbacks*

The discrete logistic equation is based on simple negative feedback: *increase* in population density *decreases* the rate of population growth, and, consequently, the population numbers themselves. In engineering terms, we can say that Verhulst introduced this feedback in order to stabilize the population, which otherwise grows exponentially. The example of DLE demonstrates that for $R < 3$ it does work, while for $R > 3$ it does not; negative feedbacks are not a common panacea, thus, and should be treated with caution.

As disciplines, cybernetics and control theory invested a lot in investigating feedback mechanisms as tools for regulating systems and keeping their dynamics up to necessary equilibrium (Riggs, 1970). As a result, there is plenty of examples of negative feedback mechanisms, both acting in nature and introduced by humans (Riggs, 1970; Jones, 1973), and the notion of negative feedbacks are typical and easily accepted.

Positive feedbacks by contrast are relatively doubtful, and too often cited in journalistic explanation of important and often cataclysmic events. As a feasible example, let us take arms races between two countries, A and B, which are afraid of each other and, thus, tend to be ready for an unexpected increase in an enemy's strength. The logic of positive feedback might infer the following common sense explanation of the race. Let us say that B provoked initial suspicion, prompting A to adopt a course of buildup. In these circumstances, B wants to be cautious, also, and so builds up its resources only to a potential sufficient to its foreseen increase in the strength of A. This registers as a threat for A, and A in turn further expands its military potential, and so on.

The consequences of the above (tit-for-tat) accelerations are catastrophic; to keep in the race, the military system of each of A and B should be able to incorporate more and more external resources. The positive feedback loop thus develops until one of the sides completely exhausts necessary resources, and either unleashes a war or calls off its "positive" reaction. In both cases, the system laws are changed. One can even try to explain the crash of the former USSR in this way, but how we can *prove* that the explanation is true?

Positive feedbacks are intuitively convenient candidates for explaining nonregular bursts or collapses in systems' behavior. From the very beginning of cybernetics and system science, they have been used in speculation about global warming, deforestation, social conflicts, economic crisis, etc., especially when the system can be considered as an *open* one, which can acquire resources from the environment to support the positive feedback loop (De Angelis *et al.*, 1986).

As we mentioned above regarding deterministic chaos, the explanations of irregular system behavior or sudden qualitative changes that are not based on chaotic regimes or positive feedbacks are usually possible; the influence of irregularly changing hidden external factors, first and foremost. Regarding the collapse of the USSR, for example, at the same amateur level as above, one can say that a totalitarian political system invoked such low labor productivity that the economy was bound to slow cessation, no matter what the part of the military industry was. Illuminating the forces behind positive feedbacks and proofs of their control over the system, thus, demands serious research, and not so many verifications can be found. As one of few thoroughly investigated examples one can consider the paper by Cochrane *et al.* (1999) who prove that accidental fires create positive feedbacks in future fire susceptibility, fuel loading, and fire intensity, and all this transforms large areas of tropical forest into scrub or savanna in a longer run.

The idea of delayed response is simpler and also appeals to common sense; parameters respond to changes, but not immediately and only after some time. For example, in the case of the logistic equation applied to mammals, we can assume that the population birth rate depends on the density at the moment of conception, and thus, the period of pregnancy becomes a delay time τ. The equation of population dynamics in this case will be

$$\mathrm{d}N(t)/\mathrm{d}t = rN(t)(1 - N(t - \tau)/K) \tag{3.35}$$

Accounting for delay results, usually, in more oscillations—during the delay time τ, the system follows "old" tendencies despite internal changes that have yet to go to work; the period of stable or decaying oscillations is usually several times bigger comparable to the delay time (Aracil *et al.*, 1997). As an example, we can consider how delay causes solutions to oscillate in the case of otherwise stable behavior: for DLE with $R = 1.3$, when solution monotonously converges to equilibrium, $X^*_{\mathrm{high}} = 1 - 1/1.3 = 0.231\ldots$ amazingly

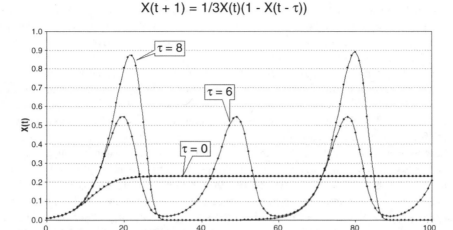

$$X(t + 1) = 1/3X(t)(1 - X(t - \tau))$$

Figure 3.8 *Solutions of the discrete logistic equation with delay for three values of* τ

fluctuates with a period of about 35 iterations when the delay is 6 and with a period of about 60 iteration when the delay is 8 iterations (Figure 3.8).

Thus far, we have talked about three mechanisms of irregular and "externally" unexpected changes in an open system—deterministic chaos, positive feedbacks, and delays. The consequences of their action can be considered differently regarding the *reversibility* of the systems they act on. Let us consider the discrete logistic equation and assume that the current value of the control parameter R entails deterministic chaotic population dynamics. If conditions change, and control parameter R decreases to the value that entails convergence to equilibrium, the system follows this change and stabilizes. It is not the case in arms races or forest fires—positive feedbacks destroy the system and the changes they cause are *irreversible*. System theory culminates in the study of irreversibility of changes and emergence and persistence of novel structures; this process is called *self-organization*.

3.1.2.2.5 Self-organization and Synergetic Behavior of the Systems. According to the *Principia Cybernetica* (2003): self-organization is an attribute of a system that changes its internal organization on its own account, neither in response to conditions in another system with which it may interact nor as a consequence of its membership in a larger "metasystem." Popular nonphysical phenomena, which are usually used as examples when explaining self-organization, are fish schooling, the honeycomb, animal's skin patterns, etc. (Murray, 1989). Alan Turing, who we will talk a lot about in the next chapters, was the pioneer in this field as well (Turing, 1952).

The first person to use the word "self-organization" in print that we know about was W. Ross Ashby in 1947 (Ashby, 1947, 1956). The notion was absorbed by general systems theory in the 1960s, when von Foerster (1960) applied this principle to a wide range of topics from biology, cybernetics, learning, and automata theory at the Biological

Computer Laboratory of the University of Illinois in Urbana-Champaign. Soon thereafter, Ilya Prigogine demonstrated that far-from-equilibrium spatial structures in open chemical systems could preserve themselves by exporting—dissipating—entropy. Prigogine called such systems "dissipative structures" (Prigogine, 1967; Nicolis and Prigogine, 1977; Prigogine, 1980), and considered their (self-) organization as triggered by "fluctuations," which force the system into far-from-equilibrium states. The system can remain in such a state by exchanging energy with the outer world. As we have seen above, sudden fluctuation can be also the result of chaotic behavior, positive feedbacks, delays, or combinations of all these reasons.

Herman Haken focused on the other characteristic of self-organization, namely, on simultaneous—*synergetic*—behavior of many system elements (Haken, 1983, 1993). He came to "synergetics" in the beginning of 1960, while working on laser theory. Haken had this idea: the interaction between many independent laser atoms and the photoelectric field they create leads to a coordination of atoms to monochromatic laser light. The laser atoms thus become "enslaved" by their own light field. His focus on cooperation between individual particles of a system instead of the dissipation of energy opened doors for nonphysical applications of the notion. Haken and followers successfully applied the synergetic approach to cognitive, social, and urban science (Haken and Portugali, 1995; 1996; Weidlich, 2000).

System theory seeks to discover the rules that make self-organization and emergence possible, and studies the forms it can take. One of the recent advances is in the direction of considering *self-organized criticality* in open systems, revealed by Per Bak and colleagues (Bak *et al.*, 1988). They have studied the behavior of dynamical systems, simulating the development of a "sandpile" when sand is poured onto a plane, continuously, from a point source. One can easily do that and then see how, after an initial period of pile growth, miniature avalanches of its slopes begin. The size of the avalanches depend on the slope of the pile. When it is low, avalanches are small, but when the slope is steep, by getting the sand wet, for example, avalanches become large and span the entire slope.

In a normal case, with sand added continuously, the pile grows until it reaches a *critical* slope. Then the pile maintains itself at this slope, and *avalanches of all sizes* occur, resulting in a power distribution of avalanche size. Bak *et al.* (1988) claim that the pile form self-organizes to a critical state; they refer to this process as self-organized criticality, which, they say, might be the underlying concept. They also propose that power law distributions of some phenomena indicate it as related to self-organizing critical systems.

Self-organized criticality, as well as the period-doubling series of bifurcations we discussed previously, is based on the present-day approach to complex systems, with focus on their *universal* properties. The theory of universality is beyond the remit of this review, but it is worth noting that it is based on an approach like that which we employ in geosimulation—studies of the collective outcome of interaction between many units, each following simple and measurable rules of behavior.

To continue learning about system theory, the reader is advised to consult more than a few recently published and fascinating books (Yates, 1987; Kauffman, 1995; Bak, 1996; Jensen, 1998).

3.1.2.2.6   From General System Theory to Urban Dynamics. The examples we considered above do not use the generality of formulations (3.1) and (3.2) we began with. They are all based on systems of one, two, at most three equations, and employ the simplest forms of nonlinearity, expressed by the *products* of variables. We were interested in external and aggregate characteristics of the components—population numbers or compounds, densities—and ignored their internal structure, say, spatial distribution of chemical reactions, or age structure of lemmings and foxes in Lotka-Volterra models.

For the cases where internal structure of the component can be ignored, cybernetics utilizes the metaphor of the *black-box*. The only information we employ in the black-box model is the balance of the *input and output flows* between model components, which defines their changes—$F_i(X_1, X_2, \ldots, X_K)$—in the end. All the models we presented before employed analytically simplest, *quadratic*, expressions for this balance, and functions $F_i(X_1, X_2, \ldots, X_K)$ were always constructed in linear and quadratic terms $a_i X_i$ and $b_{ij} X_i X_j$.

The black-box approach to model formulation is referred to in economics as *compartment* or "*stocks and flows*" models and was very popular in the 1960s and 1970s. From there it drifted to geography. The discussion will now begin to tie all of these to urban theory, which, incidentally, accepted the stock and flow view easily.

3.2 The 1960s, Geography Meets System Theory

3.2.1 Location Theory: Studies of the Equilibrium City

Geographic theory followed the mainstream of natural science. Beginning in the nineteenth century, work in *location theory* was directed toward a search for equilibrium spatial distribution of settlements, land-use, facilities, population groups, etc. The tradition was set by Johann Heinvich von Thünen's book (von Thünen, 1826). In that work, von Thünen suggested that existing urban structures provide the most effective functioning of the system. For more than 100 years, until the 1950s, geographic interpretation of urban systems was oriented toward optimality and, generally, steady structures.

Several basic geographic "laws" were discovered in this way, all regarding steady and optimal settlement patterns and land-use, and all based on understanding optimality in an economic sense. In particular, three such laws were important in shaping understanding of urban systems: central place hierarchy (Weber, 1909; Christaller, 1933, 1966; Lösch, 1940, 1954); power distribution of settlement size (Allen, 1954; Clark, 1967), which still remains an active research topic (Gabaix, 1999; Blank and Solomon, 2000; Batty, 2001b; Ioannides and Overman, 2003); and the equilibrium, decaying with distance from the urban business center, distribution of urban land rent, as popularized by Alonso (1964).

The reader is referred to Haggett *et al.* (1977) for more comprehensive historical and scientific review of location theory during the nineteenth and first half of the twentieth centuries.

It is worth noting that the revolutionary (in its dynamic view on urban processes) theory of innovation diffusion proposed by Torsten Hägerstrand was also formulated in that time, but remained unclaimed (Hägerstrand, 1952, 1967).

Models of equilibrium and optimal urban patterns deal with urban land-use and land prices. Many urban components—population, jobs, services, and transport networks—remain beyond these models' framework.

Ira Lowry's "A Model of Metropolis," from the early 1960s, constitutes *the* famous attempt to relate the main compartments of the modern city quantitatively (Lowry, 1964).

3.2.2   Pittsburgh as an Equilibrium Metropolis

Lowry's model (1964) was the first attempt at designing an "integrated" model to first try to formalize the structure of an *urban system*, and second to represent that system structure in the model. It was implemented for the city of Pittsburgh, and its success was measured according to correspondence between the model and real observations of Pittsburgh in 1958. In the model, the urban structure was specified in a simple way, with three sectors of spatial activities:

- a basic sector, including industrial, business, and administrative activities, whose clients are mostly nonlocal;
- a retail sector, dealing with the local population;
- a householder sector.

Lowry's model is static, assumes that a city is in equilibrium state, and distributions of sectors' activities are considered at a resolution of mile-square tracts (Figure 3.9); this is strongly reminiscent of cellular automata models, which became popular a decade later.

To obtain steady distribution of activities X^* over city regions, Lowry followed a standard iterative procedure: begin with some initial vector X_0 and, based on the system's formal description, build a series of approximations $X_0, X_1, X_2, \ldots$ converging to equilibrium X^*. The model starts with the real-world distribution of the *basic sector* in each tract. The rest of the land in the modeled city is considered as available for residential and retail use. The labor force, which is necessary for employment in basic industrial activity, is then allocated; so too is retail activity, allocated in proportion to market potential, given by potential customer numbers. Finally, employees of the retail sector are located in space. This first loop results in imbalanced distribution X_0 and Lowry iteratively adjusts the components by further loops that redistribute population and retail industry (Figure 3.10). It is worth noting that this iterative process is, sometimes, wrongly accepted as a dynamic one; it is not.

Lowry's model bases its distribution of population and nonbasic industry on the *gravity model*, which was very popular at that time, and had been tested empirically in several studies (Hansen, 1959; Schneider, 1959). The gravity model is, actually, a concept; if the distance between two points equals d, then any interaction between the objects located at these points is inversely proportional to some power of d. For example, Lowry assumes that the attraction of shopping opportunities for the population, at a distance d from them, is proportional to $d^{-\alpha}$, as well as the number T_{ij} of trips between two regions i and j at a distance d.

The parameter α of a gravity model represents the *frictional effect of distance*. For different cities and situations, estimates of α vary between 0.5 and 2.5, and in the case of Pittsburgh, both the gravity and generalized model with

$$T_{ij} \sim 1/(a + bd + cd^2) \tag{3.36}$$

provided quite a good "fit" with observed data, regarding distribution of population, calculation of place potential, and, finally, distribution of services.

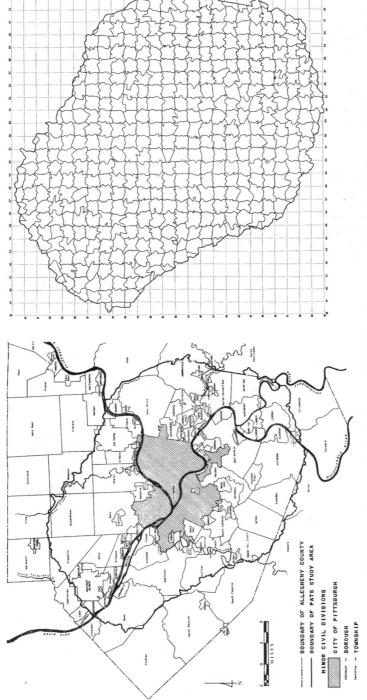

Figure 3.9 *Nonregular and close to CA division of space in a Lowry model (Lowry, 1964)*

BOUNDARY OF ALLEGHENY COUNTY
BOUNDARY OF PATS STUDY AREA
MINOR CIVIL DIVISIONS
CITY OF PITTSBURGH
BOROUGH
TOWNSHIP

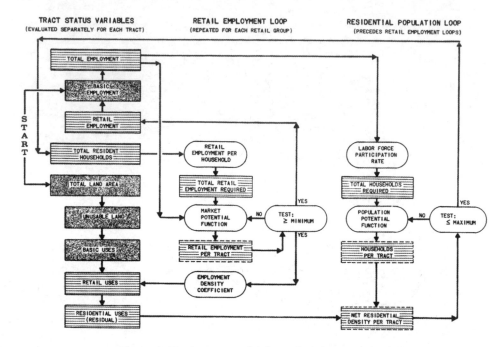

Figure 3.10   *Lowry model flow chart (Lowry, 1964)*

The results of the Pittsburgh modeling exercise were quite encouraging. The goodness-of-fit of the model was tested in two ways. First, by comparing the model and real housing and retail distributions in 1958 over four city regions, each subdivided into three zones, and, second, according to distribution of housing and retail uses depending on distance to the CBD. For most of the zones, the difference between the model and reality was at a level of 10% or so, while it reached a level of 20–30–40% for three of the zones. This correspondence afforded Lowry the right to lay down general points regarding the validity of the general assumptions (the gravity model is one such assumption), the relationship between modeling and planning, the importance of data supply, and many others, which retain much currency today, 40 years later. To mention one, "... there is no reason to anticipate eventual agreement on a single all-purpose model of metropolis.... If nothing else, differences in the relevant time scales, in the need for geographical detail, and in the available data-base must be reflected in the grand strategy of model building" (Lowry, 1964, p. 133). This almost preempts the geosimulation approach today!

The Lowry model contributed significantly to unification and implementation of the achievements of location theory in the 1960s. Its basic, static, version was further developed in many respects and applied to several cities and regions. Quite a few thorough and significant papers on the subject escaped popularity, such as models of Copenhagen (Pedersen, 1967) and St. Louis (Meyer *et al.*, 1975), the latter actually featuring raster mapping of the results, back in the mid–1970s. Books by Alan Wilson (1968) and Michael Batty (1972, 1976) review much of that genre of work, as does other material by Helly (1975) and Wilson (1979).

Now things get interesting; to animate the model, to make it come to life, we are going to throw time into the soup.

3.2.3   The Moment Before Dynamic Modeling

Lowry himself clearly understood that his model should become dynamic (Lowry, 1964, pp. 39–42). He points out that consecutive approximation of the equilibrium solution can be associated with time steps and an initial distribution X_0 can produce subsequent distributions at the next time steps X_1, X_2, etc., in a way that the scheme in Figure 3.10 demonstrates.

Urban science stood in the way of dynamic modeling, and from a present-day point of view, one can say that the old geographic "down to earth" tradition favored explicit, and, thus, simple models, focused on proximity and local interactions and aimed at simulating real urban systems. Revolutionary changes in other scientific fields and the establishment of the system paradigm did not allow for delay in the advancement of this course of work.

The beginnings were awkward; the first book with explicit introduction of urban dynamics was written by Jay Forrester, who was not, incidentally, a geographer. Forrester's interpretation of the city was as a genuine example of black-box systems. By developing a black-box model of the world economy in 1961 (Forrester, 1961), he was one of the founders of "world modeling" (Meadows *et al.*, 1972; The Executive Committee of the Club of Rome, 1973; Bruckmann, 2001). In 1969, Forrester did something similar in the context of urban systems; he developed an economically-oriented model of urban system dynamics (Forrester, 1969). We will discuss it shortly.

Not everybody followed the black-box paradigm. There was one notable outlier in the 1950s, although it enjoyed popularity only toward the end of the 1960s: the theory and dynamics model of innovation diffusion by Torsten Hägerstrand (Hägerstrand, 1952; 1967), who can be considered as a founder of geosimulation. Let us infuse a short reminder of that theory before continuing back along the mainstream.

3.2.4   The Models of Innovation Diffusion—The Forerunner of Geosimulation

The model of innovation diffusion is now quite famous, but something important lies buried in that early work and is often missed; it is this: Hägerstrand's model invokes geosimulation, and does this back in the 1950s! The simulation models he describes in his now famous paper (Hägerstrand, 1952) and in a book published in Swedish in 1953 and translated into English in 1967 (Hägerstrand, 1967) are dynamic, of high-resolution, and built in the fashion of general system theory, i.e., with consideration of phenomena as a result of collective processes and subject to what we now call self-organization. The results are carefully compared to reality at the highest resolution possible at that time (often at resolution of individuals).

These principles are formulated explicitly throughout the book. For example, "... the essential thing is not to consider the situation at t_{1700}, t_{1800}, t_{1900}, etc. We must instead consider the situation between the time t and $t + \Delta t$, where Δt can be a decade, a year, or even a shorter time, depending on the source material, and in doing so direct our attention to the way one situation gives birth to another. The accuracy of the prognosis will be the final criteria of whether or not the method of approach works well" (Hägerstrand, 1967, p. 2); "The locational relationships here under investigation are essentially horizontal man $\leftrightarrow$ man relations..." (Hägerstrand, 1967, p. 6); "The digging up of as many 'causal factors' as possible is not an economical allocation of research resources. On the contrary, it is desirable to isolate a few critical factors, which go a long way toward substantially explaining the phenomenon in question. Only after this has been done, may the rough edges of the analysis be smoothed off by the consideration of the other factors" (Hägerstrand, 1967, p. 133), and so forth.

Even technically, the models are implemented on the basis of two-dimensional grids of cells, which is very close, if not identical, to what we now call cellular automata and multi-agent systems. Hägerstrand's model was simply ahead of its time. The mainstream of urban modeling in the 1970s did not include it, and it took a decade at least to be appreciated popularly.

Hägerstrand proposes three levels of organization of socially meaningful information. He claims that spread of an "innovation" into an area populated by individuals who can accept or defeat it depends on *public information*, which comes with media or from the population as a whole—global information in modern terms; and *private information*, coming from particular members of the population—neighbors as we call them now. "No doubt, any given distribution of information is always the product of an involved mixture of public and private information," he defines (Hägerstrand, 1967, p. 139). Later in the book, "neighborhood" or "proximity" effect is added (p. 163).

The innovations considered include the rotation of crops with pasturing and the improvement of natural grazing areas, which were documented as having begun during the first decade of the twentieth century, and some others. Models are implemented on a 9 × 9 cell grid with 30 individuals in each cell. Hägerstrand's Model I is, actually, the background scenario, where individuals accept innovation at a random place and in a random order. And the conclusion is that ". . . the series of model maps created according to the rules of Model I could not repeat an observed process." In his Model II, individuals directly inform each other with respect to innovation. Namely, it is assumed that in the beginning only one person is informed about the innovation; acceptance or defeat occurs immediately and the information is forwarded further. The model is run in one-dimensional and two-dimensional versions, and demonstrates better but still not satisfactory correspondence with empirical data. Finally, Model III assumes that there is delay in acceptance and further transfer of the information. To implement that, resistance is introduced; an individual should receive information about innovation several times, and, even if accepted, the information is not forwarded further immediately, but only after some constant time interval. Without meandering to further detail, it is worth metioning that Model III does demonstrate correspondence with experimental data.

After reading "Innovation Diffusion as a Spatial Process" (Hägerstrand, 1967); the interpretation that follows is often of the following description; were this material given a popular platform in the 1950s, this body of work may well have placed geography in a leading position within system science! But, upon further inspection it might be understood that is no simple undertaking; translation of the book in 1967 attracted quite a volume of attention, certainly. But, it did not herald a revolution. Both in the 1950s and in the 1960s—even in the 1970s—a view of geographic systems in a context of collectives of individual elements was far beyond the horizon of the general state of the scientific mindset in those periods. In 1967, geography easily accepted the dynamic part of the Hägerstrand's work; it ignored a not less important part: the dynamics of innovation as *collective dynamics*.

Back to the 1960s; the ideas of system science took center stage; finally. From then on, urban modeling was synonymous with *modeling of urban dynamics*. Its pre-geosimulation stage is usually called *regional* or *"stocks and flows"* modeling, and it flourished during the 1970s.

3.3 Stocks and Flows Urban Modeling

3.3.1 Forrester's Model of Urban Dynamics

Jay Forrester (1969) was the first to popularly claim that city dynamics could not be understood within the static framework. His book *Urban Dynamics*, published in 1969, provoked a volume of reaction, mostly negative. At least partly, that was the result of one fact—the book's list of references totaled six entries, one to Kurt Lewin's *Field Theory in Social Science* (1951) and the other five to Forrester's own works on "industrial dynamics"!

Regardless of this, Forrester's book ushered in a new era of urban modeling; it summed up the basics of system science of that time regarding urban systems, and introduced the ideas to the attention of geographers. In the book, two general system chapters summarize the views of the 1960s (and might well seem overenthusiastic to the present-day reader). Following the fashion of that time, Forrester sees positive feedbacks as the main source of complex and counterintuitive behavior of natural systems in general and urban flavors in particular. He proposes that one positive feedback *dominates the system for a time* and then this dominance is shifted to another feedback, responsible for the other part of the system. The behavior of the system changes, thus, so much with such a shift that the two regimes seem unrelated. At the same time, while one of the feedbacks dominates, the system maintains resistance to the other, thus marginal, changes (Forrester, 1969).

Until recently, there was not sufficient evidence to confirm these views. Other logical chains, say change of persistent regimes through bifurcations (Portugali, 2000), can result in the same portrayal.

3.3.1.1 Computer Simulation as a Tool for Studying Complex Systems

Besides focusing on dynamics, the work represented in Forrester's book manifested another important paradigmatic shift—toward the computer (and computing) as the *main* tool for investigating the behavior of complex urban systems. Forrester was not the first to hold these views; Lowry (1964), for example, understood that not less clearly and mentions it several times in his work. Forrester worshiped computers in his first book (1961); in *Urban Dynamics* he claimed, again, that by changing the guiding policies within the system, one could see how the behavior of the actual system might be modified. From then on, computer simulations became the mainstream of urban modeling. This was synonymous with a general upsurge in the use of computing and computer simulation as a research tool at the time, something we take for granted today.

3.3.1.2 Forrester's Results and the Critique They Attracted

As we mentioned above, *Urban Dynamics* is one more application of the logic implemented in *Industrial Dynamics* (Forrester, 1961). Models of this kind were very popular during the 1960s (Meadows *et al.*, 1972), and attracted a great deal of public attention. *Urban Dynamics* attracted followers, detractors, and severe criticism. The critics attacked Forrester's lack of knowledge of urban theory and his complete reliance on former Boston mayor John F. Collins; the supporters were excited by his system dynamics approach.

Forrester himself had no doubts regarding the validity of the model and was extremely definitive in his conclusions, all of which supported conservative urban policies. The obvious one-sided skew of his views prompted much suspicion of the model itself and even modeling as a whole, and Garn and Wilson (1972) point out that the overly ambitious style of Forrester will prompt a natural idiosyncrasy among decision-makers, who might abstain, thus, from using or relating to the models at all.

Forrester's model of industrial dynamics is based on nonspatial stocks and flows, and he applies the same approach when modeling urban dynamics. To define stocks, Forrester divides housing, jobs, and the population into three classes. Each class is characterized by a different level of income, and stocks of the same level interact directly, i.e., low-income families are attracted by the availability of low-income housing and vice-versa. Other components are not related directly—i.e., low-income families are not interested in the availability of upper-income housing.

A basis on delays is important for Forrester's urban dynamics. As we discussed above, the consequences of delays are qualitative—during the delay time, the system can continue to develop in a "previous direction," and this might destabilize the system in a way that is similar to action of a positive feedback. As in many other aspects, the delays Forrester uses may be the point of the critique his work prompted; for example, the delay between change in "urban image" and the reaction of new immigrants to this change is set equal to 20 years, and can automatically imply decaying oscillations of the longer period before stabilization, just as represented in Forrester's simulations.

The city considered in Forrester's book is an abstract creature and subsequently most of the model parameters are justified only "intuitively." According to Forrester's passion, a proper model structure was more important. The parameters chosen entail convergence to a stable equilibrium, while delays made convergence nonmonotonic, with definite period of overdevelopment of a city (Figure 3.11). Changes in parameters result in changes of the steady state only. The conclusions were intuitively justified and straightforward; Forrester's modeled city is a simple system, therefore!

From a contemporary perspective, we can perhaps say that the form of *presentation* of Forrester's results engendered the large part of the backlash against the idea, and the critique had less to do with the actual results he proffered. On the surface, his favorite conclusion was that the building of lower-income housing for low-income families would only sharpen urban problems by attracting even more low-income families, further exploiting public resources without improvement. This idea directly contradicted the Great Society programs of the 1960s and the tendency of that time to invest resources in urban renewal. Forrester claimed that (according to the model) such programs only exacerbate the problem.

There were more contextual criticisms of the actual mechanics of the model also. For example, Garn and Wilson criticized Forrester's model both from general and specific standpoints (Garn and Wilson, 1972). In general, they point out that Forrester artificially separates the central city from its suburbs. When they changed some model parameters and made their own experiments with Forrester models, they found less conservative

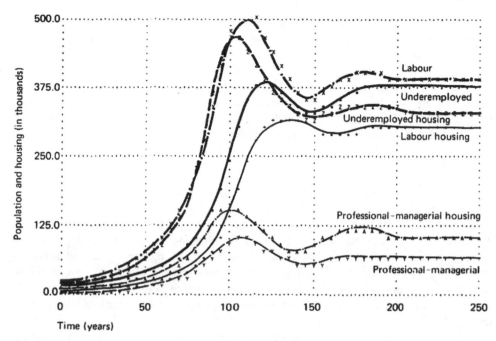

Figure 3.11 *Dynamics of population and housing in Forrester's model of urban dymanics (Forrester, 1969)*

scenarios of improving the city, and demonstrated much less urban stagnation than that suggested by Forrester. Their paper is one of the first that concludes that correct values of parameters are of the same importance as the feedbacks included. As we will see in time, the problem of parameters became one of the deepest problems of comprehensive modeling and we consider it as one of the basic reasons of its decline in two decades.

Complete ignorance of geographic theory caused a storm of criticism, which, as it often happens, threw the baby from the bath together with the water. From a present-day vantage point, the problem with the Forrester approach was that it is "too general" a view of urban systems. Explicit incorporation of geographic knowledge into the models was soon developed, but the negative attitude remained and could, perhaps, be one of the reasons behind the general retreat of geographic science from quantitative methods and modeling during the 1970s and 1980s. Many interesting stocks and flows urban models of the 1970s and 1980s remain largely unacknowledged until now. Let us review some of them.

3.3.2   Regional Models: The Mainstream of the 1960s and 1970s

One of the lessons of system theory is to incorporate as few variables as possible into a model, to reduce the volume of possibilities, and facilitate analytical investigation of the model. Applying this view to urban dynamics, we can try to understand the latter at conceptual and aggregate levels, just as we described predator-prey dynamics by a simple system of two differential equations. The other extreme approach is to ignore worries of complications and reflect the urban system in the model just as we see it, hoping that proper formulation of scenarios and computer simulations will bring us understanding of "real" urban dynamics. The reader can find comprehensive description of the state-of-the-art of the urban modeling of that period in a book by Batty (1976).

During the 1960s and 1970s both approaches were employed. The first when studying conceptual models, the second when modeling real-world cities. Let us present several typical examples, beginning at the standpoint represented by simple aggregate models and proceeding to the mainstream of regional models.

3.3.2.1   Aggregated Models of Urban Phenomena

Conceptual urban models follow the tradition of mathematical ecology and economy (Helly, 1969; Day, 1982; Dendrinos and Mullally, 1985) and are often formulated in ecological terms. Studies of *competition* between two social groups for space (Zhang, 1989) or between economic sectors (Dendrinos and Sonis, 1990; Zhang, 1993, 1994), explanation of emerging urban hierarchy (Rosser, 1994), and other interpretations are considered; see Nijkamp and Reggiani (1995) for an overview. Dynamics of one or two interacting populations are among the most elaborate models in mathematical ecology and economy, and have direct urban interpretations. Let us present a couple of examples of this kind.

The model developed by Dendrinos and Mulally (1982, 1985) assumes that with an increase in city population comes a decrease in the economic status of the city and they utilise predator-prey equations to express this formally.

$$\mathrm{d}X/\mathrm{d}t = r_X X(Y - Y_m) - aX^2$$
$$\mathrm{d}Y/\mathrm{d}t = r_Y Y(X_m - X)$$

(3.37)

where $X(t)$ denotes population size (predator), $Y(t)$ is per capita income (prey), X_m is the city's population capacity, and Y_m is per capita income.

The dynamics of solutions of Eq. (3.37) depend on the parameters r_X, r_Y, and a, which define the type of socioeconomic interactions in the city. Self-limitation of population growth, reflected by the term $-aX^2$, provides oscillating convergence to equilibrium (and not stable oscillations as in basic predator-prey models). Dendrinos and Mullally focus on the period of these oscillations and demonstrate that for 32 U.S. metropolitan areas this period varies between 20 and 150 years (Dendrinos and Mullally, 1982, 1985).

Work by Orishiomo offers a different interpretation for a similar system of equations: population is prey, while land price is a predator (Orishimo, 1987). In a recent paper by Capello and Faggian (2002), mean urban income is prey and mean urban rent is a predator, and the model parameters are estimated based on 95 time series of Italian cities from 1963 to 1996. Work by Batten demonstrates an approach by which a general two-component model (3.24) is employed for understanding evolution of urban systems as a whole (Batten, 1982).

The diffusion model is a natural candidate for representing growth and dispersion of urban population. O'Neill (1985) examined its capabilities for approximating processes of diffusion of black ghettos in Chicago in the period from 1968 to 1972. Assuming radial symmetry of the population distribution, and logistic growth in numbers, the model becomes a special case of Eq. (3.22):

$$\partial N(\rho, t)/\partial t = r(1 - N)N + D/\rho(\partial N/\partial \rho + \rho \partial^2 N/\partial \rho^2)$$

(3.38)

where $N = N(\rho, t)$ is population denisty, ρ is distance from the center of the area, r is net population growth rate, and D is a diffusion coefficient.

The agreement between the model and observations was very good. The equation that fits best to the experimental data corresponds to $r = 0.191$/year and $D = 0.128$ km^2/year:

$$\partial N(\rho, t)/\partial t = 0.191(1 - N)N + 0.128/\rho(\partial N/\partial \rho + \rho \partial^2 N/\partial \rho^2) \qquad (3.39)$$

and the comparison between model result and reality results in $R^2 = 0.79$, significant at $p < 0.001$.

Today, and for the most part, economists support the aggregate trend, and several journals repeatedly publish models devoted to different, while mostly economic, sides of urban reality. Some of them demonstrate reasonable fit to reality, just as the simple models above, thus demonstrating that complex models are not necessary for representing basic aspects of complex systems dynamics. Nonetheless, the tendency in the 1970s and 1980s was to reflect the city as a system, and not to be limited to some specific phenomena. Most of the models of that time were thus developed within the *integrated regional* framework that merges the models of Lowry (1964) and Forrester (1969).

3.3.2.2   Stocks and Flows Integrated Regional Models

Forrester's model is nonspatial. Lowry's model is coarsely spatial and is one of the first examples of models that focus on dynamics of urban systems as represented by a set of related aggregated spatial units. This view—regional modeling—came into being in the mid-1960s; it united black-box stock and flows and gravity models and became the mainstream.

Regional modeling is based on representation of geographical systems by means of "regions," described by several "stocks" (population, land, jobs, dwellings, services, etc.) and exchange among these stocks. A vector of stocks, which are actually socioeconomic and infrastructure indicators, characterizes each region. Flows between regions, say, transportation flows, are derived from the distributions of workers, employment, and the costs of traveling to work. Any aggregate models, including Forrester's one, can be considered as a degenerated case of a regional model with the city as a single region alone.

Depending on formulation of flows, the dynamics of the regional model can be very complex and can exhibit all possible regimes from stability to deterministic chaos. Dendrinos and Sonis (1990) have demonstrated this in a book that studied a model of a city divided into up to four regions, each characterized by up to two stocks. The model is formulated as a system of discrete equations, basically similar to the logistic equation we discussed before. As one might expect, they have demonstrated that, depending on parameters, the model demonstrates all possible regimes one can observe in the discrete logistic model, from convergence to steady state to pseudochaotic, fluctuations of the stocks in the regions, and even more complex regimes.

Overcoming the overgenerality of regional models relates to knowledge of parameters in the models: the narrower the intervals of their variation are, the lower the variety of potential alternatives in urban dynamics may be. This does not eclipse Forrester's claim regarding the critical importance of model structure as the representation of flows between the regions. Recalling results of work described at the beginning of this chapter, we immediately reach a second point—depending on the description of flows, the same values of parameters can result in different dynamic regimes. Geographers never really hung onto hopes that the 'ultimate' analytic expression of flows could be elaborated. Different expressions can be compared for the case of two, or a maximum of three, spatial components, but it cannot be done when regions are many in number. The description of flows in regional models, thus, should be necessarily unified in this case, and the gravity models we mentioned in the context of the Lowry model were actually elevated to a position as the standard in the 1970s.

Imitating Newton's gravitation law, flows between regions in the gravity model are proportional to the stocks concentrated in them and inversely proportional to the square, or another power of a distance, between them; for example, the flow T_{ij} of workers is

$$T_{ij} = kN_iB_j/d_{ij}^{\alpha} \tag{3.40}$$

where N_i denotes population numbers in regions i, B_j denotes a number of jobs in region j, parameter α generalizes power 2 of Newton's gravitation law, and k is a constant.

The gravity model determines specific analytical form of the functions $F_i(X_1, X_2, \ldots, X_K)$ in Eqs. (3.1)–(3.2): the distance d_{ij} between regions does not change in time, and the flows, thus, are proportional to the products of the variables characterizing regions.

Combining the simplest linear Eq. (3.3) and gravity description of flows, Eq. (3.40), we can specify a right-hand term of general model (3.1) and (3.2) as at most quadratic form:

$$F_i(X_1, X_2, \ldots, X_K) = \sum_{m=1}^{K} a_{im}X_m + \sum_{l,m=1}^{K} b_{ilm}X_lX_m \tag{3.41}$$

where coefficients b_{lm} specify interaction between region l and m, and, formally, are counterparts of coefficients k/d_{ij}^{α} in Eq. (3.40).

Weidlich (2000) considers quadratic models as sufficient for description of the urban system in *every* aspect and proposes a general mathematical approach to investigation of such systems by means of a "master equation." For obvious reasons, he bases this general claim on the examples of segregation in a city consisting of two or three regions.

Equations of type (3.41) extend the linear model and have applications in different fields; regarding urban modeling, they are sufficient to implement the idea of interacting regions. At the same time and despite seemingly simple analytics they introduce the problem of "overcomplexity" into the center of urban modeling. One should recall here that all the examples in the beginning of this chapter—predator-prey, Belousov-Zhabotinski, Lorenz models—were based on systems of equations of the same analytic form as Eq. (3.41), and none of them had more than three variables. A city consists of much more than three regions and, thus, using Eq. (3.41) we enter the world of oscillations, bifurcations, and chaos. Conceptually, it is wonderful—we become able to explain a wider spectrum of phenomena. Practically, it is terrible—in order to understand an urban system we apply the model, the special cases of which are very complex topics of thorough research and have generated numerous papers.

The problem, actually, is not related to the gravity model. The important result of general system theory is that, generally, any unified "sufficiently nonlinear" description of the relationships between the pairs of regions has the same properties—for some sets of parameters (a_{im} and b_{ilm} in the case of Eq. (3.41)) the behavior of the model solutions can be simple convergence to equilibrium; for others, the behavior is very complex.

Possible dynamic regimes can be sincerely evaluated in very restricted situations, when the overall number of parameters in the model is low (Dendrinos and Sonis, 1990).

Fears of overcomplexity were far over the horizon at the end of the 1960s and early 1970s, and many attempts at theoretical investigation of regional models and their employment in real-world situations were made at that time. Soon after the Lowry model became popular, it was extended and became dynamic. The expectation was that, based on gravity law, *spatial interaction models* and families of such models could offer a unified and practical view of cities. There was a hope that numeric values of parameters, once available, would allow the models to become operational. Dynamic models of spatial activities in the city were developed (Wilson, 1969; Putman, 1970), and these are summarized in Wilson (2000).

Among the pioneers, Peter Allen and colleagues tried to simulate urban dynamics at different spatial and time scales, including models of Brussels, economic development of North Holland, Belgian economics as a whole, etc. (Allen and Sanglier, 1979; Allen and Sanglier, 1981; Allen *et al.*, 1986). We present here, as an example, the model of an artificial city termed "Brussaville," the structure of which reflects the structure of Brussels (Engelen, 1988).

As fitting the logic of regional modeling, Brussaville is divided into 37 spatial zones. Each zone is characterized by a vector V_i, $i = 1, \ldots, 37$, which consists of "population" and socioeconomic subvectors. The population subvector, in turn, consists of two sets of components, representing properties of white-collar and blue-collar residents, and each set consists of five variables: the numbers of active residents X_i, nonactive residents N_i, population totals P_i, immigrants $M_{in,i}$, and emigrants $M_{out,i}$. The socioeconomic subvector accounts for the "quality" of each zone I_i (services, pollution, etc.), housing stock H_i, overall potential employment J_i, and actual employment of several types S_i^k—finance, heavy industry, etc.—represented by an additional index k.

The crucial factor for further analysis is the definition of the zones' interactions; for 37 regions, the potential number of interactions is $37^2 = 1369$. To keep the number of model terms within reasonable limits, the authors introduce the *attractiveness* R_i of zone i, and split residential movement into two stages. At a first stage, emigrants leave zones, while at a second stage they choose new residence according to zones' attractiveness. In this way the number of parameters becomes proportional to the number of regions. In order to account for the distances between the zones of origin and destination, an attractiveness of zone i for a migrant from zone j is given, not by the power function of the gravity model, but by the not less popular Alonso-type exponential function $R_i \cdot \exp(-\tau \cdot d_{ij})$, where d_{ij} is the distance between zones i and j.

The dynamics of stocks in the zones is described based on linear feedbacks. For instance, the dynamics of the active (X_i^k) and nonactive (N_i^k) residents of type k (say, white- or blue-collar employees) in zone i is based on the following description:

$$dX_i^k/dt = \varepsilon^k N_i^k (1 - X_i^k/R_i^k) + zMG_i$$
$$dN_i^k/dt = \lambda_i^k(X_i^k + N_i^k) - \varepsilon^k N_i^k(1 - X_i^k/R_i^k) + yMG_i \tag{3.42}$$

where ε^k is a net rate of employment for nonactive residents of type k; λ_i^k is a net growth rate of k-th population group in zone i; R_i^k is a potential for the activities of type k at i; MG_i represents a balance between in- and out-migration at i, and z and y are parameters.

The nonlinear model (3.42) can exhibit very different dynamic regimes. The only way to investigate it is by Forrester-like computer simulation of different scenarios. In one of them, for instance, the canal that crosses the city is replaced by a line of hills, and following this dramatic change, the industry concentrated along the canal begins to spread out over the outer perimeter of the city, where the crowding is less and the land is cheaper. Over the course of time, white-collar residents move in an opposite direction, from the city's outskirts to a new hill area, where they have good access to the CBD and are far from the nuisance of industry (Engelen, 1988).

A number of attempts were undertaken, trying to combine all the main components of the city in the frame of one model (Anselin and Madden, 1990; Putman, 1990; Bertuglia *et al.*, 1994). Bertuglia *et al.* (1994) present the most general formulation, which they call the integrated urban model. The following components are considered as sufficient for the description of city dynamics: housing, jobs, services, land, and transportation. The state of the components is described by numbers of population groups, housing stock, industry, and employment according to branches. Spatially, the city is divided into N zones and the flows of population between zones account for the costs of trips. The high dimension of the model is the evident problem of the above approach.

Quite a number of examples refute the general skepticism we presented above. As an example, let us consider the Amsterdam housing model (Wissen and van Rima, 1986), one of the most elaborate and accomplished regional models we are aware of. Despite the very high dimension of the model, the focus on population dynamics makes description of interactions clear and estimates of parameters reliable enough. Amsterdam is divided into twenty zones in the model; in each zone, 11 dwelling types and 24 types of households of four different sizes are distinguished. The intensity of migration and the residential choice of each family is assumed to depend on the age of the family head (according to five-year age categories) and on the number of family members (seven groups), and the corresponding experimental data are used to estimate dependencies.

As in the Brussaville model, the attractiveness of the dwelling does not depend on the individual's current occupation, and thus, the number of parameters linearly depends on the number of model variables. Nonetheless, the Amsterdam housing model includes about 1500 parameters of demographic, residential, and employment processes, but exceptional quality of Amsterdam data over the period covering 1971 to 1984 made it possible to obtain reliable estimates of parameters for each zone and provided very good fit of the simulation results and Amsterdam's population and household dynamics. For 13 of 20 zones, the R^2 statistics of correspondence between actual data and model results were higher than 0.9; for the remaining zones, excluding one, it was not less than 0.5.

Based on this correspondence, two scenarios of Amsterdam population and household dynamics for 1985 to 2000 were compared. The first reflects the plan of the central government to build new dwellings, while the second represents the tendency of the local government to decrease construction quotas in the expanding suburbs. The short period of prognosis implies close outcomes of both scenarios, which diverge at the level of 10% or less regarding most of the indicators of interest, including total population as well as population and household numbers for each zone. Today, at the beginning of the 2000s and far ahead of the original forecasts, we are not aware of published comparisons between this prognosis and reality for Amsterdam housing.

The Amsterdam residential model is relatively recent and is focused on one of the components of the urban system. Many other regional models were developed, starting from the 1970s, and the flow continues now. The various flavors of model vary from more systemic, Forrester-type, views to pure simulation approaches that include more and more components and parameters. Reviews and applications of these efforts can be found in a variety of resources (Wegener, 1994; Batty, 2001a; Guo *et al.*, 2001; Eradus *et al.*, 2002; Saysel *et al.*, 2002; Noth *et al.*, 2003; Shaw and Xin, 2003).

The successive models we quote were developed a decade and even later after stocks and flows urban modeling came into being. The goal in developing these models was modest: to maintain the balance between understanding of the process, complexity of description, and availability of data. In the beginning, the early 1970s, the few existing models were relatively pretentious; severe critique had come very soon thereafter, and, despite being overemotional, that critique has remained valid, in many cases, until the present time.

3.4   Criticisms of Comprehensive Modeling

The strongest, or at least most popular, attack on comprehensive modeling came from Douglas B. Lee (1973) and the other papers that composed the now famous issue 3 of volume 39 of the *Journal of American Association of Planning*. In his "Requiem for Large-Scale Models" Lee lays siege to the various streams of system scientists that stood ready to relegate urban geographers to a role of mere data supplier. He challenged all three heavy-weights at that time—the Forrester model, PLUM, and NBER (see Batty (1976) for more information about these models) and offered a series of criteria that a 'good' urban model should satisfy. Comparing outcomes of existing models to the two-fold goal of urban modeling—real world planning support and development of theory—Lee listed "Seven Sins of Large-Scale Models," which he regarded as standing in the way of the modeling toolkit of that time in meeting those goals. One can say that Lee foresaw the pitfalls of the system theory we talked about earlier in this chapter, even those still unpopular or uncovered at the beginning of the 1970s.

3.4.1   List of Sins

Lee's list of sins begins with *hypercomprehensiveness*, the attempt to just explain too much in a model, and *wrong-headedness*, the use of too many constraints or relationships, mechanisms that even the model-builders might not perceive or distinguish. In hindsight, we now understand clearly that additional variables and relationships make system dynamics more and more dependent on the parameters of the model, and relatively small errors can lead to completely wrong understanding of the processes.

We enter a dead loop in this way, and travel through this loop can be exhausting! On the one hand, the model exhibits *hungriness* and requires tremendous amounts of data; on the other hand comes *complicatedness*, inability of the modelers to adequately understand their own black–box creatures. More and more precise values of parameters are needed to understand what is going on in the model at all. *Grossness*, i.e., reliance on aggregate input, exaggerates the above problems even more, just because the precision of parameters related to physically nonexisting aggregates is inherently limited. All this is accompanied by the epistemological sin *tuningness* that always undermines belief in the objectivity of the model construction—the tautological loop of tuning the model until outputs conform to "reasonable" expectations.

Looking back at Lee's paper, we can only fault two of Lee's sins from our contemporary viewpoint. These are *mechanicalness*—understandable resistance to new and seemingly narrow language of computation, and, more important, *expensiveness*, the high price of data and parameter estimates. The municipal and regional GIS that came after the 1970s did cost a fortune, but they are established now and can be used for modeling.

The particular criticism of Lee's paper is weaker than the general one. He does not criticize models; he criticizes their results, based on Stonebraker's simplification of the Forrester model (Stonebraker, 1972). As for the Forrester model, Lee admits its value as a learning tool, and by that, he foresaw the future development of the Forrester views (see http://sysdyn.clexchange.org/).

Whatever, the main prescriptions of Lee remain highly relevant today: keep it simple!

3.4.2   Keep it Simple!

Lee ends his paper with guidelines for model-building, which seem to remain valid forever:

- Let the problem needing to be solved drive the methodology, not vice-versa;
- Balance theory, objectivity, and intuition;
- Keep it formally simple. Complexity will come from inside the model; this might become an indestructible obstacle too soon.

The criticisms of the 1970s had an important influence on development in the field and shaped the future path of urban modeling. Now, thirty years after Lee's manifesto, we can say that its main ideas have been implemented. Urban modeling made the critical step from engineering of aggregate black-boxes to study of collective dynamics of ensembles of interacting urban objects, the behavior of which is explicitly formalized in model structure. This paradigmatic shift was characteristic of system science as a whole.

3.5   What Next? Geosimulation of Collective Dynamics!

Geographic problems with black-box models were never specific. For nonlinear systems, whenever the population of a box is thought of as *heterogeneous*, general system theory turns the red lights on.

3.5.1   Following Trends of General Systems Science

Students of a basic course of statistics are usually taught that the average of the product of two *dependent* variables X and Y, m_{XY}, is not equal to the product of the averages $m_X m_Y$, and there is nothing specific in the "product" function; the same is true for any other nonlinear function of X and Y. This down-to-earth statement expresses the essence of the problem with the aggregate black-box models. As far as we believe in the nonlinear nature of the systems we study, the model, whatever it will be, should have products or other nonlinear functions of the variables in the right-hand part of Eqs. (3.1) and (3.2). But this immediately entails that an outcome of a population model, for example, with a birth and death rate nonlinearly depending on family income, depends on the partition of the population into age classes. The results not necessarily change significantly; following the statistical analogy above, the average of the product m_{XY} can be close to $m_X m_Y$. But the appetite of the reader, keeping in mind phenomena of bifurcations and other nonsmooth reactions of the complex systems to parameter changes, is already lost. To cut this Gordian knot we have to compare aggregate and disaggregate models, but to do that the latter should exist.

In general terms, the statement above means that an outcome of the aggregate model, which refers to "average" representations, depends on partition of population into groups. Geography recognizes this to be a problem of ecological fallacy (Wrigley, *et al.*, 1996), but that fallacy persists in modeling, due largely to the absence of alternatives. Of course, the concept of average individuals wandering around in cities is counter to common knowledge. It might be a marketer's dream, but it's just not realistic.

System science recognized the importance of population heterogeneity in the middle of the 1970s, and Herman Haken's "Synergetics" (Haken, 1983, 1993), which we talked about in Section 1.2.2.5 of this chapter, could be considered as a reaction. Synergetics did not fall out of the ether; the idea is a natural development of the old physical tradition—if we want to account for system heterogeneity and hierarchy in a straightforward way, we usually have to use integrodifferential, instead of differential, equations, or the analogous discrete-time constructions. As we mentioned above, Haken came to Synergetics in formulating the action of a laser as the interaction between many independent laser atoms and "enslaving" of the laser atoms by their own light field. The interested reader can refer to books by Portugali (2000) and Weidlich (2000) for further details of the Synergetics views and its application to urban systems.

Synergetics models demand direct and likelihood interpretation of elementary real world objects; in this way it is possible to investigate the real importance of nonlinearity; self-organized criticality is an example. Groups amid such structures are defined from the bottom-up as assemblies of model objects, and not as aggregates defined in advance. We are interested in synergetic phenomena regarding urban systems, and, thus, continue with the topic in the next two chapters. Numerous models presented there fit the criteria of geosimulation and synergetics, all at once.

3.5.2 Revolution in Urban Data

Geographic science is a complex system by itself and in some ways scientists are the most important objects acting in the system. As an outer observer, students of such systems can thus see the development of urban dynamics as collective phenomena. One might consider the development of geographic information systems (GIS) and remote sensing (RS) technologies as one of the "hidden" factors that might explain the currency of this viewpoint at this very moment.

High-resolution GIS-RS databases became institutional during the 1990s and today detailed data sets are becoming available for use. In some cases, they already contain data spanning a decade and eventually will accumulate as historical data, enabling detailed calibration of fine-scale geosimulations. The strengthening of linkages between models and GIS–RS databases has already helped in the area of usability, particularly with the communication and interpretation of results, but the need for an interactive environment for directly manipulating models remains largely unrealized in operational contexts. In Chapter 2, we introduced the general geosimulation concept, which offers vast improvements in usability over "traditional" models. In later chapters, we will explore a series of geocomputation techniques that offer the potential for a more resourceful handling of detailed data.

3.5.3   From General System Theory to Geosimulation

Based on a background of synergetics for understanding general system dynamics and on high-resolution GIS–RS databases as input suppliers at "real-world resolution," geosimulation models can be applied to urban space explicitly. In this way, geosimulation can meet the demands of geographers, demographers, and planners reasonably well. In a sense, geosimulation models are also more modest and their formalism is usually based on specific applications akin to Geographic Automata Systems—cellular automata and multiagent systems—that are beginning to become established and are currently being researched and developed with much intensity and enthusiasm.

Cellular automata (CA) geographic applications are the successor of the multicomponent regional modeling approach to modeling city *infrastructure*. Definition of cell states as land-uses of the cell as a whole entails essential constraints on CA applications to urban dynamics. First, it implies limitation of cell size; cells should be sufficiently small to keep internal homogeneity. Second, it limits, from below, the duration of model iteration—land-use does not change in days or even months. In cases where the proper cell size and time intervals are chosen, estimation of CA parameters can be performed in a straightforward manner according to high-resolution land-use maps available in a course of a time.

The CA approach regards the physical infrastructure of the city. However, it does not account explicitly for the "soft" components of the city, namely its population. Individuals' decisions and behavior are considered explicitly in models of Multiagent Systems (MAS). The general aim of MAS modeling, in this context, is to investigate the consequences of individuals' social interactions in a city environment. Human individuals are represented in MAS models by free agents that carry the economic and cultural properties of human individuals and change their location in city environments.

Chapter 4

Modeling Urban Land-use with Cellular Automata

4.1 Introduction

Cellular Automata (CA) models were introduced to geography in two stages. Initial consideration of their use as geographic tools has origins in early experiments with computer-based mapping in the late-1950s, with raster conceptualization of space. The view of space as a lattice of identical cells was adopted widely and in the 1960s and 1970s several models of urban infrastructure dynamics were formulated on this basis. These models considered cities as spatially distributed systems, and units' dynamics were defined with reference to unit characteristics as well as global factors. However, the main attractive feature of CA was ignored—interactions between neighboring land units; although, the regional modeling tradition of the 1960s and 1970s popularly introduced the idea of flows of population, goods, jobs, and information between larger intraurban areas.

Merging the two ideas—raster and regional modeling—is formally quite straightforward: take the idea of flows and apply it to cells in a raster representation of the city instead of larger "regions." Indeed, there are only superficial differences between regional models developed by Allen and Sanglier (1979) (specified as a triangular lattice of 50 urban regions), for example, and "proper" CA; the difference lies in the number of spatial units of urban partition. However, progress of the CA idea and its methodology, through geography, was delayed for ten years at least, because of the treatment of the units that comprise such models. The CA paradigm necessitates a departure from ideas of "comprehensive modeling," which we discussed in Chapter 3, and accounts for as many processes and factors as are possible (Bertuglia *et al.*, 1994). The tradition in CA

Geosimulation: Automata-based Modeling of Urban Phenomena.   I. Benenson and P. Torrens
© 2004 John Wiley & Sons, Ltd   ISBN: 0-470-84349-7

modeling, by contrast, is in limiting exploration to a few key effects and the factors that define them.

It took more than a decade after the 1970s for geography to bid farewell to comprehensive modeling. This delay seems somewhat controversial today, in hindsight. On the one hand, it was taken for granted that cities could be considered as examples of complex systems, following the classic work of Ilya Prigogine and Herman Haken (Prigogine, 1967; Haken, 1983), and geographers accepted these views (Wilson, 1979). On the other hand, it was not until the late-1980s that geographers *en masse* began to introduce these ideas into the actual practice of urban modeling. Papers by Waldo Tobler, where he implicitly applied "almost" CA views to modeling Detroit development (Tobler, 1970) and then introduced the view of the city as cellular space explicitly (Tobler, 1979), remained largely unnoticed for almost a decade. (Although his declaration of a "First Law" of geography in his 1970 paper did not go unnoticed: "everything is related to everything else, but near things are more related than far things" (Tobler, 1970, p. 236).) The idea of integrated modeling ruled supreme during the 1970s (Wilson, 1979), and it took some time for geographers to grapple with ideas from system theory, and interpret them in the understanding, exploration, and modeling of *complex urban systems*. As we discussed in Chapter 3, general system theory forces researchers to reduce the number of processes represented in models and the number of factors investigated. The "illusion" of comprehensive modeling (Bertuglia *et al.*, 1994) was not always compatible with these principles.

Nonetheless, geographers did not tolerate early failures in comprehensive models (Lee, 1973) for very long. CA entered the geographic literature as an alternative to regional modeling, and from the very beginning focused attention, predominantly, on one component of urban systems—urban land, its development, and its use. Helen Couclelis (1985, 1988), Robert Itami (1988), and Michael Phipps (1989) followed Waldo Tobler's (1979) view of urban space as a coverage of very many relatively small land units, each with its own properties. They framed problems of urban land-use dynamics in terms of CA and fostered an interest among the geographic community, interest in representation of cities by means of high-resolution grids of cells, the use of each of which depends on the use of adjacent cells (again, Tobler's "First Law"). It is important to note that, from the very start, geographers envisaged CA much beyond their mathematical prototype; again, Tobler said it eloquently in his 1970 paper, "the purpose of computing is insight, not numbers" (Tobler, 1970, p. 235).

There are many possible factors underlying the late introduction of CA concepts and tools to mainstream geographic research. The retreat of quantitative geography in the 1980s is relevant, but for the most part the slow uptake seems to have been a by-product of delay in the introduction of CA into environmental science. Ecology, biology, and other environmental sciences, which never really questioned the quantitative approach, nonetheless ignored CA ideas (Lindenmayer, 1968) for many years, turning to CA modeling only in the mid-1980s (Ermentrout and Edelstei-Keshet, 1993).

Regardless of initial motivation, geographers woke up to the automata idea in the early-1980s; today it is among the main tools for modeling land-use change. CA models are applied to a wide spectrum of land-use questions, from purely theoretical to applied and explicit examples, and the systems they are applied to range from relatively small city areas to large regions and states. CA models genially fit to the geosimulation idea and evidently correspond to fixed geographic automata, as we explored in Chapter 2. Besides direct applications to land-use dynamics, CA approaches often feature prominently in high-resolution urban modeling, which makes use of ideas about the autonomy and decision-making abilities of elementary entities. These *multiagent systems* are considered in the next chapter.

The beginning of this chapter presents a short overview of the development of the CA approach to complex systems modeling; the focus then turns to the state-of-the-art in CA modeling of urban land-use and infrastructure dynamics.

4.2   Cellular Automata as a Framework for Modeling Complex Spatial Systems

Three distinct periods of CA development can be defined, before urban applications began to appear.

4.2.1   The Invention of CA

The invention and early development of the CA framework took place in the 1950s and 1960s and is generally associated with famous names and great discoveries of the twentieth century. Mathematics, and what we now call computer science, offered up Alan Turing's "computational machine" (1936) and John von Neumann's self-reproducing artificial structures (1951). At the same time, the pioneering work of Warren McCulloch and Walter Pitts (1943) on formal neurons, and Norbert Wiener and Arthur Rosenblueth's (1946) work on artificial neural networks established a background for viewing "large" spatial systems as excitable media—a net of very many *active and interacting* discrete units. Geographic applications are understood to have been built on the basis of von Neumann views of CA, while in practice geographers generally accept the CA approach in its broadest (Wiener's) sense, as a framework for spatially distributed systems that consist of discrete and locally interacting units. The theory of Markov processes, and especially Markov fields, provided a basis for understanding the role of stochasticity in the changes of those units. Modern geographic applications combine the geometry and simplicity of the von Neumann system with cybernetics views of Wiener and Rosenblueth, on the basis of the results of Markov processes studies. However, at the beginning of 1950, urban applications of CA were still three decades away.

4.2.1.1   Formal Definition of CA

The formal definition of cellular automata (originally "cellular space") offered in von Neumann's lecture of 1951 (von Neumann, 1951) is just the same as the definitions used today, and we introduced it in Chapter 1 of this book. To remind the reader, under von Neumann's scheme a CA is defined as a one- or two-dimensional grid of identical *automata* cells. Each automata cell processes information, and proceeds in its actions based on data received from its environment and following rules that it stores or holds internally. As we defined in Chapter 1, each automaton A is defined by a set of *states* $S = \{S_1, S_2, S_3, \ldots, S_N\}$ and a set of *transition rules* T:

$$A \sim (S, T) \qquad (4.1)$$

Transition rules define an automaton's state, S_{t+1}, at time step $t + 1$ depending on its state, $S_t(S_t, S_{t+1} \in S)$, and *input*, I_t, at time step t:

$$T: (S_t, I_t) \rightarrow S_{t+1} \tag{4.2}$$

A grid of automata become cellular automata when the *set of inputs is defined by the states of neighboring cells*. The neighbors of a given automata are defined by the grid in which the automata are located; historically, neighborhood descriptions feature in the CA definition, i.e., for the automaton A belonging to a cellular automata lattice

$$A \sim (S, T, R) \tag{4.3}$$

where R denotes automata neighboring A, and defines the boundary for drawing input information I, which is necessary for the application of transition rules T.

In a one-dimensional CA, neighborhoods R typically consist of two cells, one on the left and one on the right of a target automaton; wider neighborhoods, including two or more cells on each side are also considered (Wuensche, 1998) (Figure 4.1).

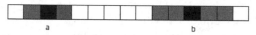

Figure 4.1 *Typical neighborhood configurations of 1D CA. (a) Neighborhood consists of two cells on the left and on the right of a given cell. (b) Neighborhood consists of two cells on each side of the given cell*

Two-dimensional CA are usually considered on a square grid, and the neighborhood consists typically of four or eight adjacent cells, which are often referred to as the von Neumann (1951) and Moore (1964) neighborhoods, respectively; wider neighborhoods are also often used, especially in applications to natural systems (White and Engelen, 1997; Yeh and Li, 2001) (Figure 4.2).

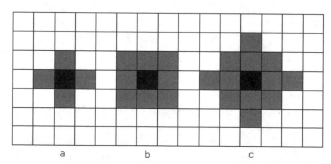

Figure 4.2 *Typical neighborhood configurations of 2D CA. (a) Von Neumann 3×3 neighborhood. (b) Moore 3×3 neighborhood. (c) Von Neumann 5×5 neighborhood*

Typically, solitary automata in CA have few states; in von Neumann studies of self-reproduction (see Section 2.1.5), the number of states reached 29. Even for few states and small-sized neighborhoods, the number of possible transitions is high. For the minimal case of one-dimensional (1D) CA, consisting of Boolean cells (the states of which are either false or true) and a neighborhood consisting of two neighbors, the number of possible transition rules equals $2^{2^3} = 256$. The power—$2^3 = 8$—in this formula is the number of combinations of states of the triple, consisting of the cell and its two neighbors. This formula can be easily generalized. If we denote the number of cell states as N and the number of cells in the neighborhood as K, then the number of transition rules equals N^{N^K}; the number grows enormously with increase in K or N (Wolfram, 1983).

4.2.1.2 Cellular Automata as a Model of the Computer

Cellular Automata in their classic sense were invented by Ulam and von Neumann in the mid-1940s. Von Neumann and Ulam were interested in exploring whether the self-reproducing features of biological systems could be reduced to purely mathematical formulations—whether the forces governing reproduction could be reduced to logical rules (Sipper, 1997). At that time, the two worked at Los Alamos National Laboratories on the atomic and, later, hydrogen bombs and Stanislaw Ulam, together with Edward Teller, signed the patent application for the latter. Mathematical folklore attributes the CA idea to Ulam, who was very well known for his exceptional mathematical imagination and avoidance of writing. Although, there is debate about the origins of the idea: "One can say that the 'cellular' comes from Ulam and the 'automata' comes from von Neumann" (Rucker, 1999, p. 69). By 1943 Ulam suggested the idea of cellular space, where each cell is an independent automaton, interacting with adjacent cells (and published the ideas much later in 1952 (Ulam, 1952)), and shared the idea with von Neumann. The common view, now, is that Ulam's idea was also a secondary one, and was based on a paper by Alan Turing (Turing, 1936), where he demonstrated that a simple automaton, later termed a "Turing machine," can simulate any discrete *recursive* function. Regardless of the origins, CA came into being amid a soup of very talented intellects.

Having been responsible for researching some of the most critical defense projects of World War II, Ulam and von Neumann did not care too much about publishing their theoretical thoughts (most of the papers by von Neumann on CA were completed and published after his death, in the 1960s (Taub, 1961; Burks, 1966)). The first paper by von Neumann, "The General and Logical Theory of Automata," introducing what are now known as cellular automata, was published in 1951 (von Neumann, 1951, 1961), and discussed the problem of designing a self-reproducing machine (see Section 2.1.5).

During the years of World War II, von Neumann put the idea of CA aside, and concentrated his activity on construction of the Electronic Numerical Integrator and Calculator—ENIAC, which was one of the first digital computers. The initial goal of ENIAC was rapid calculation for artillery tables, but that use faded when ENIAC came into being in 1946. Instead, the physicists and mathematicians of the Los Alamos team used it to solve equations describing explosion of hydrogen bombs.

4.2.1.3 Turing Machine

The theoretical interests of von Neumann lay far beyond his applied activities. Almost from the very start of the ENIAC project, he considered the computer as a "universal calculator"—a tool that is able to implement any formal algorithm. The idea of *computational universality* belongs to Turing, the other great mathematician of the twentieth century, who proposed the very possibility of a machine, that can implement any algorithm, in a purely theoretical paper published in 1936 (Turing, 1936). In this paper, he proposed what is called now the "Turing machine"—a device which has a "head" that can read and write symbols on "tape." The tape is divided into "squares" and is potentially infinite in both directions. The head observes one square at each moment and interprets symbols in a cell as instructions. Depending on the instruction in a square and the symbols (instructions) written in a limited number of previously observed squares, the head can write another symbol into a square it observes and then either halt its activities or move one square to the right or to the left along the tape.

It is evident that the Turing machine is an automaton and that it is based on recursion—the next action of the head depends on what was previously written on the tape. A 1936 paper by Turing (1936) offers explanations as to why the proposed machine is able to reproduce *every* recursive function.

Turing himself clearly understood the applied value of the machine he invented. During World War II, he applied his genius to deciphering coded Nazi communications. A similar device was subsequently applied to deciphering of Japanese communications in the Pacific Theater. Following his ideas, the group he worked with built the first digital computer—"Colossus"—about half a year before ENIAC, and this fact remained unknown to the public for a long time. The extreme secrecy and low budget of the British defense ministry could not compete with US government support for ENIAC and the popularity and public activity of the project leader, John von Neumann. Consequently, the ENIAC, and not the Colossus, influenced further developments of computers in the twentieth century (Britannica, 1982).

Turing and von Neumann were like-minded in their foresight of the future of computers. In 1946 they independently suggested the next principal step in logical design for digital computers—to store programs in the same way as data and to build the computer as a two-component scheme, consisting of memory of addresses, containing instructions and numbers, and a processor—Turing's "head" (Stern, 1980). The proposal was inspired by another discovery in the field of automata—neural networks.

4.2.1.4   Neuron Networks

Cells in the Turing machine are inert. They serve as frames for symbols, while only the head is responsible for interpreting symbols and performing consecutive actions. Living matter, however, consists of *active* cells. To represent that, Warren McCulloch and Walter Pitts suggested a formal model of the nervous cell in 1943—the neuron—and of a network of neurons exchanging signals (McCulloch and Pitts, 1943). The formal neuron reflects the basic features of the nervous cell: it is activated when signals input to it exceed an activation threshold. Just as Turing discovered that his machine can simulate any recursive function, McCulloch and Pitts demonstrated that nets comprising the simplified neural units could represent the logical functions comprised of AND, OR, NOT, and the quantifiers $\forall$ (*for all*) and $\exists$ (*exists*) and are thereby sufficient for expressing logical formulas of formal theories (McCulloch and Pitts, 1943).

The goal of McCulloch and Pitts was to formalize a representation of the neural system and, thus, neuron synapses are supposed to be connected in an arbitrary way and do not follow rectangular or other regular grid connections as CA usually do. From the beginning, their research was oriented toward networks and, thus, connections between neurons are considered as having their own property: *synaptic weight*. Neural network theory also addressed another important aspect of CA transition rules—*the order* in which these rules are applied, and in which solitary neuron automata change their states. Both Turing and von Neumann considered *sequential* updating of cells, when at each time tick, only one cell changes its state, and this change is immediately "known" to the other cells. A network of McCulloch and Pitts' neurons functions *synchronously*, that is, each neuron changes state at each tick of the time scale, and changes on the base of the *same* pattern of previous states; all neurons are then simultaneously exposed to the new states, i.e., the network exposes the new pattern.

4.2.1.5 *Self-reproducing Machines and Computational Universality*

Besides the very idea of a universal computation machine, Turing's work provided the basis for von Neumann's thoughts on the nature of live matter. Turing himself also addressed this issue and in 1946 proposed the now well-known criteria for distinguishing human and machine intelligence according to their reasoning (Turing, 1950) and these criteria[1] remain indomitable today. Turing thought of machines that could think. Von Neumann concentrated on the other basic property of living matter—its ability to reproduce itself—and tried to construct machines with these abilities.

Around 1945, and following Turing's ideas, von Neumann formulated the idea that a self-reproducing machine requires three parts: a universal constructor, a tape that keeps the program of construction, and a tape copier. The machine reproduces itself when the constructor builds a new universal constructor and tape copier, and the copier copies the tape, which of course has the instructions for the universal constructor on it. To work out the details, von Neumann took a suggestion from Ulam, who also knew about the Turing machine, and designed a self-reproducing machine as a two-dimensional cellular automata (von Neumann, 1951), the cells of which are connected to its four orthogonal neighbors (Figure 4.2a), and, just as the cells of the Turing tape, are updated sequentially. Von Neumann was able to prove that CA of about 200 000 cells, each with 29 possible states, governed by quite complicated rules, meet all the requirements of self-reproduction.

Von Neumann based his model on Turing's ideas, while proving that the CA model he constructed could, in turn, simulate a Turing machine. In modern terms we can say that von Neumann's CA were both self-reproducing and *computationally universal*, that is able to reproduce any recursive function, just as the Turing machine can.

4.2.1.6 *Feedbacks in Neuron Networks and Excitable Media*

Studies of neuron networks progressed in parallel with the development of computer science we presented above. The view of neuron networks as *excitable spatial media*

---

[1] To determine if a computer program possesses intelligence, Turing proposed "the imitation game," which is understood today as played by a human (A), a machine (B), and human interrogator (C); the interrogator is allowed to put questions to A and B via a terminal. The object of the game, for the interrogator, is to determine which of the other two is human and which is machine. If the machine can "fool" the interrogator, it is regarded as being intelligent.

prompted investigation of their spatial dynamics; until then, the investigation of spatial dynamics was monopolized by use of diffusion equations and their generalizations. The diffusion equation assumes *continuous* space, and this is the novelty of a 1946 paper by Norbert Wiener and Arthur Rosenblueth (Wiener and Rosenblueth, 1946)—considering a discrete neuron network as a spatial system. Following ideas of the time, they assumed that parameters of the neuron network depend on its global state; that is, the system as a whole feeds back its elementary components—the neurons.

Very soon after, Wiener published his famous Cybernetics book, where he developed this view in depth (Wiener, 1948/1961). Cybernetics greatly influenced the science of that century, but appeared too heavily formal for scientists outside mathematics and physics. This might be one of the reasons that modern high-resolution models in ecology, economics, social science, traffic studies, and geography all fall back on a very general understanding of space and relationships between spatial units, and are usually framed—or named—as cellular automata, despite often being more reminiscent of Wiener's excitable media.

4.2.1.7 Markov Processes and Markov Fields

Initially, the laws describing von Neumann self-replicated CA were deterministic, as were the laws governing the neuron networks of McCulloch and Pitts and the excitable media formulated by Wiener and Rosenblueth. Immediately after those ideas were proposed, it became clear that each cellular automaton can also be considered in the context of a stochastic system, whereby state transition rules are based on *probabilities* that an automata cell will change its state from S_i at a moment t in time, to S_j at moment $t + 1$. Discrete *stochastic processes* may be characterized by this condition—the state of a variable describing the process at moment $t + 1$ is completely determined by its state at moment t; in other words, the "next" state is determined by the "previous" state alone. These stochastic processes were among the favorite objects in mathematical statistics in the twentieth century. The Russian mathematician Andrey Markov introduced them in a 1907 paper, published in the *St Petersburg Academy of Science Journal*, and these *Markov processes* were subsequently studied in much depth, long before the beginning of the CA epoch (Sheynin, 1988).

Formal definition of Markov processes is very close to that of CA. Just as in Eqs. (4.1) and (4.2), the classic Markov process is considered in discrete time and characterized by variables that can be in one of N states from $S = \{S_1, S_2, \ldots, S_N\}$. The set T of transition rules is substituted by a matrix of *transition probabilities* P, and this is reflective of the stochastic nature of the process:

$$
\boldsymbol{P} = \| \, p_{ij} \, \| = \left\|
\begin{matrix}
p_{1,1} & p_{1,2} & p_{1,3} & \cdots & p_{1,N} \\
p_{2,1} & p_{2,2} & p_{2,3} & \cdots & p_{2,N} \\
\cdots & \cdots & \cdots & \cdots & \cdots \\
p_{N-1,1} & & & & \\
p_{N,1} & p_{N,2} & p_{N,3} & \cdots & p_{N,N}
\end{matrix}
\right\|
\tag{4.4}
$$

where p_{ij} is the *conditional probability* that the state of a cell at moment $t + 1$ will be S_j, given it is S_i at moment t:

$$
\text{Prob}(S_i \rightarrow S_j) = p_{ij}
\tag{4.5}
$$

The Markov process as a whole is given by a set of states S and a transition matrix P. By definition, in order to always be "in one of the states," for each i, the condition $\Sigma_j p_{ij} = 1$ should hold.

Classic Markov processes assume that p_{ij} are constant, and, not surprisingly, their investigation follows the same line we presented for linear difference equations in Chapter 3. The basic result is also similar; with time, the population of Markov units "almost always" reaches *unique equilibrium distribution* of states. For degenerate situations, characterized by additional analytic relationships between values of p_{ij}, the limit distribution can be cyclic or even more complex. The theory of Markov processes provides tools for estimating equilibrium distribution and the process of divergence, just like the theory of stability of solutions of difference equations we discussed in Chapter 3.

According to the definitions (4.1) and (4.2), CA satisfy the basic tenets of the Markov process—the next state depends only on the previous state. The switch from deterministic to stochastic CA and interpretation of rules T as providing *probabilities* of transformations is also easy. What really distinguishes probabilistic CA from classic objects of Markov processes theory is the dependence of transition probabilities p_{ij} on the states of neighbors. In CA, these probabilities *are not* constant, but depend on the states of neighbors. That is, for the cell C in state S_i:

$$\mathrm{Prob}_C(S_i \to S_j) = p_{ij}(N(C)) \tag{4.6}$$

where $N(C)$ denotes C's neighbors.

Probability theory incorporates study of the processes defined in Eq. (4.6), which are called "Markov fields." The problems of Markov fields theory are quite close to those studied by the theory of CA: what the most probable patterns of cell states are, what the conditions of convergence to them (if ever) are, whether more complex dynamic regimes beyond quasistationary distribution exist in a system, etc. (Guttorp, 1995).

It is worth noting that the probabilities p_{ij} and their dependence on the state of neighbors can be determined experimentally, on the base of comparison between cell states at t and at $t + 1$. This approach is often employed quite enthusiastically in land-use studies based on remote sensing data (see Section 4.4 of this chapter).

4.2.1.8   Early Investigations of CA

Early pioneering work on CA, and associated ideas, suggested future paths of inquiry for CA studies; notions of self-reproduction and the universality of computation were foremost among these threads of research until the mid-1970s. (For example, see Burks (1970) for a collection of essays on important problems addressed by cellular automata during this period.) The original self-reproducing CA of von Neumann were simplified several times. Codd (1968) introduced an eight-state machine and Banks (1970) provided the simplest known self-reproducing machine with four states per cell. Banks (1970) has also described the simplest known computationally universal (but not self-reproducing) 2D CA, with three states per cell. Toward the mid-1970s, it was clear, following much of this work, that relatively simple CA constructs could support and produce spatiotemporal patterns as complex as that which a recursive function could produce!

4.2.2   CA and Complex Systems Theory

As we discussed in Chapter 3, the 1960s and 1970s played host to the rise of general system theory. Far-from-equilibrium and self-organizing systems (Prigogine, 1967; Haken, 1983) became hot topics in natural science. Systems of nonlinear differential equations have been applied to socioeconomic systems of all levels, from the world as a whole (Forrester, 1961; Meadows *et al.*, 1972) to regions and cities (Forrester, 1969; Allen and Sanglier, 1981).

Revival of interest in CA can be related to an apparent dearth of simple and sensible examples of complex self-organizing systems in the literature. Deficiencies of illustrative examples inspired interest in things like John Horton Conway Game of Life the CA model he developed in 1970 that demonstrated, with unusual clarity, many basic principles of complex systems. Together with Lorenz's chaotic system and discrete logistic equation (see Chapter 3), the Game of Life brought ideas about complexity to life, popularizing the notion far beyond the domain of physicists and mathematicians, and inspired common curiosity about the complex dynamics of spatial patterns.

4.2.2.1   *The Game of Life—A Complex System Governed by Simple Rules*

During the 1960s, public interest in CA hovered less over mathematic publications, and gradually decayed. Revival in interest came at the beginning of 1970, amid the popularity prompted by Martin Gardner's presentation of John Horton Conway's model of "Life" in the October 1970 issue of *Scientific American* (Gardner, 1970, 1971). Conway's initial motivation was to design a simple set of rules to study the microscopic spatial dynamics of population (Berlekamp *et al.*, 1982). Aware of the computational universality of CA and their ability to generate complex spatial structures, Conway looked for rules that were simple, but generated population dynamics that were not easily predicted or expected. After a great deal of experimentation, Conway settled upon a set of rules for a 2D "population" of cells in a CA model; these cells could be in one of two states—dead (0) or alive (1). According to the three rules of the Game of Life, a cell remains alive, dies, or is reincarnated, depending on the number of live neighbors within a 3x3 Moore neighborhood (Figure 4.3).

- Rule 1: Survival—a live cell with exactly two or three live neighbors stays alive.
- Rule 2: Birth—a dead cell with exactly three live neighbors becomes a live cell.
- Rule 3: Death owing to overcrowding or loneliness—in all other cases, a cell dies, or if already dead, will stay dead.

These rules should be applied to all cells simultaneously.

Surprisingly, the simple rules of the Game of Life support fantastic variation in patterns of growth in a simulation. Because the rules are so simple, the model can be reproduced quite easily, and an amateur can get a taste of the complexity of the model (Berlekamp *et al.*, 1982). There are countless Internet sites devoted to "Life" (Summers, 2000; Silver, 2003); we recommend www.math.com/students/wonders/life/life.html for general introduction, while a good stand-alone version of the model can be obtained Online at psoup.math.wisc.edu/Life32.html.

The Game of Life is a nice toy, but its significance goes far beyond mathematical or visual elegance. It is computationally universal in the sense that any Turing machine can

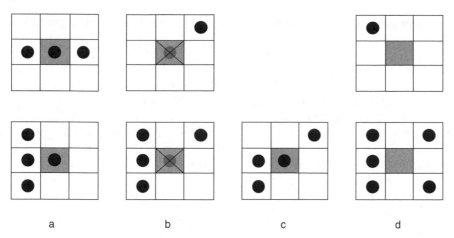

a   b   c   d

Figure 4.3   *Illustration of the Game of Life rules. An alive cell in the center of a 3 × 3 Moore neighborhood: (a) survives if it has two or three neighbors; (b) dies from overcrowding if it has four or more neighbors or from loneliness if it has only one neighbor. A dead cell in the center of a 3 × 3 Moore neighborhood (c) is born anew if it has exactly three neighbors; (d) remains dead otherwise*

be interpreted in its terms (Smith, 1976). Studies of the formal properties of the Game of Life abound (Hemmingsson, 1995; Ninagawa *et al.*, 1998; Evans, 2003), and have inspired connections between CA and complex spatial systems, the latter being previously dominated by models based on continuous—differential—equations. Its significance in broadcasting awareness of the abilities of CA within the scientific community is no less important; from the mid-1970s, it became almost unnecessary to re-introduce CA to the reader when applying CA to specific situations. Succinctly stated, the Game of Life introduced CA as an interdisciplinary tool for representing complex spatial systems and investigating their dynamics.

4.2.2.2   Patterns of CA Dynamics

System theory is concerned with persistent dynamic regimes and the ways in which a system converges to them. After the introduction of CA to complex systems studies, investigation into the limits of systems' spatial patterns began. Podkolzin in 1976 (see Culik *et al.*, 1990) did pioneering work, studying the spatial patterns for evolving CA and early work in this area was undertaken by Willson (1978, 1981). The topic was popularized by the work of Steven Wolfram (1983, 1984a, b). Wolfram's contributions mark what we might call the modern stage of CA studies, which focuses on analysis of the *space-time patterns* of CA evolution.

Study of the Game of Life had already demonstrated at that stage that very simple CA rules can yield complicated self-organizing dynamic patterns. And this is particularly important when considered in comparison with chaotic regimes of nonspatial systems, say Lorenz's system or dynamics of the solutions of discrete logistic equations (see Chapter 3).

The discovery of attributes of chaotic dynamics was the dominant question pursued in CA research at that time; Wolfram was the first to shape the research field, with extensive numerical analysis of the growth patterns of one-dimensional CA.

Wolfram's research regards the simplest of CA, which are defined in one-dimensional space and the cells of which can be in one of only two states. The cell neighborhood is specified as the cell itself and the cells immediately to the left and to the right of it (Figure 4.1a). As we mentioned above, for such simple CA, there exist 256 possible sets of local rules, and Wolfram studied limiting patterns for all of them (Wolfram, 1983, 1984a, b). (He later continued this investigation with further extensive analysis: see Wolfram, 2002.) Based on numeric experiments, he demonstrated that the limiting configuration of CA does not depend on initial states; it is defined by the transition rule. To describe the limiting patterns and convergence to those patterns, Wolfram defined several statistical parameters, such as:

- average density of cells in state 1;
- average density of sequences of n adjacent sites with the identical value of 0 or 1;
- average density of full 1-triangles of base length n, in a space-time 2D pattern.

Based on these measures, he proposed an empirical classification of rules, depending on the limiting pattern the rule generates. The classification identifies four classes, which correspond to qualitatively different modes of system dynamics (Table 4.1):

Table 4.1   *Wolfram's classification of 1D CA behavior*

Class	CA dynamics evolves towards	Type of system dynamics
I	Spatially stable pattern—each cell reaches the stable value of "0" or "1"	Limit points
II	Sequence of stable or periodic structures—each cell changes its states according to the fixed finite sequence of "0"-s and "1"-s	Limit cycles
III	Chaotic aperiodic behavior—the sequence of cell state is not periodic, but the spatial patterns repeat themselves in time	Chaotic (strange) attractors
IV	Complicated localized structures, which are sometimes long-lived and are more complex than those of class III	Attractors unspecified

Typical patterns produced by the rules of each class are presented in Figure 4.4; created using tools at http://course.cs.york.ac.uk/nsc/applets/CellularAutomata/index1d.html.

It is important to note that the rules leading to complex dynamics look very "regular." For example, consider these simple rules: the next cell state is the sum modulo 2 of the current states of two neighbors, or the sum modulo 2 of the current states of the neighbors and the cell itself. Despite their external simplicity these "totalistic" CA, in which the local rule depends on the sum of the states of the neighbors (and which belong to class III and IV), are computationally universal (Culik *et al.*, 1990).

Wuensche (1998) investigated two-state CA with neighborhoods of width K higher than 3 ($K > 3$) and demonstrated that four classes remain relevant. He also notes that in the

Figure 4.4 *Typical dynamics of CA for each of four Wolfram's classes: (a) Class I: Stable pattern, converging to all 0-s. (b) Class II: Cyclic pattern. (c) Class III: Chaotic aperiodic pattern. (d) Class IV: Complicated pattern, which is more complex than those of class 3 (created using tools at http://course.cs.york.ac.uk/nsc/applets/CellularAutomata/index1d.html)*

case of $K > 3$ many rules entail limiting patterns that contain both limit points and short limit cycles, though one or the other may predominate. He suggests, thus, that classes I and II may usefully be combined. Wuensche (1998) proposes characterizing CA by entropy and more complex "signatures" and, based on this characterization, concludes that CA of class IV can be considered as a transition form between the "ordered" CA of class I and II and the chaotic CA of class IV, an idea initially proposed by Langton (1992). Wolfram's classification could thus be readjusted as follows (Wuensche, 1998):

Ordered dynamics (classes 1 and 2) $\rightarrow$ Complex dynamics (class 4)
$\rightarrow$ Chaotic dynamics (class 3)

With growth of the number of states N and neighborhood size K, the higher fraction of rule-sets from the N^{N^K} rule-sets possible entail behavior characteristic of classes III and IV (Wolfram, 2002).

Wolfram's approach to CA classification has a problem: one cannot decide what the resulting pattern is, based on the formulation of the rule alone. In other words, in deciding which class (I to IV) CA with a given set of rules belongs to, it is necessary to investigate the dynamics of the CA with these specific rules. It was later proven that this disadvantage is inherent to the Wolfram classification scheme, and, thus, the class membership of a given rule is *undecidable* (Bandini *et al.*, 2001). Many other classifications of CA according to state and rule sets have been proposed (Culik *et al.*, 1990; Kayama *et al.*, 1993; Braga *et al.*, 1995; Cattaneo *et al.*, 1997; de Salesa *et al.*, 1997; Domain and Gutowitz, 1997; Cattaneo *et al.*, 1999; Oliveira *et al.*, 2001), while Wolfram's classification remains most popular.

Let us mention the classification introduced by Christopher Langton, which is particularly interesting for geographers. Under his scheme, CA, in which the set of states includes an "inactive" state, are considered; a cell in this state cannot change (Langton, 1986, 1992). Langton's classification does not depend on the number of states (N) or size of the neighborhood (K), but on the fraction of neighborhood configurations (denoted by λ), which *do not* lead a cell to become inactive in the next time tick. Langton's classification is also "undecidable;" to estimate λ, one should test all possible configurations of the cell's neighborhoods, which number N^K. It is clear that with increase in N and K, the number of configurations grows enormously, and Langton uses Monte Carlo approaches to investigate the "most probable" limit pattern of the CA.

Langton has demonstrated the relationship between λ and the limit pattern of CA. With growth in λ, the typical limit pattern of the CA passes from Wolfram class I to II, then to IV and, finally, to class III, i.e., Wolfram's classification is thus reordered. For λ close to zero, the limit pattern is always the same—eventually all cells are in an inactive state. With growth of λ, persistent stable or cyclical patterns appear when λ reaches 0.2 and prevail until λ grows to about 0.3. For λ over 0.3, limit patterns become complex and unpredictable behaviors appear. These complex regimes remain typical until λ reaches 0.5 and for higher λ, over about 0.5, chaotic limit patterns prevail. The boundary between stable and chaotic regimes, i.e. $\lambda \sim 0.5$, corresponds to the most complex behavior, which can be investigated online at alife.santafe.edu/cgi-bin/caweb/lambda.cgi.

Studies of 2D and higher-dimensional CA (Packard, 1985; Gerling, 1990b; a; Gora and Boyarsky, 1990; Magnier *et al.*, 1997; Gravner and Griffeath, 1998) show that it is possible to classify two-dimensional CA along the same lines as one-dimensional CA, and specifically, Wolfram's classification remains valid while 3D CA behave in a more complex fashion. Two- and one-dimensional CA do show marked differences in several respects (Golze, 1976; Gravner and Griffeath, 1998), while all of these seem of minor value for urban applications, where the four classes of the Wolfram classification are perhaps more relevant. In general, analysis of CA limiting patterns makes it clear that the Game of Life is not an esoteric example, and extended classification of 1D and 2D CA can be found in Wolfram (2002). Actually, many sets of (externally) very simple rules produce extremely complicated patterns; the implications regarding urban applications are evident.

Before going deeper into geographic applications of CA, let us briefly review results of CA studies where assumptions deviate from the classic ones. As we noted above, geographers often interpret CA in a context that is much wider than that in which CA were initially defined and, thus, these results might be important for geographic applications.

4.2.3 Variations of Classic CA

A number of CA characteristics are commonly varied in application: interpretation of cell states, input information and neighborhoods, for example. Just as with any abstract skeleton, basic CA are evidently insufficient for representing real-world systems, for which regular grid-based neighborhoods of identical size and shape, discrete and clearly distinguished states, independence of cells' time-scale, predefined order of change of cell states, etc., are usually overconstraining. To investigate urban reality with CA models, we have to distinguish between superficial effects, caused by the formal framework, and effects that do not depend on details of formalization and, thus, can be attributed to the modeled system. Significant effort has been invested into investigation of the consequences of "deviation from the standard" in definitions for CA dynamics.

4.2.3.1 *Variations in Grid Geometry and Neighborhood Relationships*

Neighborhood relationships between cells of von Neumann CA are defined via the adjacency of the cells in a uniform discrete grid, and, historically, this view governs most theoretical and applied CA models. However, uniform neighborhoods are not at all necessary. Neuron networks, for example, with nonuniform neighborhoods, seem more relevant for real-world geographic applications, and graph interpretation of CA, with nodes representing cells and edges connecting between two neighbors is sufficient for both regular and irregular neighborhoods. We pointed out these variations when introducing Geographic Automata Systems in Chapter 2.

In general, we can say that extensions of CA models toward a non-square grid and beyond von Neumann or Moore neighborhoods do not introduce significant effects. That has been demonstrated for CA over triangular and hexagonal grids (Gerling, 1990b; Eloranta, 1997), on "Cayley" graphs, in which each node is connected to the same number of neighbors (Machi and Mignosi, 1993, Roka, 1994, 1999) as well as on less regular graphs (Chua and Yang, 1988); see Schonfisch (1997) and O'Sullivan (2001) for reviews.

This is *not* the case, however, when the neighborhood structure is allowed to change in time. CA defined on a regular grid or irregular graph are nonetheless both "static" in the sense that the neighborhood relations between cells do not change with time. It is possible to consider CA where these relations do change. The most important class of CA possessing this property is called L-systems and was introduced by Lindenmayer in 1968 (Lindenmayer, 1968). L-systems formalize processes of tissue growth by introducing new nodes between existing ones. Models of this kind are applied mostly for modeling growing biological structures, especially plants, and successfully simulate complicated forms of leaves and tissues (Harary and Gupta, 1997; Stauffer and Sipper, 1998). We do not consider L-systems and other similar views (Silva and Martins, 2003) here in more detail; Ferdinando Semboloni's model (2000b) implements the idea of dividing units in an urban context, assuming that land percels—just as cell tissue—can be subdivided during growth.

4.2.3.2 *Synchronous and Asynchronous CA*

The order of update of cell states is of crucial importance for both the general theory of CA and for their applications, and we have introduced that explicitly as a property of Geographic Automata Systems in Chapter 2. As we noted, two polar formalizations of update schemes were introduced at the very beginning of CA studies; cells of von Neumann self-reproducing automata are updated sequentially, according to a predefined order, while neurons in neuron networks are all updated on a simultaneous basis. During the 1970s, the paradigm of parallel update took center stage; cell states are updated synchronously in the examples of the Game of Life and the numeric investigations that led to Wolfram's and Langton's classification schemes. The differences in dynamic behaviors of CA with synchronous and asynchronous update have come into focus relatively recently, after it was demonstrated that the most famous patterns of the Game of Life vanish if update is asynchronous (Ingerson and Buvel, 1984; Schonfisch and de Roos, 1999).

Asynchronous update presumes that cells are evaluated one after another and that the results are immediately available to the other cells. As opposed to unique processes of synchronous update, however, different asynchronous processes, depending on how the order the cells are retrieved for updating, can be considered. Two asynchronous update schemes are particularly popular in simulating real processes.

- At a tick in time, the cell to be updated is selected according to its characteristic or randomly.
- The update of each single cell is governed by the internal time of the cell itself. The probability that the cell will be updated at a given moment is a function of this 'waiting' time; exponential distribution of the waiting time is usually employed.

The dynamic outcomes of different update methods have been compared statistically for broad classes of CA. The general result is that the behavior of asynchronous CA is always *simpler* than that of synchronous CA. The effects can also depend on the order in which the cells are considered. For example, Ingerson and Buvel (1984) compared synchronous and asynchronous updating schemes for one-dimensional Boolean automata and demonstrated that, for rules that entail simple behavior of classes I or II, terminal patterns do not depend on the update scheme employed. By contrast, complex patterns of class III or IV usually degenerate under asynchronous update (also see Rajewsky and Schreckenberg, 1997; Schonfisch and de Roos, 1999, regarding the influence of updating schemes on Wolfram classification).

Intensive analysis of limiting patterns during the 1980s and 1990s marked a second period of established research with CA, and positioned CA as the standard model of complex spatially distributed systems. Toward the 1990s, CA theory reached a sufficient condition for the proposition of basic geographic questions of the nature: To what extent do the laws of local interaction between urban spatial units influence global city dynamics? As we noted above, during the 1980s, geographers adopted CA as a concept (Tobler, 1979; Couclelis, 1985; 1988; Itami, 1988; Phipps, 1989), but it took a further decade for the first well-known CA simulation of urban dynamics—a model of Cincinnati, developed by Roger White and Guy Engelen (White and Engelen, 1993)—to appear.

4.3   Urban Cellular Automata

4.3.1   Introduction

As we discussed in Chapter 3, in terms of simulation, geographers' attention in the period from the mid-1960s to the mid-1980s was almost exclusively focused on regional models.

Despite serious warnings and general critique, the hope that regional models would serve as a qualitative and even a quantitative panacea for understanding the dynamics of geographic systems had perpetuated for some time. Toward the end of the 1980s these hopes evaporated. Two main reasons, it would now seem, included an unjustified complexity in model design, even when minimal number of flows between the regions is accounted for, and the lack of the experimental data necessary to "anchor" the researcher in a sea of possible dynamic regimes. Toward the mid-1980s, a common understanding that something simpler should come instead had set in (Klosterman, 1994; Lee, 1994); CA offered such an opportunity.

As we have demonstrated in the previous sections of this chapter, CA matured as a research tool toward the end of the 1980s. First, the formal background of CA had been established by that stage and they became the standard device of theoretical computer science, physics, chemistry, and neural networks theory. Second, over the late-1980s, CA became widely accepted in ecology and their potency and convenience for interpreting common-sense understanding of local determinacy of many environmental phenomena was demonstrated (Hogeweg, 1988). It is important to note here that several important attempts to attract geographers' interest and incorporate the models, which were very close—ideologically—to CA, into geographic research had been made at earlier stages. Nonetheless, despite several influential publications in the 1970s, some of which are frequently referred to now (Tobler, 1970, 1979) and many that remain largely unreferenced (Latrop and Hamburg, 1965; Chapin and Weiss, 1968), the introduction of CA into mainstream geography had to wait until the end of the 1980s. Transition to urban CA research began with models that were based on a raster representation of urban space, but did not account for neighborhood relationships.

4.3.2   Raster but *not* Cellular Automata Models

Geography never really lagged behind advances in computing. Only a decade after ENIAC, toward the end of the 1950s, vector and raster computer maps were introduced (Creighton *et al.*, 1959; Tobler, 1959). Following that, raster computer maps (which were relatively "easy," computationally) and the idea of cell space as a basis for description of urban dynamics were almost immediately accepted. The coarse representation of urban data by means of multiple raster layers (we refer the reader to Steinitz and Rogers's book (1970)—Figure 4.5) is actually a prototype of the raster GIS that were to follow a few years later. Raster computer "maps of factors" provided the basis for modeling changes in urban cells as a function of these factors (Figure 4.6).

Raster models possess all the features of CA but one; they do, however, miss the most important feature. Just as in standard CA:

- the city is represented by means of cellular space;
- each cell is characterized by its state;
- models are dynamic, and the state of each cell is updated at each time-step.

However, a very significant feature of CA is ignored, namely, *the dependence of cell state on the states of neighboring cells*. Looking back, it is hard to understand why it took geographers a further 20 years to embrace this feature.

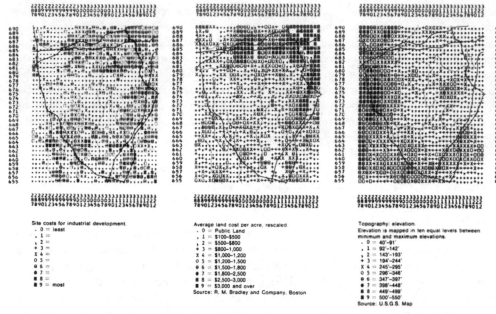

Site costs for industrial development.
. 0 = least
, 1 =
, 2 =
+ 3 =
X 4 =
0 5 =
● 6 =
● 7 =
■ 8 =
■ 9 = most

Average land cost per acre, rescaled.
. 0 = Public Land
, 1 = $100–$500
, 2 = $500–$800
+ 3 = $800–1,000
X 4 = $1,000–1,200
0 5 = $1,200–1,500
● 6 = $1,500–1,800
● 7 = $1,800–2,500
■ 8 = $2,500–3,000
■ 9 = $3,000 and over
Source: R. M. Bradley and Company, Boston

Topography: elevation.
Elevation is mapped in ten equal levels between
minimum and maximum elevations.
. 0 = 40'–91'
, 1 = 92'–142'
, 2 = 143'–193'
+ 3 = 194'–244'
X 4 = 245'–295'
0 5 = 296'–346'
● 6 = 347'–397'
● 7 = 398'–448'
■ 8 = 449'–499'
■ 9 = 500'–550'
Source: U.S.G.S. Map

Figure 4.5   *Raster layer presented on nongraphic display; source Steinitz and Rogers (1970)*

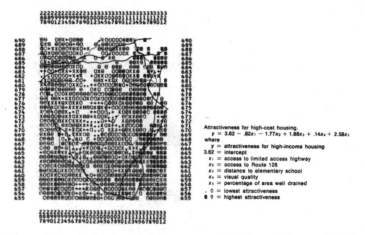

Attractiveness for high-cost housing.
$y = 3.62 - .82x_1 - 1.77x_2 + 1.88x_3 + .14x_4 + 2.58x_5$
where
y = attractiveness for high-income housing
3.62 = intercept
x_1 = access to limited access highway
x_2 = access to Route 128
x_3 = distance to elementary school
x_4 = visual quality
x_5 = percentage of area well drained
. 0 = lowest attractiveness
■ 9 = highest attractiveness

Figure 4.6   *Derived raster layer; source Steinitz and Rogers (1970)*

One of the best examples of raster modeling is the model built to simulate urban development in Greensboro, North Carolina (population 200 000) during the period 1948–1960, developed by Stuart Chapin and collaborators (Donnelly *et al.*, 1964; Chapin and Weiss, 1965, 1968).

The Greensboro simulation and two other raster models that we refer to in this section display a level of foresight ahead of the "constrained cellular automata" approach that was explicitly introduced 25 years later (White and Engelen, 1993, 1994). Namely, the overall changes in built area, population, land-uses, etc., over the study region, are all considered as *externally defined* constraints. These "limits of growth" are usually known for the area as a whole and for long periods, say ten years. The idea is to distribute these changes more or less uniformly over the interval of simulation, say, to set parameters in such a way that during a year 1/10 of the changes happen and then try to simulate *spatial allocation* of these changes.

The computing power at hand in the 1960s placed restrictions on the resolution of the maps, and to cover the area of the Greensboro, raster cells were specified as 300×300 m^2 (9 ha) in size. A nine hectare cell is too large for uniform land-use and to overcome the problem it was further considered as consisting of 3×3 virtual 100×100 m^2 "ninths," these ninths having unique land-use attributes. The model considers urban development in terms of the transformation of land from nonresidential into residential land-use; in terms of ninths, residential use can vary from zero to nine at a grand cell. Seven factors determining utility of a cell for residential use are considered. Three of them are taken as "major" (endogenous) factors and are controlled internally: (1) the land is marginal, and still not in urban use, (2) accessibility to work areas, (3) assessed value. Four "secondary" factors are considered as controlled by public officials (i.e., are exogenous): (1) travel distance to the main street, (2) distance to the nearest available elementary school, (3) residential amenity, (4) availability of wastewater facilities (Chapin and Weiss, 1962).

The model imitates allocation of residential development; this process is considered in two stages.

First, the *potential* of each cell for development (in numbers of ninths) is calculated as a linear function of seven utility factors. The Greensboro model clearly distinguishes between the potential of the cell and its actual *state*, and, thus, explicitly applies economic geographic theory of that time (Lowry, 1964). Cell potentials are then normalized and considered as *probabilities of allocating* new dwellings.

Second, externally defined demand for residential use per time-step is distributed, proportionally to potentials, among cells. The land-use of a cell can change if it has unused nonresidential ninths or if it is *adjacent* to a cell in which residential development is already initiated. One can relate this rule to both CA as well as to Hägerstrand's principle of innovation diffusion (Hägerstrand, 1952, 1967), which we discussed in Chapter 3.

The Greensboro simulation begins with a map of initial conditions in 1948 and makes four three-year iterations to reach the spatial pattern of land-use in 1960 (Figure 4.7a). Two scenarios are compared. A rough scenario (Figure 4.7b) is based on overall changes but does not distinguish between different types of residential use and the likelihood represented in this simulation is limited. Another (refined) scenario generates a visually realistic map (Figure 4.7c), accounts for variation in land value, and considers maximum possible residential density of a cell as a function of land value; it also excludes zones where residential use is impossible.

A clear consideration of variance in model outcomes is among many important methodological novelties of the Greensboro model. To achieve that, 50 model runs with identical values of parameters were executed for each scenario, and 50 outputs were generated and ordered by overall deviation between simulated and actual land-use maps of 1960. The variance of 50 outputs is a variance of the model, and not of reality, and, thus, the reality should be compared to the average of these outputs, and not to the most fitted. This average was taken as the "median simulated distribution," for which overall deviation from the actual case is 26th in a list of 50 deviations. For the refined scenario, the correspondence between this median simulation (Figure 4.7c) and reality (Figure 4.7a)

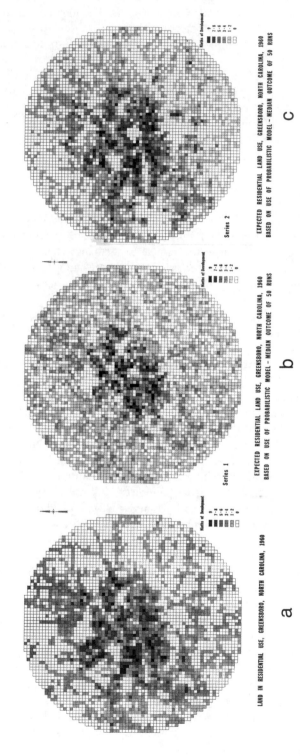

Figure 4.7 *Greensboro's land-use map versus simulation results; source Chapin and Weiss (1968). (a) Actual land-use in 1960. (b) Simulated pattern of 1960 according to a rough scenario. (c) Simulated pattern of 1960 according to a detailed scenario*

was very good indeed—simulated residential use of more than 80% of the cells differ from the actual in only two or less of the ninths.

Now, 40 years later, it is evident that the Greensboro model was ahead of its time. Many of its ideas should be (and have been) "rediscovered."

- Focus on the allocation of externally defined development.
- Separation between cells' potential and locations where change does occur.
- Two-level hierarchy of urban space.
- Separation of the factors of land-use change into endogenous, determined by land unit properties and location, and exogenous, controlled by public authorities.
- Investigation of the variance of model results.
- Comparison of the actual pattern with a median, and not the best fit, outcome.

The cellular view of urban space and a mechanism of allocation that was close to diffusion make the Greensboro model very close to CA. Disappointingly, it took geography 25 years more to make the last and "evident" step, to assume the development potential of the cell to be dependent on the land-uses of neighboring cells.

The work by Chapin and co-authors is an outstanding, but not lonely, example of the raster models of the mid-1960s. Latrop and Hamburg (1965) used a cellular grid at resolution of several hundred meters when modeling allocation of activities in the Buffalo metropolitan area. In estimating potential of a cell for change, they followed Alonso (1964) and concentrated on one factor only—travel distance to the city center. Based on distance, a potential surface is calculated and a new unit of activity is located at a point of maximum potential. New activities in the model are consequently located, one by one, and the potential surface is recalculated after each act of allocation. Just as in the Greensboro model, we can point to novel principles that would be rediscovered in urban CA two decades later.

- Use of cells' potential for change.
- Allocation of externally defined development at a cell of maximum potential.
- Application of asynchronous updating rules.

The other notable achievement of the 1960s was "The Systems Analysis Model of Urbanization and Change," devised at the Harvard School of Design during a spring 1968 course (Steinitz and Rogers, 1970). This model seems to be the most elaborated use of raster modeling for planning purposes at that time. Its goal was investigation of development scenarios for a 45×45 km^2 area to the Southwest of the Boston region, with Newton at the Northeast corner (Figure 4.8a). Twenty guests of the seminar— academic and public planners, engineers, and computer experts—together with 22 students on the course, carefully collected data on highway networks, land cost, pollution, vegetation density, topography, and designed a computer model based on 1×1 km^2 grid representation of the area. The Boston model introduces its own important innovation: the *cell potential is multidimensional* and its components include potential for several land-uses: high-, medium-, and low-cost housing, regional and town recreation, commercial use, etc.

The book that describes the Boston model (Steinitz and Rogers, 1970) does not mention the Greensboro or Buffalo models we discussed, but implements an approach that is very close. Just as with the Greensboro model, cell potentials in the Boston model are

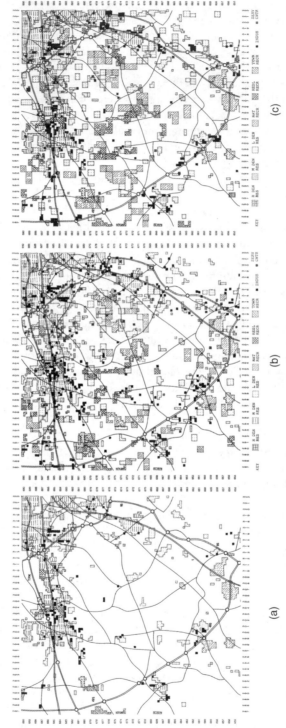

Figure 4.8 *Initial land-use map and the scenario outcomes for a Boston model. (a) Initial land-use map. (b) Current development tendencies applied to the next 25 years. (c) Planning scenario applied, based on zoning and extension of recreation areas, source, Steinitz and Roger (1970)*

estimated as a linear function of environmental factors (Figures 4.5, 4.6). It is worth noting that long before the onset of general disappointment with comprehensive models, the participants of the seminar clearly declared that, despite the fact that relationships can be nonlinear, this complication does not make sense until specific analytical representation is justified theoretically.

As similar to the Buffalo model (Latrop and Hamburg, 1965), externally defined changes are allocated in the Boston model over cells having maximum potential. Two scenarios of regional development for 25 years are considered at five-year resolution. The first scenario is based on the tendency of urban development, estimated according to data about Boston growth during the period prior to model development. The second scenario implements the planning tendencies of that time, considering concentration of industrial development in specially established zones and expansion of continuous recreation areas. The outcomes of these scenarios (Figures 4.8b and 4.8c) diverge significantly toward the end of the 25-year simulation period. Regrettably, we are not aware of attempts to compare the results to current conditions of urbanization in Massachusetts.

To conclude, raster models of the 1960s anticipated most of the conceptual and technical ideas of high-resolution land-use models, rediscovered in later CA simulations. The logic of raster models was very close to those of CA, in all but one aspect—they ignored dependence of cell potential on the state of neighboring cells. Waldo Tobler made this last step in his paper of 1979, which is cited today most frequently in reports on urban CA simulation.

4.3.3   The Beginning of Urban Cellular Automata

A short paper by Tobler (1979) looks superficial to a student of mathematics; for geographers, it was of significant value. Tobler begins with raster layers, as a self-evident representation of geographic space, and influencing factors and formulates three possible analytical formulations of spatial dynamics:

A historical or autoregression model

$$g_{ij}(t + \Delta t) = F(g_{ij}(t), g_{ij}(t - \Delta t), g_{ij}(t - 2\Delta t), \dots, g_{ij}(t - k\Delta t)) \tag{4.7}$$

A multivariate model

$$g_{ij}(t + \Delta t) = F(u_{ij}(t), v_{ij}(t), w_{ij}(t), \dots, z_{ij}(t)) \tag{4.8}$$

A geographical model

$$g_{ij}(t + \Delta t) = F(g_{ij}(t), g_{i-1,j}(t), g_{i+1,j}(t), g_{i,j-1}(t), g_{i,j+1}(t)) \tag{4.9}$$

Indices i and j in the formulae above stand for location of the cell on a rectangular grid, t represents time, Δt is a time interval, $g(t)$ is land-use or another characteristic of a cell, and $u(t)$, $v(t)$, etc., stand for the factors influencing urban development.

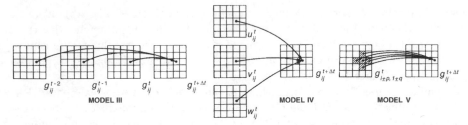

Figure 4.9   Tobler's historic model (III), multivariate model (IV), and geographic model (V); source, Tobler (1979)

Two of the three models were not new. The autoregression (or historical, in Tobler's terms) model (Figure 4.9a) is nothing but a special case of Markov processes (and we consider applications of these models in land-use modeling shortly). One can easily trace the multivariate model (Figure 4.9b) to the raster models of the 1960s that we reviewed already. New dimensions, however, are introduced via the *geographical* model. In today's terms, it is a cellular automata model, and Tobler's graphics (Figure 4.9c) are evidently based on a von Neumann neighborhood. Tobler's reference to Codd's book on cellular automata (Codd, 1968) illustrates the decade delay between invention of CA and their geographic applications.

To make the last step from raster to CA models, Tobler considers *several* land uses— residential, commercial, industrial, public, and agriculture—as cell states. From then on, these settings became standard. Tobler also specifies that the size and the form of neighbor- hoods can be important for geographic applications and recalls the Game of Life (but does not refer to any source) as an example of complex outcomes of simple rules of CA dynamics.

One important formal novelty of the Tobler paper is the launch of a linear transition function as a simple version of a general formulation (4.9). The analytic expression he proposes for the linear transition function is specified as follows:

$$g_{ij}(t + \Delta t) = F(g_{ij}(t), g_{i-1,j}(t), g_{i+1,j}(t), g_{i,j-1}(t), g_{i,j+1}(t))$$
$$= \sum_{p \in [-1,1], q \in [-1,1], p+q \leq 1} w_{pq} g_{i+p,j+q}(t) \qquad (4.10)$$

where w_{pq} denote "weights of influence" of the neighboring cells on the central one; this form became characteristic of most later CA geographic models (Wagner, 1997).

It is worth noting that the conceptual paper of 1979 followed his paper of 1970, where the CA approach, in a slightly obscured form, was employed for modeling spatial development of Detroit population. Tobler considers population $P_{ij}(t)$ of 1.5×1.5 miles cells and implements the "diffusion of urban area" idea in a way that we definitely regard as a CA model today. In parallel to Eq. (4.10), dynamics of the Detroit population distribution are described analytically as

$$\sum_{p \in [-2,2], q \in [-2,2], p+q \leq 2} w_{pq} P_{i+p,j+q}(t) \qquad (4.11)$$

where $w_{pq} = a_{pq} + b_{pq} \Delta t$, and coefficients A_{pq} and B_{pq} are estimated on the basis of Detroit population maps of consecutive periods; one can compare this formulation to principles implemented in raster models.

Were the paper noticed then with the popularity it enjoys today, Tobler's 1970 publication could have provided a crucial link at that time, connecting raster and CA

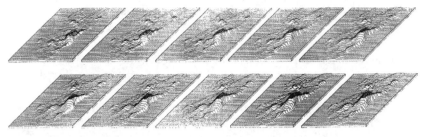

Figure 4.10 *Growth of Detroit city in Tobler's simulation; source, Tobler (1970)*

models. That did not happen, however, and despite exciting illustrations for the end of the 1960s (Figure 4.10) and lucid formulation of "almost CA" ideas, Tobler's papers had to wait several more years before the interest of geographers in CA joined them with accumulated momentum.

Helen Couclelis (1985) was the first to recall Tobler's work in the context of connecting raster models and CA, and made the claim that the achievements of CA and system theories can be combined and applied to investigation of urban and other geographic systems. Nakajima also published a paper in French, using a CA approach for simulation of urban growth, some years before (Nakajima, 1977). Added to this list, by the end of the 1980s, Robert Itami (1988), Michael Phipps (1989), Arnaldo Cecchini, and Filippo Viola (1990) were among the few authors to introduce CA, as an approach, to the geographic public; these developments paved the way for acceptance of CA as a modeling tool, capable of substituting regional models. Around the same time, a book by Peter Albin (1975) introduced CA and MAS as a tool for investigating complex socioeconomic systems.

It is important to note that Tobler's papers, and all the following geographic CA papers, considered CA far beyond von Neumann's definition, as the latter is given, for example, in Codd (1968). "Cellular geography" does include the basic characteristics of CA: partition of geographic space into cells, discrete system time, and transition functions that determine cell state at a next time moment as a function of the state of neighboring cells. At the same time, not less basic operational constraints of "mathematical" CA such as a finite set of states, identical form of the neighborhood, deterministic nature of transition rules that directly determine next state of the cell on the basis of the previous ones, were definitely dismissed or ignored. One can say that, from the very beginning, the notion "Cellular Automata" was used in geography in a very broad sense and not as a rigid formal scheme. The Game of Life, referred to in the first works on geographic CA (Albin, 1975; Tobler, 1979; Couclelis, 1985; Phipps, 1989), was always used for demonstrating complex outcomes of a simple set of transition rules, but never as an example of a geographic system.

This tradition of "free" (as in liberty; not as in beer) extension of the CA framework in geographic applications is continued today. One can easily recognize characteristics of neural networks, excitable media, and Markov fields as well as specifically geographic nonlocal motivations in the urban models of the last decade. Specifically, we have the following extensions.

- Cell neighborhoods are not necessarily identical and can vary in size and shape.
- Characteristics of cell states can be of any kind—nominal, ordinal, or continuous. Cell states can be fuzzy and can be characterized by several variables simultaneously.

- Both deterministic, stochastic, and fuzzy transition rules are employed.
- Transition rules can be given by equations, "if-then" and more complex predicates, tables of the probabilities of possible transformations, etc.
- Factors at above-neighborhood level of urban hierarchy, first and foremost accessibility of urban networks, and characteristics of a city as a whole, are employed.

In the middle of 1990, several years after the conceptual papers on geographic CA were published, the first of a new batch of CA models of urban dynamics began to appear.

4.3.4   Constrained Cellular Automata

The wider application of CA to modeling urban dynamics dates to 1993, when Roger White and Guy Engelen (White and Engelen, 1993) introduced their "constrained CA model of land-use dynamics." This model became the mainstream CA application in geography during the 1990s. The approach of White and Engelen merges the cellular space models of the 1960s with Tobler's geographic model (Tobler, 1979) and implements the assumption that the potential of a land cell to undergo a certain land-use transformation depends on the states of cells' neighborhood. The latter underwent "geographic" treatment; White and Engelen claim that, at resolution of homogeneous land units, cells beyond a circle of adjacent cells influence the given cell. Based on this, the neighborhood is extended from a standard 3×3 field to 113 cells at a distance of six or less cell units from the center (Figure 4.11).

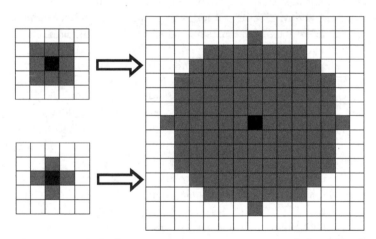

Figure 4.11   *From Moore's and von Neuman's 3 × 3 to an extended neighborhood, consisting of cells at Euclidean distance less or equal to 6 from a central cell; source White and Engelen (1993)*

Constrained CA follow the basic principle of the Greensboro model of Chapin and Weiss (1968): the numbers n_i of cells that must have use S_i, $i = 1, \ldots, N$, at time step t is considered as an external parameter that determines the amount of the overall changes. For each land-use i the model determines the cells for which the potential for transformation into S_i is the highest, and distributes n_i among these cells. The rule of allocation incorporates three stages.

Stage 1: The *potentials* $p_{c,i}$ of transition from the current state into S_i, $i = 1, \ldots, N$, are estimated for each cell.

Stage 2: For each cell, obtained potentials are sorted in decreasing order.

Stage 3: An externally defined amount n_i of land that must be in S_i use is distributed over the cells c, for which the potential $P_{C,i}$ is the highest.

The essence of the constrained CA model is in recalculating the potential of transition into S_i, based on the state of the cell and of the neighboring cells. The exact formula for the potential has evolved during the decade over which the model has been used and developed (compare White and Engelen, 1993, 1994 with Engelen *et al.*, 2002), but its main component that describes the influence of the neighbors has not. To formalize neighbors' influence on a cell C, White and Engelen follow Tobler's (1979) proposal and combine neighbors' effects linearly:

$$P_{C,i} = \left(1 + \Sigma_{\text{over all cells within C's neighborhood}} w_{d,j,i}\right) \cdot \varepsilon \qquad (4.12)$$

where d is the distance between C and the neighboring cell, and $w_{d,i,j}$ is the 'weight' representing the effect of cells in state S_j at distance d, on the potential of cells' transformation into state S_i.

The factor ε in Eq. (4.12) makes the model stochastic, and the analytical expression for ε is chosen in a form of $\varepsilon = 1 + (-\ln(R))^\alpha$, where R is uniformly distributed on (0, 1), and parameter α controls the frequency of large fluctuations. With increase in α, distribution of ε becomes more and more skewed toward 1, and the frequency of large deviations thus decreases (Figure 4.12).

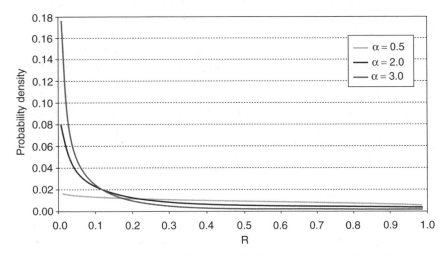

Figure 4.12 *Distribution of the stochastic multiplier of White and Engelen's transition function for different values of parameter*

There is one more conceptual shift from standard CA, "hidden" in the form of dependence of weights $w_{d,j,i}$ on distance d from the central cell. White and Engelen (1993) broke the tradition of *monotonic decay* of neighbor's influence with distance. Instead, they used weights $w_{d,j,i}$ to represent the balance of "attractiveness" and "repel" forces acting between different land-uses. As a sum of two opposite forces, the weight $w_{d,j,i}$ is not necessarily monotonic and can even be negative. Let us consider, for example, the influence of commercial land-use of a neighboring cell on the transition of a given cell into commercial use (White and Engelen, 1993). As a balance between two tendencies— one that is not too close, and the other not too far from existing commerce—this influence is supposed to grow initially with d, and then decrease (Figure 4.13).

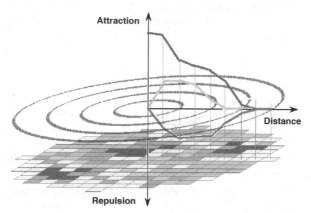

Figure 4.13   *Nonmonotonous dependence of weighting function on distance; source Engelen et al. (2002) (See also color section.)*

The introduction of a strict order of possible transitions represents another step beyond the classic approach. Land-use in the model is only allowed to change in the following sequence: Vacant → Housing → Industry → Commerce. This irreversibility does not contradict the basics of CA, but is rarely implemented in nongeographical applications.

Reproduction of realistic urban land-use patterns was a specific goal of the modeling exercise. Based on data for Cincinnati, White and Engelen did obtain visually realistic patterns. To characterize patterns quantitatively, they followed the fashion of 1990s, and used fractal dimension (Fotheringham *et al.*, 1989). Analysis of the patterns demonstrated that they are bifractal, the fractal dimension of the inner and outer city rings differ. The patterns of real-world cities—Atlanta, Cincinnati, Houston, and Milwaukee (White *et al.*, 1993) and Berlin (Frankhouser, 1991)—were then tested, and were also found to demonstrate bi- and multifractality as well.

The framework of constrained CA is well-suited to simulating real-world city dynamics. White and Engelen did that in 1997, again for Cincinnati (White *et al.*, 1997). Conceptually, the difference between the basic model of 1993 (White and Engelen, 1993) and that of 1997 (White *et al.*, 1997) lies in the introduction of a road network and other long-living infrastructural elements (water bodies, railways) as additional, spatial, constraints for urban land-use development. The idea of inclusion of above-neighborhood structures into transition functions was independently formulated at that time in several other papers (Xie, 1996; Batty and Xie, 1997; Phipps and Langlois, 1997) and from then on became a standard in urban CA modeling. Besides making simulations realistic, account of long-living spatial constraints made it possible to assess the influence of infrastructure elements on urban evolution.

Formal accounts of spatial constraints differ in different applications of constrained CA. Here, we present the latest formulation, employed in the generic model that Engelen and White propose for European cities (Engelen *et al.*, 2002; Barredo *et al.*, 2003):

$$V_{C,i} = P_{C,i}(1 + T_{C,i})^{\sigma 1}(1 + Z_{C,i})^{\sigma 2}(1 + A_{C,i})^{\sigma 3} \tag{4.13}$$

where $V_{C,i}$ is overall potential of the transition of cell C to S_i; $P_{C,i}$ is given by Eq. (4.12); $T_{C,i}$ reflects the effect of suitability, $Z_{C,i}$ of zoning, and $A_{C,i}$ the effect of accessibility of a cell C for land-use S_i; and each σ_1, σ_2, σ_3 is either unity or zero, depending on which of the factors are taken into consideration. Effects of spatial constraints should be specified further, and accessibility $A_{C,i}$, for example, is calculated as

$$A_{C,i} = 1/(1 + D/a_i) \tag{4.14}$$

where D is the distance from the cell C to the nearest cell of the transportation network, and a_i reflects the importance of the network for land-use S_i.

The initial urban pattern of the Cincinnati model was set according to an 1840 map of the city and the model was run 50 time-steps ahead (at a rate of $\sim$2.5 years per step) at resolution 250×250 m^2 per cell, until the year 1966 was reached. Comparison between the simulated and actual patterns can be performed in several ways, from comparing integral characteristics (fractal dimension, for example) to pixel-by-pixel comparisons of the simulated and actual patterns. In the case of Cincinnati, the estimates of fractal dimension were compared and did confirm good fit (Figure 4.14).

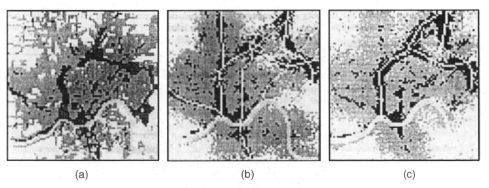

(a)   (b)   (c)

Figure 4.14   *Results of a Cincinnati simulation versus the real urban pattern; source White et al. (1997). (a) Cincinnati, 1960. (b) Iteration 50 of the simulation. (c) Iteration 50 of the simulation, run with the network as a constraint*

Detailed comparison between the model and reality was not performed, but pixel-by-pixel comparison between different simulated patterns was employed in studying the sensitivity of the model. This study revealed the results to be important for general understanding of the abilities of constrained CA models: within the reasonable range of parameters' variations, locations of *wrongly simulated* cells in the Cincinnati model were mostly the same! One can easily assert that these cells represent areas in which development proceeded as contrary to the general scheme represented by model. The

CA model can be considered, thus, as a tool for recognizing these areas, estimating consequences of their restructuring, etc. One more encouraging general results of the Cincinnati model regards the very ability of constrained urban CA to predict urban development. Contrary to what one can assume, stochastic variations, given by ε, do not necessarily increase variance of the model trajectories in time. Global constraints imposed on the system can dampen stochastic fluctuations and the system, thus, behaves more determinedly than one can expect.

It is worth noting that a year before White and Engelen published their own simulation of the Cincinnati development, Back *et al.* (1996) employed constrained CA defined by White and Engelen (1993) to maps scanned from that same paper and other publications of the Cincinnati model (White *et al.*, 1993). Despite low quality of the initial data obtained in this way, the simulations of Back *et al.* (1996) did produce realistic patterns of Cincinnati development, which were similar to those (White *et al.*, 1997) published a year later! This example remains an encouraging exception in the history of urban modeling.

In the methodological milieu, the work by Back *et al.* (1996) points to a weakness in constrained CA that remains unresolved today, namely, that there are "too many" free parameters of the linear transition functions (4.12). We are not aware of any systematic investigation in this regard, while recent simulations of real urban dynamics with constrained CA are often based on smaller neighborhood and transition functions with less parameters (Li and Yeh, 2000; Li and Siu, 2001; Yeh and Li, 2001, 2002).

Ferdinando Semboloni (2000a) has recently proposed an interesting generalization of the constrained CA model for a 3D city. His simple transition function of form (3.3.2.1) includes the effects of the "number of floors" k:

$$P_{C,i,k} = \left(1 + \Sigma_{\text{over all cells within C's neighborhood}} W_{d,j,i}\right) \cdot \varepsilon - B_k - F_{i,k} \qquad (4.15)$$

where B_k is a building cost for floor k and $F_{i,k}$ is a cost related to performing activity i at a floor k. Smart illustrations (Figure 4.15) generated in an initial version of the model point to the necessity of further investigation of this generalization.

Successful applications of constrained CA at the city level entail the extension of model applications, and White, Engelen, and their co-authors have applied it to several

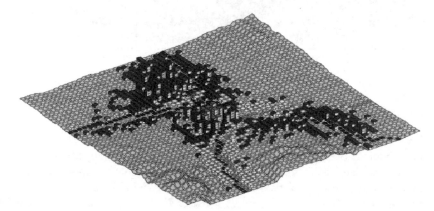

Figure 4.15 *Illustration of the results of Semboloni's modification of constrained CA into a 3D model; source Semboloni (2000a)*

situations, mostly at the Maastricht Research Institute for Knowledge Systems (RIKS) (see www.riks.nl). Examples include the island of Saint Lucia (in the West Indies), Jamaica, the coolest city in Europe (Dublin), and parts of the (and the entire) Netherlands (White and Engelen, 1994; Engelen *et al.*, 1995; Engelen *et al.*, 1997; White and Engelen, 1997; White *et al.*, 1997; White and Engelen, 2000; Engelen *et al.*, 2002). The concept of incorporating a local CA model within a regional model is considered later in this chapter. Over the last decade, the RIKS institution has also developed software for urban and regional modeling, the core of which is a constrained CA model.

The constrained CA model developed by White and Engelen was the first straightforward urban CA model. Just as in linear regression, the linear transition function (4.12) provides the simplest approximation of a transition rule, a step the researcher can hardly avoid approaching in a novel situation. It may serve as an anchor for evaluating the models with more complex ones, which:

- employ nonlinear transition rules;
- vary representation of external constraints;
- vary representation of urban space.

Let us consider models that employ these extensions.

4.3.5 Fuzzy Urbanization

Constrained CA follow the idea of discreteness quite rigorously; some reservations always remain regarding consistency of the discrete framework and our always-partial ability to comprehend what is going on over the area of a city. One is often unsure regarding the use of land units; transitions can be considered as a reaction to continuous accumulation of factors and they always take quite some time. To resolve doubt, fuzzy views of land-use and of transformation as a reaction to continuously accumulating potential were introduced.

Fuzzy set theory provides a convenient framework for expressing partial and vague knowledge of cell states. Briefly put, fuzzy set theory works like this: if X is a collection of land units representing an urban area, then a fuzzy land-use set U consists of pairs

$$U = \{x, \mu_U(x) \in [0, 1] | x \in X\} \tag{4.16}$$

where $\mu_U(x)$ is a 'membership' function, which represents the fuzzy state of urbanization of a cell characterized by parameter x. $\mu_U(x) = 0$ means that a cell is not developed, $\mu_U(x) = 1$ that it is fully developed, and values between 0 and 1 imply intermediate urbanization. The fuzzy set approach demands a clear algorithm measuring the degree of urban development; for example, if p is population density at a cell, $\mu_U(p)$ can be given as

$$\mu_U(p) = \begin{cases} 0 & \text{for } p \leq p_{min} \\ (p - p_{min})/(p_{max} - p_{min}) & \text{for } p_{min} \leq p \leq p_{max} \\ 1 & \text{for } p_{max} \leq p \end{cases} \tag{4.17}$$

Liu and Phinn (2001, 2003) offer a detailed introduction to the use of fuzzy sets in CA models.

Fuzzy CA models operate with the characteristics and membership functions of cells and their transition rules describe laws for updating characteristics based on membership functions. Fulong Wu (1997) considers transition rules based on membership functions dealing with intentionally simple geographic indices, averaging the state of a neighborhood; see Wagner (1997) for review of these indices regarding raster GIS. Employing fuzzy set CA for modeling development of the Chinese city of Guangzhou, Wu describes transitions of woodland (D) and cultivated lands (C) into urban (U) uses, with water (W) and transport (T) uses counted as spatial constraints by means of indices-based rules, where indices are calculated over a 5×5 Moore neighborhood:

Uniformity Un of current usage S of the central cell : $Un = N_S/N$
Agglomeration Ag of urban use : $Ag = N_U/N$ (4.18)
Access Ac to a neighborhood : $Ac = N_T/N$
Capacity Ca of a central cell for transition to urban use : $Ca = (N_D + N_C)/(N - N_W)$

N_S is the number of cells in S-usage, and N is the overall number of cells in the neighborhood. An analogy between these rules and the rules of totalistic CA that we mentioned in Section 2.2.2 of this chapter is apparent.

Membership functions are based on piecewise linear transformation of indices (4.17). For example, the capacity membership function, $f_{capacity}$, is calculated as

$$f_{capacity}(Ca) = \begin{array}{ll} 0 & \text{if } Ca \leq 0.2 \\ (Ca - 0.2)/0.8 & \text{if } 0.2 < Ca < 0.8 \\ 1 & \text{if } 0.8 \leq Ca \end{array} \qquad (4.19)$$

The land transition rules are then simulated on the basis of four membership functions $f_{capacity}, f_{agglomeration}, f_{access}$, and $f_{uniform}$. For an urban cell in, say, a "cultivated land" state (C), the values of these four functions for possible state transitions $C \rightarrow C$ and $C \rightarrow U$ are calculated and the transition that has the highest membership function is employed.

Direct accounting of global factors and accessibility of the transportation network as the coefficients of membership functions, as Eq. (4.13), is a necessary step in employing CA for real-world land-use modeling. Examples of applying the model to real-world scenarios in Guangzhou were approached, with measures of accessibility of urban networks and other globally defined constraints included as parameters in the membership functions (Wu, 1998c; Wu and Webster, 1998). The heuristic assumptions did not differ much from reality; the Guangzhou simulation of development provided rather likely patterns, which react to variation in parameters of the membership functions by significant variations in land patterns after 20 and 40 years of development.

Separation of the actual state of the cell and its urbanization "potential" and direct modeling of both of them is a next step following fuzzy views of cell state.

4.3.6   Urbanization Potential as a Self-existing Characteristic of a Cell

The idea of potential is broadly used in physics and turned out to be useful in theoretical and applied models of urban dynamics, and Allen and Sanglier introduced potential for development in their models at an early stage of urban modeling (Allen and Sanglier, 1981). In the previous sections, potential was utilized as an intermediate characteristic, calculated by an "external observer" in order to compare land units regarding possible changes before they occur. Batty and Xie (1997) suggested that "an observer" not only sees the land-uses, but can also remember the previous estimates of potential, which influence land changes together with the actual land-uses. In this interpretation, potential becomes a self-existing characteristic of a cell, and Batty and Xie apply potential as an additional variable, which is assumed to be continuous and vary on [0, 1].

The idea of potential entered urban modeling together with the idea that three standard levels of the urban hierarchy—land unit, neighborhood, and city as a whole—are not sufficient for representing and formalizing urban dynamics. This was formulated explicitly by Batty and Xie (1994), who introduced an additional above-neighborhood level and referred to it as an "interaction field" of a land cell. They also proposed one more level above the interaction field, but it was not adopted in recent CA models.

It is worth noting that, implicitly, the idea of above-neighborhood, yet nonglobal, factors, influencing cell transitions was always active in CA models. First and foremost, it is expressed by the dependence of cell transitions on distance to urban networks, applied in most of the models we discuss in this chapter. The wider neighborhoods of constrained cellular automata we considered in Section 3.4 can also be seen as a compromise, aimed at combining minimal 3×3 neighborhoods and the wider interaction field into one acting factor.

Taken together, the ideas of potential and above-neighborhood influence introduce considerable functionality to urban modeling. In a theoretical sense, they help in explaining emergence of secondary urban centers (see below); in applied studies they reflect the influence of urban networks, as well as existing or emerging urban zones of different types: industry, dwellings, services, etc.

4.3.6.1 From Monocentric to Polycentric City Representations

Michael Batty and Yichun Xie (Batty and Xie, 1997) used the idea of potential as an independent characteristic of a land cell in an abstract model aimed at investigating the role of *positive feedback* in land-use transformations, when an increase in urban use within a neighborhood forces transition of yet nonurban cells to urban uses.

The most interesting instance of the model is when the potential $P_C(t)$ of the cell C evolves on the basis of itself:

$$P_C(t+1) = (1 - \delta)P_C(t) + \beta\Sigma_{B \in N(C)}P_B(t)/\Sigma_{B \in N(C)}1 + \lambda\varepsilon(t) \qquad (4.20)$$

A physics background is easily identified in the models (4.20): δ can be considered as the rate of time-decay in potential, β as a rate of diffusion of potential around the cell, and $\Sigma_{B \in N(C)}1$ is a number of cells in the neighborhood. The stochastic variation of potential is defined by amplitude λ, given distribution of an error ε is standardized normal. Positive feedback is expressed with positive β; the higher the urbanization potential of neighbors, the higher the potential of urbanization for C itself.

Explanation of polycentricity is similar to that of the dissipative structures of Prigogine (1967) that we discussed in Chapter 3; random fluctuations provide potential seeds of urban development far from existing centers, and then positive feedbacks enforce their further development. To implement this sort of functionality, the amplitude of the random fluctuations should be large, and this is confirmed in numerical experiments. For a given positive value of β, low λ results in a monocentric steady state of the city, while increase in λ breaks this tendency and the city's steady state becomes polycentric.

Wu (1998a) tries to explain polycentric urbanization based on positive feedback between the potential of the cell C to urbanize and the rate of urbanization over the neighborhood, given by the number $r_C(t)$ of neighbors over the 3×3 Moore neighborhood $N(C)$ of C that undergo transition to urban states S during the last iteration, if we denote the state of a cell B at moment t as S_B (t), then, $r_C(t) = \Sigma_{B \in N(C)} S_B(t) - \Sigma_{B \in N(C)} S_B(t-1)$.

Let us note that, in this way, the standard CA framework is extended and the cell transitions at $t+1$ become dependent on the state of the neighborhood both at moments t and $t-1$. Positive dependence of potential on a rate of increase can result in polycentricity. If the potential increases, when neighborhood cells were urbanized just at the previous time-step, then the rate of urbanization there increases and the seeds of urbanization will self-accelerate and the city becomes polycentric. Wu (1998a) compared several representations of potential $P_C(t)$, of a cell C including the following.

1. Potential as dependent on the rate of urbanization:

$$P_C(t) = \alpha + \beta r_C(t) + \varepsilon \tag{4.21}$$

2. Potential as dependent on the rate of urbanization with C's neighborhood and, in addition, on the average density of the urban cells $D_C(t)$ within an above-neighborhood area around cell C:

$$P_C(t) = \alpha + \beta r_C(t) + \gamma D_C(t) + \varepsilon \tag{4.22}$$

In both formulae, ε represents random variation.

After the potential $P_C(t)$ of the cell C is estimated, the probability p_U that the cell will change state from nonurban into urban is calculated as

$$p_U = \exp(P_C(t))/(1 + \exp(P_C(t))) \tag{4.23}$$

According to experiments, positive feedback between potential, and the rate of neighborhood urbanization, formally expressed by $\beta > 0$, is insufficient for emergence of new subcenters: in the case of Eq. (4.21), the city aggregates around the main center only. The situation changes when additional—negative—feedback between potential and the density of urbanized area in the neighborhood is assumed. Namely, when the number of urban cells in the main or other center becomes "too high," the transition potential of the nonurbanized cells around decreases and marginal centers begin to develop relatively faster. Formally, the model (4.22) in this case should have $\beta > 0$ and $\gamma < 0$, and, indeed, for $\alpha = -2.35$, $\beta = 1.27$, $\gamma = -0.15$, the growth of the initial center is blocked, and new peripheral centers emerge and develop with time (Figure 4.16).

It is worth noting that, in probabilistic CA, probabilities of transition between cell states have the same meaning as potentials. Phipps and Langlois (1997) studied such a model and demonstrated stable polycentric patterns in the model, where probability of transition

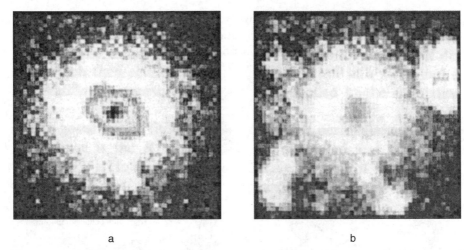

a   b

Figure 4.16 *Pattern generated in two versions of Wu's fuzzy land-use model; source Wu (1998). (a) Monocentric. (b) Polycentric*

between nonurban and urban states depends on the fraction of already urbanized cells in the neighborhood and distance to the main city center.

The explanations of mono- and polycentric structures above do not exploit any specific view of urban systems, and we mention them here and not in the general system (Chapter 3) just because they were inspired by urban dynamics and are not applied to anything but urban systems. Ferdinando Semboloni (1997) investigates this problem in a more realistic context. He considers ecological-economic explanation of the emergence of polycentric settlement patterns assuming that urban development is determined by population migrations. Specifically, he considers five states of urban cells: empty (M), white-collar dwellings (W), blue-collar dwellings (B), base activities (E), and services (S), and assumes that the population of dwelling cells (W + B) is kept proportional to the number of infrastructure cells (E + S):

$$(W + B) = \alpha(E + S) \tag{4.24}$$

the city population structure does not change;

$$B = kW \tag{4.25}$$

and the number of service cells is determined by the city's population:

$$S = \beta(i_w W + i_B B) \tag{4.26}$$

The proportion of white- and blue-collar dwellings (k), between population, and dwellings and services (α), population/services ratio (β), and income capacities i_W, i_B of white and blue-collar citizens are established externally.

The model is driven by an externally defined population growth rate, which determines growth in numbers of dwellings, basic activity, and service cells, according to

Eqs. (4.24)–(4.26). The model also follows optimization principles when formalizing allocation of service and, while demanding further analysis, never results in growth of one center and decline of the others, but in distribution of growing population and accompanying activities and services between the existing settlement centers (Figure 4.17).

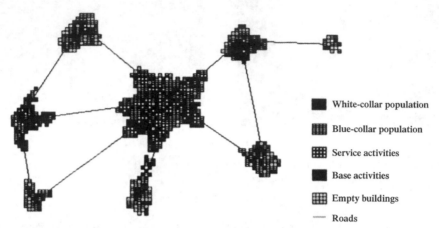

White-collar population

Blue-collar population

Service activities

Base activities

Empty buildings

Roads

Figure 4.17   *Typical settlement pattern generated in the Semboloni model; source Semboloni (1997)*

A lot more work has been undertaken regarding theoretical and experimental study of the mono-polycentric dilemma in urban development (Anas *et al.*, 1998). The models above present several likely, but still too general, dynamic mechanisms, which account for both self-regulation and external forces. Further and more detailed model-based studies are evidently necessary to understand the applicability of the models to urban reality; applications of CA models to real-world cities usually consider mono- or polycentric structure of the city they imitate, and take that as a given.

4.3.6.2   Real-World Applications of Potential-Based Models

Tight-coupling between potential-based CA and raster GIS, where maps of factors are easily available, is especially convenient for modeling real-world problems. In this way, the "geographic" and "multifactor" models as defined by Tobler (1979) are naturally merged. Yeh and Li employ this approach in modeling the rapidly growing Dongguan area, located between Hong Kong and Guangzhou (Figure 4.18) (Yeh and Li, 1998; Li and Yeh, 2000; Yeh and Li, 2001, 2002).

The Dongguan area had 1.9% annual population growth rate during 1988–1993, and, thus, is a wonderful area for modeling and forecasting. To simulate its development, Yeh and Li employ the constrained CA of White and Engelen (1997), but base it on smaller neighborhoods; instead of cells within a circle of radius six, they employ a circle of only two-cell radius. The latter results in 20 neighboring cells instead of 113. The model follows an urban–non-urban (agriculture) dichotomy of land-use and concentrates on transition rules; the latter account for the combination of global and local factors (Yeh and Li, 1998; Li and Yeh, 2000; Yeh and Li, 2001, 2002).

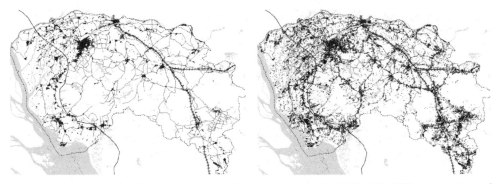

Figure 4.18   *Map of urbanization of the Dongguan area: (a) 1988, (b) 1993; source Li and Yeh (2000)*

The principal feature of the model is an assumption that the potential of transition of a cell C from agricultural into urban use accumulates in time:

$$P_C(t+1) = P_C(t) + \Delta P_C \tag{4.27}$$

The increase in potential ΔP_C is defined in the model by the characteristics of the neighborhood and depends on external factors given by GIS layers of the area, namely

$$\Delta P_C = NBH_{Urban}f(Form)g(Env)R_\varepsilon \tag{4.28}$$

Let us specify four factors included in Eq. (4.28). Generally, two first components reflect the influence of nearby urban cells; a third component represents the influence of environmental factors, and R_ε represents random variation. The first factor—NBH_{Urban}—simply reflects the density of the urban cells within the neighborhood $N(C)$ of C, and is calculated as a number of cells in urban state divided by the total number of cells in $N(C)$. The influence of the *form* of emerging urban structure on a cell C is given by a second factor:

$$f(Form) = \exp(-\sqrt{((w_R^2 d_R^2 + w_r^2 d_r^2)/(w_R^2 + w_r^2))}) \tag{4.29}$$

Here d_R stands for distance from C to a main urban center and d_r to the closest subcenter ($w_R + w_r = 1$); main and subcenters are defined *a priori* and do not change in time. The negative exponent in Eq. (4.29) entails positive feedback between the potential of urban transformation and the closeness to a center; the closer C is to the main or subcenter, the higher the values of the form-factor and the potential of transformation from agriculture into urban use are. Coefficients w_R and w_r function as a switch between mono- and polycentric development: high w_R enforces monocentric trends, while high w_r forces polycentric tendencies in urban development.

The third component of Eq. (4.28) reflects the influence of external factors, given as

$$g(Env) = \Sigma_i w_i (1 - E_i)^k \tag{4.30}$$

where each E_i represents a specific exogenous factor, such as soil quality or slope, normalized to a (0, 1) interval, with 0 as the "best" and 1 as the "worst" values for urban development. In addition, $\Sigma_i w_i = 1$, and the integer value of k varies between 1 and 5 in model scenarios.

The analytic form of the random component follows White and Engelen (1993):

$$R_\varepsilon = 1 + (-\ln\gamma)^\alpha \tag{4.31}$$

with γ uniformly distributed on (0, 1). An approximation of Eq. (4.31), given by

$$R_\varepsilon \approx 1 + \alpha(\gamma - 0.5)/0.5 \tag{4.32}$$

is also employed.

Following the logic of constrained CA, the Dongguan model focuses on allocation of externally defined surplus of developed cells at each time-step. However, as different from the standard practice of allocating cells of the *highest* potential (White and Engelen, 1997), the rule of allocation in the Dongguan model is closer to the "flows" tradition of regional modeling, as *proportional to cell's potential*.

In numerical experiments, variation in parameters w_R and w_r and α of Eqs. (4.27)–(4.32) generate different patterns of urban development. The variance of random error, defined by α, determines whether an evolved city pattern is compact (demonstrated with $\alpha = 0$) or dispersed (demonstrated with $\alpha = 1, 5, 10$). A monocentric city is obtained for $w_R = 1$, $w_r = 0$, while for the polar case $w_R = 0$, $w_r = 1$, each local center develops independently and the city is polycentric.

Correspondence between modeled scenarios and real conditions was measured at the global level by means of measures for suitability loss and compactness, determined as a mean of area-perimeter ratio $\sqrt{(S/P)}$ over urban patches. According to these measures, model results are close to reality for highly dispersed ($\alpha = 5$) and very highly dispersed development ($\alpha = 10$) (Figure 4.19).

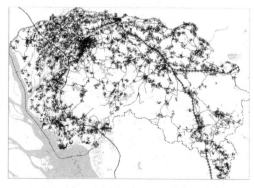

Figure 4.19   *The outcomes of the very highly dispersed development scenario for Dongguan; source Li and Yeh (2000)*

Simulation in the model is accompanied by calculations of the *cost of development*, and this factor is used for comparison between hypothesized development programs. In proposed scenarios of compact development, the evaluated cost of development falls by 50% and on this basis the authors argue that the current, noncompact, development is not optimal at all.

Ward *et al.* (2000) employ a similar approach, simulating urbanization of the Gold Coast area of Australia, near Brisbane. This urban area grew by 32% over the modeled period (1988–1992). As usual, the authors distinguish between urban and nonurban uses; the rate of growth in an urban area is assumed as constant and is interpreted as a probability of "birth" of a new urban cell (a step toward multiagent systems; see Chapter 5). The area considered is close to 100 sq. km and is modeled at 50 m resolution.

In the model, new urban cells are located within neighborhoods of existing urban cells, and the potential for their allocation is defined in a fashion similar to Eq. (4.28) as a product of components determined by different endogenous and exogenous factors. These factors include existence of a nearby "transportation" cell, weighted influence of planning limitations at different hierarchical levels in a form close to Eq. (4.30), and distance to the nearest urban cell.

The authors succeed in translating their knowledge of Gold Coast development plans for 1988–1995 into model parameters. Initial conditions were introduced based on 1988 data and it was assumed that the transportation network did not change from 1988 to 1992. According to pixel-by-pixel comparisons, correspondence between simulated and actual patterns for 1995 over the simulated area of the Gold Coast was at a level of 60–70 %.

This approach, which interprets transition functions as a product of factors that vary within a [0, 1] interval—as in Eq. (4.29)—reflects a "necessary" approach to urban change. Under this approach, transformation of a cell from nonurban to urban state is highly probable when *each* of the factors is close to unity (where 1 supports transition fully). An additive approach or one that is based on the maximum among several factors can be considered, by comparison, as expressive of a "sufficient" view of urban development; change becomes highly possible if at least one of the criteria is sufficiently supportive. Models based on fuzzy sets, which we considered in Section 3.5 of this chapter, can be considered as an example of this view.

The choice to follow a "necessary" or "sufficient" view of transition is a matter that depends on the particular tastes of the researcher, while the multiplicative form of the transition function seems more popular. A paper published by Sui and Zeng (2001) presents an example of the sufficient approach. To model the development of the Chinese City of Longhua at a resolution 180×180 m per pixel, they represent cell C potential P_C as

$$P_{C,k} = \alpha_1^* \, elevation_C + \alpha_2^* \, slope_C + \alpha_3^* \, accessibility_C + \alpha_4^* N_k + \alpha_5^* S_C + \varepsilon \qquad (4.33)$$

where $P_{C,k}$ is the probability (potential) that a cell C would change its use from 'nonused' to an urban land-use k of one of four types: built-up area, forest, agriculture, or planned urban area. The *elevation_C* and *slope_C* are derived from a digital elevation model (DEM) for the study area, *accessibility_C* is measured by distance from C to the nearest major road or highway, and ε is a random error. Two unusual factors are included in the transition function: N_k, the index of clustering of cells of usage k over the entire area, and S_C, the shape index of the urban patch that includes C, calculated as usual as

$$S_C = patch\ perimeter/(4 \times \sqrt{(patch\ area)}) \qquad (4.34)$$

Coefficients $\alpha_1 - \alpha_5$ are estimated by means of multiple regression built on the basis of remote sensing data for 1988 and 1992. The estimates obtained are $\alpha_1 = 0.014$, $\alpha_2 = 0.001$, $\alpha_3 = 0.012$, $\alpha_4 = 0.019$, and $\alpha_5 = 0.031$, with highly significant $R^2 = 0.81$. A map from 1988 was used to establish initial conditions, and data from 1992, 1994, and 1996 were used for model evaluation. The model does a good job of following global and local tendencies of Longhua development. Differences between global characteristics of predicted and actual maps for 1994 and 1996 are below 5%, while pixel-by-pixel comparison results in 15–25 % difference. With these figures, it was judged that the simulation would likely predict the area's development until the year 2010.

Bell et al. (2000) also employ additive transition functions in their computer environment for urban modeling, which is used to test possible scenarios of urban development in the city of Adelaide, Australia. The model characterizes urban cells by means of the fraction of area available for dwelling construction and, unusually, accounts for cells' release class—this parameter reflects the possibility of using dwelling area for construction; it is estimated based on a combination of information on ownership, zoning, and stage of development. The possibility of exploiting a given piece of land now and in the future is of major importance for planning purposes, but the paper published by Bell et al. (2000) is the only one we are aware of that makes use of this characteristic. The model considers eight release classes, from a highest 1 to 8, and assumes that the higher the release class, the higher the probability of allocation of urban development in the cell. In an extreme version of the model, allocation follows a strict sequence of priorities from 1 to 8 and cells of the next release class are not employed until cells of the current class allow for allocation.

The authors also claim that the elements of a planning environment should be as simple as possible in order to be best understood by nonexpert users. Consequently, the potential $P_{C,r}$ of transition of the cell C of release class r into urban state in the model is close to that proposed by Langlois and Phipps (1997): a linear combination of only three components:

$$P_{C,r} = w_c A_c + w_d A_d + w_r A_r \qquad (4.35)$$

where A_c is accessibility, A_d is adjacency (neighbors' interface), and A_r is land unit release class score; weights w_c, w_d, and w_r are set by the user.

The logic of release classes entails a sequential mode of demand allocation in the model; consequently, after the potentials are calculated, the allocation of demand begins from the most suitable cell. The model is accompanied by a 3D visualization, which makes it more attractive for general consumption (Figure 4.20); several scenarios of Adelaide development are tested and compared.

Intermediate, neither multiplicative nor additive, ways of combining different factors are proposed by Besussi et al. (1998). They employ a sequential view of CA update, whereby land-uses diffuse at a first substep; the result serves as an initial condition for

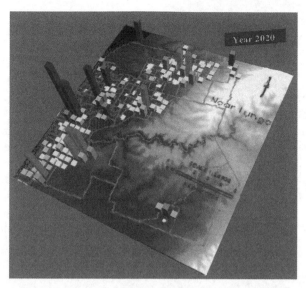

Figure 4.20   *Projected difference in population distribution (2020) in Adelaide for alternative scenarios with different forms of neighborhood influence; source Bell et al. (2000)*

application of rules describing morphological growth, then growth of the transportation network is simulated; so the model proceeds. An initial version of the model has been applied to the Veneto region of the Italian North-West, and produces visually reasonable patterns, while we are not aware of further developments.

To conclude, the results of explicit implementation of CA models are encouraging. First, different real-world cities and regions are likely simulated based on similar factors: form and level of aggregation of land-use, accessibility of the road network, and local topographic conditions. Despite different analytic forms of transition function, the results of the simulations are of similar likelihood and all allow for reasonable speculations regarding area development for ten or more years ahead in time. We can, therefore, assume that a surrogate, but "likely" transition function might be sufficient for correct modeling, provided factors are chosen properly and experimental data are available to estimate coefficients. An encouraging conclusion from the above applications is that the transition functions, which are found sufficient for likelihood simulations, are *simple* and depend on *few parameters*. We can hope, thus, that potential-based constrained CA could become an engineering tool for urban and regional modeling.

4.3.7 Urbanization as a Diffusion Process

CA model rules express a relationship between the state of cells and the characteristics of their neighborhood, and this relationship can be considered in both directions. The CA tradition is "cell-centered" and assumes that "the cell state is influenced by the neighborhood." There is an opposing view, of cells as affecting their neighbors, that has roots in views of urban dynamics as a diffusion process (Hägerstrand, 1952). As we mentioned above, this notion was characteristic of some of the pioneering urban raster models (Chapin and Weiss, 1968); it is worth examining recent developments in that area.

4.3.7.1   Spatial Ecology of the Population of Urban Cells

The famous Game of Life was designed as a model of spatial population dynamics. Its rules are superficial, but it has some relevance to urban modeling—the idea of a population of urban cells, which undergo demographic processes. In Chapter 5, we present an agent-based formulation of such a model of spatodemographic processes (Durrett and Levin, 1994; Durrett, 1999). To employ this idea in an urban context, Batty and Xie (1994) consider population of urban cells, and these cells can be in urban and nonurban states, and assume that each cell can give birth to new cells or die. Given the death rate is δ (survival rate $1 - \delta$) and the birth rate is β, the numbers of newborn (B), dead (D), and overall number of active cells (N) in a population is determined as follows:

$$B(t) = \beta N(t), D(t) = \delta N(t) \tag{4.36}$$

$$N(t+1) = N(t) - D(t) + B(t) = (1 - \delta + \beta)N(t) = rN(t) \tag{4.37}$$

where $r = 1 - \delta + \beta$ is a global rate of population growth.

To describe the process of locating newly born cells, Batty and Xie (1994) extend a standard physical view of diffusion. As we illustrated in Chapter 3, the diffusion view assumes that the offspring of urban cell C are located at positions nearest to a parent cell C neighborhood, $N(C)$. Batty and Xie (1994) consider wider areas, the *interaction field* $F(C)$, which is bigger than the neighborhood $N(C)$ and represents an intermediate urban scale between the neighborhood level and that of the city as a whole. This intermediate scale, explicitly or implicitly, is employed in many recent urban CA models (Clarke *et al.*, 1997; Engelen *et al.*, 2002; Wu and David, 2002).

Batty and Xie (1994) employ the idea of interaction field in its simplest form: they assume that a probability p of urbanization decreases with distance d from the parent cell according to either a power decay function:

$$p(d) \sim d^{-a} \tag{4.38a}$$

or a linear decay function:

$$p(d) \sim (a - d) \tag{4.38b}$$

The model also incorporates negative feedback regulation in a simple way: if the maximum possible density within a 100×100-cell interaction field $F(C)$ is reached and an offspring cannot be located, then the birth action is foregone.

Batty and Xie (1994) applied their model to description of the urbanization processes in a 20×20 km^2 area around the city of Amherst (Buffalo metropolitan area); they represented the city by means of a 600×600 grid of cells. Data have been recorded for the Amherst metropolis over a very long period, 110 years (1880–1990), and maps of the urbanized area for the years of 1880, 1935, 1945, 1960, 1980, and 1990 are available.

These data about the spatial expansion of Amherst were translated into a growth rate r of the population of urban cells for twenty-year intervals; the obtained estimates of r (per year) vary from 1.01 to 1.03 for most of these intervals. Initial distribution of urban land in 1880 within the modeled area is represented by 259 small settlements.

Two urbanization scenarios were examined for the region, representing the period from 1880 to 1990. In the first scenario, distance from "parent cells" was specified so as not to influence offspring location. In the second scenario, a linear decay of the probability of location given by Eq. (4.38b) was used. The second scenario generated better correspondence with actual data; the R^2 pixel-by-pixel correspondence between actual and simulated land-use maps remained at a level of 0.85 for all periods, save that covering 1961–1980, for which R^2 was at a value close to 0.5.

Yichun Xie has refined this model further. In a 1996 paper, he considered residential and commercial land-uses instead of Boolean urban/nonurban uses, with the supposition that the "survival" of an urban cell decreases with increase in its "age," i.e., with time passed since the last change of use (Xie, 1996). In addition, it is assumed that the probability of transformation into commercial use increases with increase in population density within the neighborhood; distinctions are made between birth and death parameters of residential and commercial urban cells.

Specified in this way, the model accounts for the low level of "fit" during the period from 1961 to 1980. Namely, R^2 matches between simulated cities and actual maps of residential use (the use which dominates most of the cells) in Amherst remain at a high level (0.75) for that period, while the R^2 value drops below 0.1 for commercial use. Low correspondence between modeled and actual distributions of commercial land-use is not an exception related only to the period from 1961 to 1980. The correspondence always remains at a level of 0.3, and an explanation might be that the diffusion description employed (Batty and Xie, 1994; Xie, 1996) is sufficient for residential, but not for commercial, uses, the latter following a more complex behavior characterized by far-distant moves, for example.

4.3.7.2   Spread of Urban Spatial Patterns

Keith Clarke and co-authors (Clarke *et al.*, 1997; Clarke and Gaydos, 1998; Candau *et al.*, 2000) have further explored a diffusion-based view of urban development in their CA models, not only assuming unitary cells, but also diffusion of more complex urban entities as a whole. Beginning in the mid-1990s with simulations of Santa Barbara in California, their models have now been applied to other regions of the United States, as well as cities elsewhere in the world (Leao *et al.*, 2001; Silva and Clarke, 2002). The goal of the models developed by Clarke and co-authors is similar to that of White and Engelen (1993): to build a general and simple tool for high-resolution simulation of urban growth. A general heuristic CA model has been built by Clarke and colleagues, called SLEUTH (*S*lope, *L*and cover, *E*xclusion, *U*rban, *T*ransportation, and *H*illshade).

The input requirements for SLEUTH are intentionally restricted. Given layers describing slope, land cover, areas excluded from use, existing urbanized areas, and transportation networks, changes in urban form are implemented in four substeps in the model, all describing different forms of spread and diffusion in space (compare to Besussi *et al.*, 1998).

1. Spontaneous urbanization.
2. Generation of new diffusing centers.
3. Diffusion at the edges of urbanized areas.
4. Road-influenced diffusion.

At each substep, different local patterns are recognized, their growth is simulated, and the results are visualized on a background image denoting hillshade. A model run proceeds as follows.

Substep 1: Spontaneous Urbanization

Every nonurbanized cell C that does not belong to the exclusion layer (e.g., water or swamps) can be urbanized. The potential of the cell C to be transformed into urban state is determined in SLEUTH by the slope at C. Probability of becoming urbanized is set to zero if the slope is 21% or above and is inversely proportional to the slope for lower values.

Substep 2: Generation of New Diffusing Centers

Each cell C spontaneously urbanized at substep 1 can become a "spreading" center. This can happen when C has two or more nonurbanized adjacent cells within a 3×3 Moore neighborhood (Figure 4.21); if so, there is a fixed probability (a parameter of the model) that C becomes a new spreading center and two of its neighbors are urbanized.

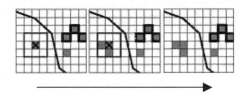

Figure 4.21   *Spreading center: a cell with two or more nonurbanized cells within a 3 × 3 Moore neighborhood generates two more urban cells within the neighborhood; source Candau et al. (2000)*

Substep 3: Diffusion at the Edges of Urbanized Areas

Existing spreading centers, as well as new ones generated in substep 2 and established earlier, can grow on their edges. Subject to exclusions and limitations of slope, the edge cell has a fixed probability (a parameter of the model) to become urban if it has three or more urbanized neighbors within its 3×3 Moore neighborhood (Figure 4.22).

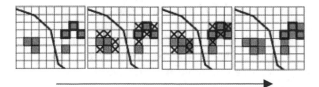

Figure 4.22   *Edge growth; source Candau et al. (2000)*

Substep 4: Road-Influenced Diffusion

This last substep is determined by the existing transportation infrastructure and the most recent urbanization simulated under substeps 2 and 3. It is more complicated than the first three steps and describes further development inspired by an urban cell close to a road. This development is "transported" along the road until it reaches a location where it can "paste" development. This process consists of three additional stages (Figure 4.23).

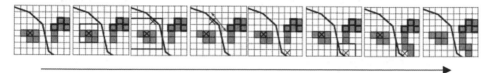

Figure 4.23   *Road-induced growth; source Candau et al. (2000)*

a. Select a cell close to the road to generate a new spreading center.

Each previously urbanized cell is approached with given probability (a parameter of the model) and the existence of a road is sought in its neighborhood of a given radius (also a parameter of a model). If a road is found, a transported center is placed at the closest road cell.

b. Simulate transport of the spreading center along the road.

The temporary center established at stage (a) conducts a walk of fixed length (a parameter of the model) along the road in a randomly selected direction.

c. Anchor spreading center at destination.

If exclusions and slope at the terminal location of a center allow for cell urbanization, two adjacent cells are randomly picked among those neighboring the terminal location and set to urban state; otherwise, the center vanishes.

   The rules above are too simplistic for representing growth of real urban areas if the model parameters remain fixed during the modeled period. To overcome this problem and to make the model capable of simulating real-world cities, Clarke *et al.* (1997) introduce a *positive* feedback relation between the rate of urban cluster growth and the parameters employed at each of the four stages of SLEUTH. Namely, when the cluster growth rate exceeds a given threshold, the probability of spontaneous urbanization and all diffusion rates are accelerated—multiplied by a factor greater than one. To prevent explosive growth, acceleration is controlled, and the multiplier is decreased linearly with aging of a cluster. In a similar way, when the cluster growth rate decreases below the other threshold, the growth and diffusion rates are slowed down even more—multiplied by a factor less than one. Once again, to prevent collapse, these parameters increase linearly with the age of a cluster.

Description of positive reaction to changes is really a nonstandard view of system development: the best reason to accept and further study it is good fit between SLEUTH and real-world urban systems. Nonetheless, the divergence usually accumulates in time, and the authors overcome this problem by reestimating model parameters each ten to 15 years, based on available urban maps; this "tuning" is, evidently, problematic.

Similar to constrained CA, the SLEUTH model has several applications. There have been several examples of Santa Barbara simulation that have been published (Candau *et al.*, 2000; Herold, 2002; Goldstein *et al.*, 2004). Elsewhere in the United States, there are applications to the San Francisco Bay area for the period covering 1900–1990 and Baltimore for an even longer period (Clarke *et al.*, 1997; Clarke and Gaydos, 1998). In Europe, Silva *et al.* (2002) applied SLEUTH to the Portuguese cities of Lisbon and Porto, and obtained satisfactory approximations for a 50-year period; in Brazil, SLEUTH has been used to estimate land demand for waste disposal in the rapidly growing city of Porto Alegre (Leao *et al.*, 2001).

Further developing the idea of growth rates depending on the age of urban cluster, Clarke (1997) and Candau *et al.* (2000) consider urban cells and clusters of urban cells as *Deltatrons,* which act as self-existing urban entities. Initially, each cell is a Deltatron, associated with specific urban state. But, differing from cells, Deltatrons remain as self-existing entities when growing on their edges. The rules of Deltatron evolution are similar to rules 1–4 of the SLEUTH model, but depend also on the age of the cell, that is, on the time passed from last change in deltatron cell states (compare to Xie, 1996, presented in Section 3.7.1 of this chapter). At each iteration, Deltatron cells age by one unit of time, and, reaching a threshold age, a cell "dies" and can then be recruited as a seed for a new Deltatron.

The Deltatron model has also been applied at a regional level (Hester, 1998), the most extensive being the Mid-Atlantic Integrated Assessment (MAIA) study area. MAIA includes seven states on the eastern coast of the United States: Delaware, Maryland, North Carolina, New York, Pennsylvania, Virginia, West Virginia, as well as the District of Columbia, and is designated by the Environmental Protection Agency for the implementation of research, monitoring, and assessment of ecological conditions there. Using a GIS database, an extensive calibration of the Deltatron model for the MAIA area was performed on a per-pixel basis, based on land-use maps of 1950, 1970, 1980, and 1990 and transportation maps of 1950, 1970, and 1980 (Clarke *et al.*, 1997). To simulate urban growth and land cover change in the future of the MAIA region, 1992 data are taken as initial conditions and the model is run for 50 Monte Carlo repetitions until the year 2050. The map in Figure 4.24 presents the most probable (present most often over the 50 repetitions) land class for the year 2050, predicted by SLEUTH on the base of 1992 coverage, demonstrating excessive growth around already established urban areas.

To conclude, let us note that the Deltatron model *differs conceptually* from constrained CA. It can be considered as another candidate engineering tool for high-resolution urban modeling, while it has fewer advocates and implementations than constrained CA. The Deltatron model has a simple and clear structure, limited number of parameters, and can be calibrated straightforwardly on the basis of existing urban data. At the same time, its mechanism of self-enforcing is problematic from a theoretical point of view and was not studied in depth. More research is necessary to verify its applicability.

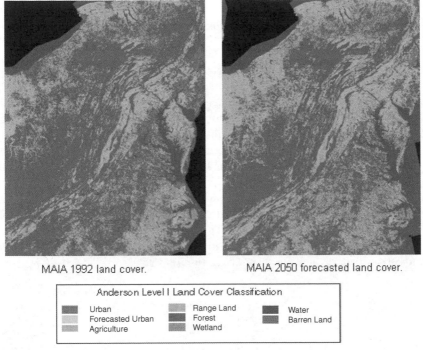

MAIA 1992 land cover.   MAIA 2050 forecasted land cover.

Anderson Level I Land Cover Classification

Urban	Range Land	Water
Forecasted Urban	Forest	Barren Land
Agriculture	Wetland	

Figure 4.24   *Map of land-uses in the Mid-Atlantic Integrated Assessment Area (MAIA) in 1992 and a SLEUTH model prediction for 2050. MAIA includes: Delaware, Maryland, North Carolina, New York, Pennsylvania, Virginia, West Virginia, and the District of Columbia; source Candau et al. (2000) (See also color section.)*

4.3.8   From Fixed Cells to Varying Urban Entities

As we already mentioned in Section 3.2 of this chapter, basing CA on a regular grid, which is characteristic of standard CA models, is superficial—*any* land partition can serve for implementation of CA rules. One can go further and note that continuity of partition is also unnecessary; in Section 2 of this chapter, we discussed the network interpretation of CA—nodes as cells, and edges as denoting neighborhood relations.

Wriggling free of a reliance on regular grids has immediate advantages: network CA offer additional degrees of freedom for establishing new urban units; new units can have specific shape and orientation, and relationships between neighbors can be *altered* during a model run. It is not surprising then, that geographic CA, which are *not* based on a regular grid, have begun to appear.

4.3.8.1   *Infrastructure Objects as Self-existing Urban Entities*

A model by Erickson and Lloyd-Jones (1997) was the first in this *milieu*. Instead of basing cells on a regular grid, the model works with *land objects* of four kinds: street objects, road objects, house objects, and garden objects; all are allocated in the *continuous* space of a growing city. The elementary objects have attributes and can be grouped into complex objects. For example, a house object has an attribute of height, necessary amount of garden space around it, and an azimuth of façade.

The differences between the CA and object-based model become clear when transition rules are considered. Namely, the rules of object location do not concern transitions between predetermined states of cells, but determine the type and location of new objects based on correlation between the kind, location, and orientation of new and existing ones (Erickson and Lloyd-Jones, 1997). For example, a new street object can be attached at the end of an emerging street, while a new house object can be located in such a way that it faces the street at a certain distance from the latter. It is clear that regular grid structure is nothing but a drawback for this approach.

The specific rules investigated by Erickson and Lloyd-Jones (1997) reflect the authors' views of settlement development of several types: an English village, the city of Delhi, South America's "barriada," etc. We will not wander into details of these heuristic rules here, since no regular investigation of the model has been performed. Nonetheless, the emerging urban patterns talk for themselves (Figure 4.25).

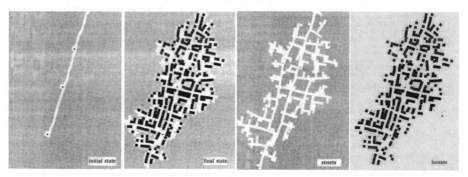

Figure 4.25   *Results from an object-based model accounting for buildings and road segments; the example of urban pattern emerging from three houses initially located along one road segment; source Erickson and Lloyd-Jones (1997)*

It is worth noting that encapsulation of the properties within an urban object model (Erickson and Lloyd-Jones, 1997) removes some more of the "inconveniencies" of the standard CA framework. Height as a property of a house object, for example, makes possible simulation of realistic three-dimensional urban patterns (Figure 4.26).

The next step is to enable change of location for urban objects, allowing objects to make location decisions by themselves or by means of simulated humans that manage them. This is a step along the way to urban agents—see Chapter 5—and full implementation of the Geographic Automata Systems we introduced in Chapter 2.

4.3.8.2   Changing Urban Partition

Ferdinando Semboloni (2000b) made the next step toward object-based modeling of urban infrastructure development. He considers urban partitions and networks, the structure of

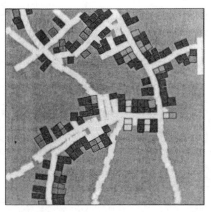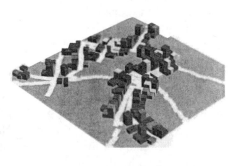

Figure 4.26   *An example of emerging urban pattern in an object-based model accounting for buildings and road segments, presented in 2D and 3D. Differing gray levels represent house types; source Erickson and Lloyd-Jones (1997)*

which may change; urban parcels can be redivided, for example. Besides clear reference to object ideas (Erickson and Lloyd-Jones, 1997), his model is also reminiscent of Lindenmayer's (1968) approach to modeling growing tissues of leafing organisms, which we mentioned in Section 2.3.1 of this chapter.

To combine an object-based view and the ability to change objects' shape, Semboloni represents urban infrastructure by means of two layers: one of *restructuring* urban partition and one of *changing* road network. Restructuring partition cannot have a regular grid as a base. Instead, coverage of Voronoi polygons (Halls *et al.*, 2001; Benenson *et al.*, 2002) is introduced, and land units are considered as neighbors if they have a common boundary. The road network is also based on this coverage; its links connect polygon centroids and are generated following the transformations of "empty" polygons of Voronoi partition into "urban" ones.

As we discussed already (see Batty, 1998, for example), Semboloni assumes that the rules that govern land repartition depend on the state of neighborhoods of two hierarchical levels. First, polygons adjacent to land unit C are considered as a neighborhood of the first order $N_1(C)$; second, $N_1(C)$ *and* polygons adjacent to polygons of $N_1(C)$ are considered, which are the neighborhood of the second order $N_2(C)$.

The units of land partition can be either Nonurban (*N*) or be in three Built states: Housing (*H*), Service (*S*), and Unoccupied (*U*); possible transitions are limited to $N \rightarrow H$, $H \leftrightarrow U$, and $U \leftrightarrow S$. Initial land-use of each unit is always *N*.

The rule for land unit transition is the most interesting of these rules. It is applied to the *pair* of adjacent units C_1 and C_2, when the *potential of division* $Y(C_1, C_2)$ passes a threshold. Potential of division $Y(C_1, C_2)$ depends on *units' potentials* P_{C1} and P_{C2}, which accumulate in time. Unit C begins accumulation of potential when one of the units in its neighborhood $N_1(C)$ turns into built states (*H*, *S*, or *U*) or a road link appears within the $N_2(C)$. The process of accumulation is described as

$$P_C(t + 1) = P_C(t) + (1 - \omega)\Sigma_{B \in N1(C)}b_B(t) + \omega\Sigma_{B \in N2(C)}r_B(t) \tag{4.39}$$

where $r_C(t) = 1$ when land unit C contains a road link in $N_2(C)$ ($r_C(t) = 0$ otherwise), $b_C(t) = 1$ if C is an urban unit ($b_C(t) = 0$ otherwise), and ω denotes the importance of roads for unit potential.

The potential of division $Y(C_1, C_2)$ is specified simply as a weighted average of units' potentials:

$$Y(C_1, C_2) = P_{C1}/n_{C1} + P_{C2}/n_{C2} \qquad (4.40)$$

where n_C is the number of neighbors in the first order neighborhood $N_1(C)$ of C. To simulate repartition when $Y(C_1, C_2)$ passes the threshold, the centroid of a new unit is established in the middle of the link connecting centroids of Voronoi polygons C_1 and C_2. The coverage is then locally rebuilt, accounting for the new centroid. The generation of a new unit is abandoned if either C_1 or C_2 are in Built states H, S, or U, or when the area of one of them is below a threshold.

Unit potential also serves as a switch for changing its state. First, the Nonurban unit C is "awakened" when one of the units within $N_1(C)$ becomes Built. Unit C then begins to accumulate potential $P^C(t)$, and when it passes the threshold, C changes its state from N into H. As standard, further development of housing unit C is in return to unoccupied state ($H \rightarrow U$) after a fixed number of time steps; then, a new loop of accumulating potential is entered, transforming into housing ($U \rightarrow H$), etc. Under certain conditions, however, unoccupied housing units can transform into services ($U \rightarrow S$). This development happens when the number of housing units within the neighborhood C of the third type, $N_3(C)$, defined by the closest nodes of the road network, is above the threshold. Like the housing units, service units return to unoccupied state ($S \rightarrow U$) after some fixed number of time steps, and then enter the new development loop.

The road network expands between centroids of cells, not necessarily adjacent ones. The criterion of extension of a link is based on the number of 'built' units in the neighborhood $N_2(C)$ of the unit the road link ends. The road network branches when the link crosses the unit that has insufficient number of links surrounding it, but the number of service units within $N_2(C)$ is high.

Visually, the growth of urban patterns in the Semboloni model closely mimics real urban growth (Figure 4.27). Qualitative correspondence between the model and real urban patterns has been tested according to global characteristics: fractal dimension, concentration of services, density of built units; all fit well to urban reality.

As with the object-based approach the idea of repartitioning seems very attractive, but as with many of the models above, it demands further investigation.

4.4 From Markov Models to Urban Cellular Automata

Transition rules of CA models can be "tested" straightforwardly. Namely, if we know land maps for a simulated area at time unit t and $t + 1$, we are able to estimate the probabilities of transition between states, numerically, by juxtaposing the state of a cell at $t + 1$ versus states of neighbors at t. In this way, we imitate the logic of "Markov field" models, as presented in Section 2.1.4.

Formally, Markov processes are more general than standard CA; in an urban context, they consider land-use of the cell at $t + 1$ as dependent on states of *any number* of units of

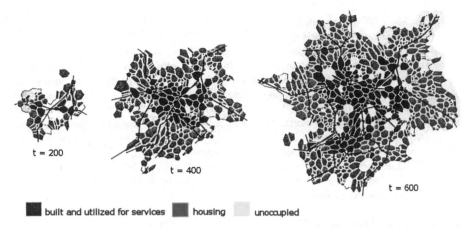

t = 200

t = 400

t = 600

■ built and utilized for services ■ housing unoccupied

Figure 4.27 *Built cells after 200, 400, and 600 iterations. Lines above the polygons represent roads; source Semboloni (2000b)*

the system at *t*, but, as we discussed above, urban CA take this extension for granted. In various urban CA models, cell transitions depend not only on the state of the cell and its neighbors, but also on nearby cells representing roads and rivers, clusters of cells, etc. At the same time, the one-step logic of Markov process is, in fact, preserved in almost all urban CA models we consider: cell state at time moment *t* remains dependent on the state of (many) elements of the urban system at *t* − 1. Markov field models are, thus, a successor of raster models on the one hand, and on the other, are very close to present-day implementations of urban CA.

The simplest Markov model ignores neighbors' influence and considers cell state at *t* + 1 as a function of the cell state at *t* alone. Raster models of the 1970s, as discussed in Section 3.2, extend this view, and consider the next state of a cell as a function of the previous state *and* external factors (Steinitz and Rogers, 1970). One might expect that introduction of urban CA by the mid-1990s would have vanquished out ignorance of this but it did not happen. The situation has remained paradoxical until today. A Markov standpoint is most general and seems ideal for the "unbiased" researcher. In its framework, changes in cell states can be related to factors at the individual, local, intermediate, and global levels and, thus, the relative importance of factors can be investigated and tested statistically. In practice, however, many researchers, usually starting from remotely-sensed images of land-use, do not follow this imperative, and still blindly follow traditions of raster modeling, the latter ignoring neighboring units when estimating transition probabilities.

CA research is not less rude regarding Markov models. The standard way of tuning CA models is in varying coefficients of theoretically established transition rules and making comparisons between simulated and observed land-use patterns. Why not justify transition rules empirically, by direct comparison between states of cells and neighbors for two consecutive distributions of land-uses?

The lack of connection between CA and Markov models may be explained by their view of the "content" of a land cell. Under CA, the cell is viewed as a homogeneous unit, or, at least, as a unit for which the *main use* determines future development. For researchers that apply Markov models to remotely-sensed data, this assumption is not always self-evident at all. Until the mid-1990s, pixels of standard satellite images were 100×100 m^2 or even bigger in size, and this experimental unit is evidently heterogeneous in reality. Moreover, if we prefer to think in terms of homogeneous land units, then, for most landscapes, the 100×100 m^2 area contains a homogeneous land unit *together* with its homogeneous neighbors. There is no need to account for the other 100×100 m^2 cells nearby, then.

CA and Markov modeling approaches began to slowly converge during the 1990s, following enhancements in land-use data on the one hand, and the tendency of CA models to relate between high-resolution CA and regional models on the other. Let us consider that in more detail.

4.4.1   From Remotely Sensed Images to Markov Models of Land-use Change

Most Markov land-use models are based on a combination of historical maps and remotely-sensed images. This is a tradition that was launched in the early 1970s (Bourne, 1971; Bell, 1974) and continues today. The typical Markov model, like the typical CA model, considers several discrete and easily recognized land-uses and accomodates estimation of the matrix of transition probabilities between them $||p_{ij}||$ (which we introduced in Section 2.1.7 of this chapter). As we have discussed, the main insight from Markov processes theory is the convergence of land-use fractions over the entire area toward stable ones, and the matrix of transition probabilities $||p_{ij}||$ completely defines this process.

One technical note is important before proceeding to models themselves. In cases where diagonal elements p_{ii} of $||p_{ij}||$ are close to unity, that is the land unit does not usually change at all, the convergence process is slow and it is hard to recognize what is actually changing in the area. Longsdon *et al.* (1996) propose a simple and effective way to estimate the most probable *changes* of land units by constructing a map of conditional probabilities $q_{ij} = p_{ij}/\Sigma_{i \neq j}p_{ij}$ of "real" transformations $S_i \rightarrow S_j$, which result in change of cell states. If a cell C is in state i and has high value of q_{ij}, then if C changes, its next state will most probably be j. Mapping q_{ij} makes it possible to recognize units that will change most probably as well as the most probable next states for each cell.

As a typical example of Markov model, let us consider research by Jahan (1986), in which she distinguishes five land-uses: open-vacant, residential, public service–recreational, industrial, and commercial. Based on land-use maps for Guelph city and the Guelph region, and for Kitchener city and the Waterloo region of Ontario for the years 1955, 1966, and 1980 she constructs 5×5 matrices of transition probabilities for periods covering 1955–1966 and 1966–1980 for each area. The matrices demonstrate that most land-uses remain the same during that period and, thus, diagonal elements of the matrices $||p_{ii}||$ remain at a level of 0.95–0.99. Jahan (1986) calculates the steady distributions for each region and investigates the time necessary to approach steady state and the sensitivity of obtained equilibrium land-use fractions to changes in p_{ij}. In another situation, Boerner *et al.* (1996) constructed a matrix of transition probabilities between three land-use types—natural vegetation, agriculture, and urban/suburban—for an area of central Ohio that was 242 km^2 in size, for the periods covering 1940–1957, 1957–1971, and 1971–1988. In this case, the probabilities that land-uses are preserved are relatively low, between 0.5 and 0.9, and, thus, land-use fractions stabilize quickly. Both pieces of research evidently resemble the raster models of the 1960s that we discussed in the beginning of this chapter.

The tradition of Markov models demands statistical confirmation of the relationships between land-use changes and environmental factors; they often demonstrate that these dependencies are strong, and thus confirm incorporation of global and midscale factors into urban CA models. Gobin *et al.* (2002), for example, demonstrate these dependences for a 40-km^2 area of Ikem in southeastern Nigeria, analyzed at very high resolution of 4×4 m^2 per pixel. They consider influence of the accessibility of a location, given by distance to road networks and main settlements, on land transition. Five of these distances were identified as significantly influencing transitions between agricultural land-uses, namely distance to the main river, to the market, to the settlement, to the water stream, and to the main road. The dependences of transition probabilities p_{ij} of the Markov model on the accessibility explain prediction of 95% of the changes revealed by interpreting aerial photographs and about 80% of changes observed at visited sample plots. In the same fashion, LaGro and DeGloria (1992) analyzed land-use/land cover change in 15 minor civil divisions in Ulster County, New York, covering a period from 1968 to 1985. Besides the significance of accessibility, measured as distance to a highway and to urban centers, they revealed that population density and the main landscape factors—elevation, slope, and suitability of soils for agriculture—determine land-use changes. Similar factors were revealed by Hathout (2002) when studying transformation of land from agricultural to urban use in a period covering 1960–1989, in the East (32 km^2) and West St Paul (50 km^2) municipalities adjacent to the city of Winnipeg.

By assuming that transition probabilities p_{ij} are not constant, but linearly depend on environmental factors, and tuning the coefficients of this dependency, good correspondence between actual and predicted land-uses is achieved. Lopez *et al.* (2001) did that for the vicinity of fast-growing Morelia city in Mexico for a period covering 1960–1995 (and built a computer system for locating potential land-use changes on this basis). Brown *et al.* (2002) developed a regression model for a vast area covering parts of Michigan, Minnesota, and Wisconsin in America's Upper Midwest. However, the approach employs a "hidden" assumption that *the same* dependences are preserved in time. When tested, it illustrates a problem, and this problem is revealed in the research of Weng (2002), which investigates stability of transition probabilities for an area of rapid land-use change over the Zhujiang Delta coastal region (15 000 km^2) during the period 1989–1997. The probabilities p_{ij} themselves and the dependence of p_{ij} on environmental factors change significantly in each of three three-year time-spans in that period. On this basis, the authors conclude that the land change process has not yet been fully established in the Zhujiang Delta. It is worth noting that the process of land-use change can be non-Markov.

Remotely-sensed data make it possible to delve even deeper into the hidden assumptions underlying Markov and CA models and to test whether dependence of the future state of a land unit on its previous state really holds, *taken for granted* in both models. The alternative to Markov, CA, or other functional approaches is the (disappointing) view that temporal relationships are superficial and the next state of a land unit is, say, randomly selected from the set of possible states. Research of this kind can hardly be a source of inspiration. Let us imagine that temporal dependence is proven; who is actually interested in these sorts of self-evident facts? And what should be done if temporal in dependence is defeated? We are aware of one only source (Bell, 1974) where the dependence of the next state of a land unit on the previous state was tested statistically—on the basis of 100×100 m resolution data of land-use change on the 36 km^2 area of San Juan Island in Washington's Puget Sound over the period 1949–1971. A χ^2 test was used to verify whether the changes were compatible with a Markov hypothesis regarding six land-uses: water, forest, grass, agriculture, housing, commercial and industrial activity. Disappointingly, the statistical examination in this case demonstrated that past land-use is *not helpful* in predicting future use, given the present pattern!

However, this is the only example we are aware of; the dearth of research in this topic does not really allow for any conclusion. One can hope that the result is esoteric, while it is worth keeping in mind that Markov dependencies can be, at least partially, statistical artifacts. The secure way to resolve this issue would involve data-sharing between research efforts, and data analysis from different statistical viewpoints.

Putting doubt to the side, let us consider one more evident disadvantage of the standard Markov process (2.7.1): no self-regulation mechanisms are included in the model. At a common-sense level, it is hardly possible that transition probabilities p_{ij} are preserved when *fractions* of land-use *essentially change*. Indeed, for an area that states of natural land shortage and sparseness, measures aimed at preserving "natural" land-use are applied and, thus, corresponding transition probabilities drop to zero.

Classic Markov models ignore these feedbacks, while they are assumed in the CA framework. In constrained CA, they are imposed implicitly and explicitly in other cases. The idea of neighborhood-dependence and, thus, extension of Markov models toward CA, become unavoidable if we consider land-use changes as part of *landscape system dynamics*.

4.4.2   The Link Between Markov and Cellular Automata Models

CA models start with neighborhood dependencies and modify transition rules to include nonlocal and environmental factors. Markov models begin with estimates of transition probabilities and make them dependent on neighbors' states if such dependence is discovered. Until now, attempts to start from one of the viewpoints and to reach the other one have not been numerous.

The first one is provided by Turner (1988), who studies changes in land-use patterns over six 4.5×4.5 km^2 areas in Piedmont County in Georgia during the period 1942–1980. Land-use data were obtained from historical aerial photographs and digitized into a matrix based on a 100×100 m^2 grid with five land-use categories distinguished: urban, crop, abandoned cropland, pasture, and forest.

To simulate observed changes, Turner modified a standard Markov model in a way that is very close to the constrained CA that we introduced in Section 3.3.2 of this chapter. Namely, she assumed the 'likelihood' of cell C transformation from category i into j to be a product $n_j \times p_{ij}$, where n_j is a number of cells in state j in the Moore 3×3 neighborhood of C and the transition that occurs is one of maximum likelihood. To illustrate this, let us consider modification for the case of two states, "1" and "2", and the cell C in state 1. If $p_{11} = 0.7$ and $p_{12} = 0.3$, then, ignoring neighborhood influence, during one time-step, the model will on average generate "1" with probability 0.7 and "2" with probability 0.3. If

C's neighborhood consists (besides C) of $n_1 = 2$ cells in state 1 and $n_2 = 6$ cells in state 2, then the likelihood that C will preserve its state (1) equals $n_1 \times p_{11} = 0.7 \times 2 = 1.4$ and the likelihood that the state of C will change into 2 equals $n_2 \times p_{12} = 0.3 \times 6 = 1.8$. Turner assumes, thus, that the next state of C should be 2.

The mechanism above is applied Turner (1988). At the beginning of the period, the numbers n_{ij} of all possible transitions $S_i \rightarrow S_j$ observed sequentially are established. The overall number of transitions $N = \Sigma_{ij} n_{ij}$ should be implemented during the period and this is done in N ticks. At each tick, the likelihood of each possible transition is calculated for each cell and the values of the likelihood are placed in order. If, for example, for the cell C in state 1, transition $1 \rightarrow 2$ is more likely than $1 \rightarrow 1$, then the state of C is changed into 2 and the number n_{12} of $1 \rightarrow 2$ transitions and the overall number of transitions N are decreased by one. Upon the next tick, *all values of likelihood* are estimated anew, based on transformations performed before then, and the most likely one is applied again.

Three descriptors were used to compare simulated and actual land-use patterns: (1) the probabilities that the nearest neighbor of a cell of type i is of type j, (2) amount of edge areas between land-uses, and (3) the number of patches of given size for each land-use. The model simulated area of land-use and patch sizes extremely well; the actual and simulated fraction of each land-use do not differ more than several percents for each of five distinguished types. At the same time, the complexity of the shapes of some land-uses was not captured fully, and the simulated patterns were more aggregated than the actual patterns.

The maximum likelihood mechanism that Turner proposed was introduced at what we can identify as the very beginning period of geographic CA modeling and is conceptually very close to the constrained CA of White and Engelen (1993), proposed five years later. In terms of our understanding today, these are just constrained CA with employment of an *asynchronous* (sequential) transition rule. However, the Turner model (Turner, 1988) was ahead of its time and, maybe for this reason, is the only example that we are aware of in which asynchronous CA are employed for explicit description of land-use changes. The very good correspondence that was obtained in her model provides strong support for further investigation of abilities of asynchronous CA for description of land-use dynamics.

Statistical confirmation of the influence of neighborhood on cell land-use changes is of paramount importance, and several pieces of research do demonstrate that. McMillen (1989) uses the multinomial logit model and demonstrates that transitions between vacant, agricultural, and residential land-uses in fringe areas of Chicago depend on property size, distance from Chicago and nearby smaller towns, and characteristics of the quarter in which the property is located. De Almeida et al. (2003) demonstrate that distance to roads, industrial areas, as well as existence of subsidized dwellings and services, in the zone that a land unit is affiliated with, all influence land transitions in the Brazilian city of Bauru. Wu and Yeh (1997) regard factors of land-use change in Guangzhou, China and use logistic regression to represent transition from nonurban into urban land-use as a function of several groups of factors: characteristics of a land unit itself, fraction of neighbors of urban use, transport accessibility, potentials of employment, population, and investment. Significant influence of the factors of each kind is found. Cheng and Masser (2003) extend the spectrum of accounted factors even further; proximity to center of the city, rivers, bridges, and roads, and characteristics of the neighborhood, they include binary variables representing intention of the planner to include the land unit into urban use or increase the dwelling density there. The study is performed over an area of the Chinese city of Wuhan and is based on comparison of 10×10 m resolution SPOT images in 1993 and 2001. Linear and logistic regressions of two binary variables, distinguishing between land units of urban/nonurban land-use and of low and high dwelling-density demonstrate the priority of urban road infrastructure and distance to developed area. The influence of "planning" variables is low, and the authors conclude that master planning lost significance for the city over the period that was examined. The next step, explicit modeling of human reaction to environmental factors (Irwin and Geoghegan, 2001; Irwin and Bockstael, 2002), is considered in Chapter 5.

Arai and Akiyama offer another advance toward employing empirical Markov relationships in CA models (Arai and Akiyama, 2004). Their research in this field involves investigation of a 10 km^2 suburban district of Tokyo at resolutions of 10×10 m and 100×100 m. The research is based on data from 1979, 1984, 1989, and 1994, and nine land-uses are distinguished: (1) woods, (2) paddy field, (3) dry field, (4) vacant land, (5) industry, (6) low-story residential, (7) high-story residential, (8) commercial, and (9) public, or other uses. Land transition probabilities are assumed dependent on land-use in the 5×5 Moore neighborhood of a target cell and accessibility of the cell to transportation networks: nearest railway station and nearest highway. This is specified as follows:

$$p_{ij} = \Sigma_m a_{ijm} x_m + b_{ij1} d_1 + b_{ij2} d_2 + c_{ij} \tag{4.41}$$

where x_m is the number of cells belong to the m-th land-use category within the neighborhood of a given cell, d_1 denotes the distance from the cell to the nearest railway station and d_2 to the nearest main road, and a_{ijm}, b_{ij1}, b_{ij2}, and c_{ij} are empirically estimated coefficients.

Implementation of these transition rules in a sequential manner provides encouraging approximation of observed situations for the area. Taking data from 1974 as initial conditions, model predictions generated for 1984 and 1994 are correct for 70–95 % of 250 cells that changed their states.

The procedure adopted by Arai and Akiyama (2004) combines Markov and CA approaches over a relatively small area. Combination of Markov and CA models at a regional level, which has been performed in several recent works, seems to be an emerging standard of regional modeling.

4.5   Integration of CA and Markov Approaches at a Regional Level

Several attempts at construction of high-resolution regional models that integrate Markov and CA approaches were performed in 1990 (Clark *et al.*, 1990), and research in this area has been developing intensely (Wegener, 2001). The main difference between approaches is in the way they employ high-level constraints on intraregional development. One extreme is presented by "flat" models, in which probabilities of land unit changes are established directly in each region (Klosterman, 2001; Kwartler and Bernard, 2001). The other view, "hierarchical" models, actually constitutes the bulk of urban models, each applied at a specific hierarchical level of partition. High-level models describe interregional relationships, while intraregional models delegate the overall changes generated by the higher-level model to models working at the level of land parcels. It is worth noting that models of both types employ similar factors and demand similar number of parameters to be determined.

High-resolution models of regional development differ in the extent to which they rely on a flat or hierarchical structure. Anyway, while recent tendencies seem to favor the hierarchical approach (Verburg *et al.*, 1999; Lambin *et al.*, 2000; Wu and David, 2002), no standard is established; to illustrate this point, we present examples of both flat and hierarchical model styles.

4.5.1   Flat Merging of Markov and CA Models

According to the Markov view, all factors are equal candidates in influencing land-use changes, and those that are ultimately selected should be chosen by statistical analysis. This view is implemented in the California model of Landis and Zhang (Landis and Zhang, 1998a, b; Landis, 2001). To estimate the influence of potential factors on transition probabilities, they utilize a logit model:

$$P_{Ck} = \exp(\beta_{k0} + \beta_{k1}X_{C1} + \cdots + \beta_{km}X_{Cm})/$$

$$\sum_{i=1}^{K} \exp(\beta_{i0} + \beta_{i1}X_{C1} + \cdots + \beta_{im}X_{Cm}) \qquad (4.42)$$

where P_{Ck} is a probability that a land unit C will be changed (developed or redeveloped) into land-use k; X_{Cj} are variables that can influence this transformation; and β_{Cj} are regression coefficients to be estimated. Some of the variables X_{Cj} represent characteristics of land unit C itself (and this reflects the Markov basis of the model), some are defined by C's neighborhood, and some by external and environmental factors. The structure of the model is independent of the partition of space, and was applied to real partitions— parcels—in an initial version, but was substituted by a 100×100 m^2 grid at a later stage (Landis, 2001).

Seven groups of factors X_{Cj} are considered as having potential influence on land-use transitions.

- *Community level:* employment change, household change, total households, employment, job/householder ratio.
- *Accessibility of a cell:* distance to San Francisco, San Jose, freeway interchange, nearest station for the local train, and to areas for which essential urban services are committed.
- *Physical state of the cell:* slope.
- *Policy constraints (dependent on community):* consideration of whether the unit is prime agricultural land.
- *Cell neighborhood:* fractions of uses—residential, commercial, industrial, public, transportation, and vacant—within a 200 m radius of the cell (one can compare this definition to that utilized in constrained CA and similar models (White and Engelen, 1993; Yeh and Li, 2001)).
- *Cell's externalities:* distance to the nearest cell where commercial, industrial, or public uses prevail.

Altogether, a logit model (4.42) with 27 featured factors is calibrated for eight counties in California, separately for each one, for the period covering 1985–1995. In 1985, two cell-states, developed and undeveloped, are distinguished, while for 1995 the states are as follows: single-family residential, multifamily residential, industrial, commercial, and "other." Testing of Eq. (4.42) demonstrates that the factors' influence vary from county to county, and about 65% of the coefficients of eight constructed logit models are significant at 0.05 (Landis and Zhang, 1998b). Only some factors, as, for instance, the distance to freeway interchange for transformation from undeveloped into single-family use, are important all over. Neighborhood's influence is significant in about 60% of cases.

To implement the regional quotas of development, Landis and Zhang (1998a,b) introduce "bidding" for land against the demand for each use; this is similar to processes proposed a decade before (Turner, 1988). All possible transitions are ordered according to their transition probabilities and listed against demand for each land-use type. The transitions are then implemented in land cells according to the order of the likelihood, until demand is satisfied.

Based on data for 1985, prediction for 1995 was performed, and its fitness regarding overall changes was very good, varying from 92% for Contra Costa County to 99% for Napa County. The goodness-of-fit for specific land-use changes is essentially lower and mostly varies between 20% and 60% (Landis and Zhang, 1998b; Landis, 2001).

The number of regions in partition of the area is always problematic in a flat model. More regions mean more parameters, while, probably, better model fit. In the above California model, the regions are simply administrative units, with no guarantee of uniformity of processes represented within each. If there is no basis to assume that land-use transition probabilities differ in different areas, then the partition can be arbitrary, and Wu and Martin (2002) do ignore municipal partition, simulating population allocation over a huge 300×300 km^2 area of Southeast England, an area that comprises about thirty million people.

The model follows logics of constrained CA and utilizes government population projections to the year 2020 (about 10% surplus), available at different levels, from the entire country to counties, as constraints. The area is then represented by a 250×250 grid, for which initial population density and distributions of developed and vacant (rural) lands are estimated based on a postcode directory. To simulate further development, development potential is assumed to be proportional to accessibility, the latter estimated based on commuting time to major London rail termini, to motorway junctions, and to principal settlements. Accessibility is also nonlinearly transformed in order to increase potential of "better" land cells and decrease potential for the "worst" ones. Finally, the projected population growth is distributed between land cells, proportionally to potential.

The legitimacy of the approach is demonstrated by good fit of 1997 population distribution simulated with data for 1991. Based on this correspondence, the authors compare two 2020-scenario population surfaces: one simulated on the basis of county-level projection, and the other on the basis of global projection, which does not specify numbers for each county. The latter scenario is essentially governed by the evolution of the accessibility surface and thus exhibits relatively stronger development in areas that are closer to London than the former, as well as greater contrast between urban and rural areas. Possible variance of population distribution, caused by inadequacy of county-level projections, potential influence of self-organizing mechanisms, etc. can be estimated in this way.

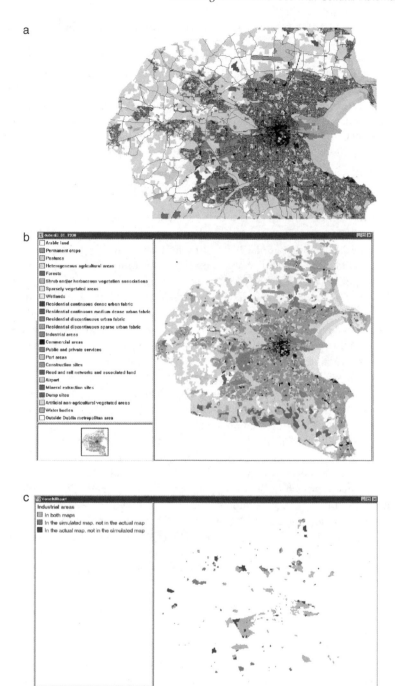

Figure 4.28 *Land-use map of funky funky Dublin in 1988: (a) actual land-use, (b) simulated land-use, (c) comparison between simulated and actual location of industry; source Engelen et al. (2002) (See also color section.)*

4.5.2   Hierarchy of Inter-regional Distribution and CA Allocation

The idea of hierarchy of models, which combine regional and local development, has grown in popularity recently. The hope is that a hierarchy of *simple* models—each acting at its own level of spatial and, maybe, temporal resolution, and accounting for its own factors—is more productive than a relatively more complex model that accounts for the same factors at one level of resolution, as in the above section (also, see Pijanowski *et al.*, 1997; Wu and David, 2002). The principle of hierarchical models is straightforward; lower-resolution models are used for calculating constraints, say, demand for land of specific type, which are delegated to detailed models. High-resolution models deliver integral characteristics of the regions to the upper levels, say, fraction of land vacancy.

The Research Institute of Natural Systems (RIKS), directed by Guy Engelen, as well as other research groups, has developed several models of this kind during the last decade (Engelen *et al.*, 1997; White *et al.*, 1997; White and Engelen, 2000; Engelen *et al.*, 2002; Barredo *et al.*, 2003). The number of levels of hierarchy, employed in the explicit regional models of RIKS never exceeds three: local, regional, and global (national). In most cases, only the first two scales are employed. For example, the Netherlands "land explorer" model (White and Engelen, 2000) divides the country into 40 economic regions, and the country growth tendencies in terms of population and jobs are distributed between regions on a gravity model basis (White, 1977). Allocation of regional quotas is further modeled at a 500 m resolution with constrained CA that account for accessibility, zoning, and suitability of the land unit (White and Engelen, 2000; Engelen *et al.*, 2002), as we presented in Section 3.4. The regional model is based on the multiplicative form of demand for certain activity in a region, and including average cell characteristics as multipliers connects CA and regional models. Three average characteristics of cells of the given region are included: mean density of each type of activity per cell, mean cell potential for land-use corresponding to a certain activity, and the mean suitability of the cell for the land-use corresponding to each type of activity. Several encouraging applications of this approach have been published (Engelen *et al.*, 2002) (Figure 4.28).

There are many other recent developments in the field of combined regional-cellular automata modeling (Berry *et al.*, 1996; Hester, 1998; de Koning *et al.*, 1999; Shabazian and Johnston, 2000; Allen, 2001; Klosterman, 2001; Wang and Zhang, 2001; Pijanowski *et al.*, 2002; Soares-Filho *et al.*, 2002). All yield hope that skepticism surrounding urban models in the last two decades (Lee, 1973; 1994) can be overcome.

4.6   Conclusions

As a field of study, research into, and with, geographic CA has passed several stages. An initial boom in the 1960s was obscured for some years and progress was delayed until the end of the 1980s. Tobler's formulation of geographic CA models and the idea of linear transition rules (Tobler, 1979) had to wait for 15 years until constrained CA and the idea of potential as a characteristic of cells were clearly formulated (White and Engelen, 1993). Nonetheless, this hiatus in research activity ended during the 1990s, and now the field is fully established. The CA view of land-use dynamics provides new impetus and offers models and methodologies that work in actual planning and assessment at urban and regional levels.

Modern geographic modeling follows the idea of constrained CA and various facets of this idea have a number of well-shaped implementations, sharing several extensions of standard mathematical CA:

- use of "potentials of change" in addition to "state" as characteristics of a cell;
- stochasticity of transitions;
- neighborhoods of several orders are employed, thus reflecting different levels of urban organization;
- factors at above-neighborhood level, usually representing accessibility of a land unit, are employed;
- cells and neighborhoods of geographic CA are not necessarily identical and the partition of urban space in the models can change during simulation;
- layers of raster and vector GIS are exploited extensively for initialization, calibration, and evaluation of models.

At the same time, urban CA models often look overcomplicated and, partially by this reason, their formal investigation is far from being satisfactory. Parameters of theoretical models are often intentionally tuned to demonstrate phenomena and model sensitivity is hardly ever investigated. In many instances, model setup is partially reported, if ever, and the model results cannot be repeated. Methods of model calibration are still under-developed, and model tuning is still the "art" in these contexts.

Several novel ideas, all developed in the context of substitution of fixed cells by interacting *urban objects* have been proposed recently. These objects—detached infra-structure units representing houses, street segments, and open spaces—encapsulate geographic characteristics, and are sufficient for adjusting properties of neighbors by varying properties or even shape. Elementary urban infrastructure objects can also unite into clusters, as Deltatrons, possessing their own properties. This revives general system ideas of emergence and self-organization regarding urban land-use dynamics.

Most generally, recent interpretations of urban cellular automata consider CA as spatially discrete systems, consisting of many simple elements, each reacting to local environment and global states of the system. This understanding merges classic frame-works—classic von Neumann cellular automata, McCullouch's and Pitts' neural net-works, Wiener's excitable media, and Markov's random fields. The Geographic Automata Systems idea that we discuss in Chapter 2 develops this view further and incorporates human urban objects, thus providing a general framework for portraying and modeling all components of urban systems. This full geosimulation view is applied in multiagent systems, and the discussion proceeds with MAS in the next chapter.

Chapter 5

Modeling Urban Dynamics with Multiagent Systems

5.1 Introduction

Fundamentally, agents are automata that fully fit to the general idea discussed in Chapters 1 and 2. As with any automata, agents are capable of processing information and exchanging it with other automata. Individual automata of cellular automata can replicate any units of excited media, from cells of a neuron network to elements of urban infrastructure that are usually thought of as inanimate. Inspiration for multiagent systems comes from humans. Agents extend the automata framework with attributes borrowed largely from studies about *behavior* and multiagent urban models focus on the global outcomes of the collective interaction of automata, which is explicitly or implicitly representative of humans. Urban agents live and work in urban systems—developer agents construct new buildings, car agents move in traffic, business agents operate in the city and provide services to customer agents, land-use agents buy, sell, and change the activity on parcels and lots. The outcomes of agents' behavior relate to all aspects of urban dynamics—emergence and decline of urban places, results of elections, traffic flows and jams, establishment and decline of business centers and so on.

Broadly speaking, the term agent is a concept, and there is no universal set of components to demarcate what is agent-based and what is not. "The notion of an agent is meant to be a tool for analyzing systems, not an absolute characterization that divides the world into agents and nonagents" (Russell and Norvig, 1995, p. 33). The construction and study of an agent-based system can be considered in the context of Artificial Life, and is perhaps more art than science (Kohler, 2000). Geosimulation adopts the notion of agents in a narrow sense, as elementary units of Geographic Automata Systems,

Geosimulation: *Automata-based Modeling of Urban Phenomena*. I. Benenson and P. Torrens
© 2004 John Wiley & Sons, Ltd ISBN: 0-470-84349-7

for which behavior is expressed by means of state transition, neighborhood, and location rules.

5.2.   MAS as a Tool for Modeling Complex Human-driven Systems

5.2.1   Agents as "Intellectual" Automata

This book is devoted to modeling; our understanding of "agent" is, by definition, applied to a formal entity, which can be simulated by means of a computer. Consequently, an *agent is an automaton*. The notion comes from the literal meaning of the word—in Latin, *agere* means "to do," and various characterizations of agents for modeling purposes use this interpretation. For example, Maes (1995b) defines an agent as "A system that tries to *fulfill a set of goals*. An agent is situated in the environment; it can sense the environment through its sensors and act upon the environment using its actuators." Among many characterizations (Hayes-Roth, 1995; Maes, 1995b; Russell and Norvig, 1995; Wooldridge and Jennings, 1995), we prefer that of Franklin and Graesser (1996), which is based on intuitive understanding of an agent as being "able to do" or "one who acts, or who can act" autonomously. More specifically:

"*An autonomous agent* (1) *is a system* [we would say '*automata*'] *situated within and a part of an environment*; (2) *that senses that environment and acts on it, over time;* (3) *in pursuit of its own agenda, and* (4) *so as to effect what it senses in the future.*"

But, are not all automata like that? At a conceptual level, the cellular automata models we discuss in Chapter 4 provide a definitively negative answer to this question. Indeed, the cells of CA hardly satisfy two of the characteristics in the definition above, but not others: CA cells do not have their own agenda and definitely do not sense the future. They also do not "act" by themselves. We are going to avoid philosophic discussions here, but our assertion is that an automated agent should have some sort of intellectual ability, in order to be able "to act," "to pursue its agenda," and, especially, "to sense the future." In urban studies, one immediately approaches humans as prototypes of computer agents; directly or indirectly, humans stand behind all the types of agents we consider below.

5.2.2   Multiagent Systems as Collections of Bounded Agents

A *Multiagent System* (MAS) is a community of agents, situated in an environment. "Community" refers to the relationships between individual agents in the system, and these may be specified in a variety of ways, from simply *reactive* to *cooperative*. Generally, a set of relations are specified within a community, linking agents to other agents and objects in their environment (Ferber, 1999). "Environments" are the spaces that house agents and support their activities. MAS are used, in modeling contexts, as an experimental medium for running agent-based simulations. "The researcher employs a multiagent system as if it were a miniature laboratory, moving individuals around, changing their behavior and modifying the environmental conditions" (Ferber, 1999, p. 37).

The rules of a MAS model run simulations, invoking the various functions of its constituent components and interaction among the objects within the system. Going further, the interactions among agents and between agents and their environments can evolve toward, say, coadaptation in a MAS (Kohler, 2000), and these phenomena can be investigated in simulation, but this direction of research is beyond the present state-of-the-art for urban geosimulation.

5.2.3   Why do we Need Agents in Urban Models?

Regional and CA models ignore agency and, specifically, goal-oriented behavior of elementary model components and their ability to forecast. There is no wonder. Regional modeling has roots in the cybernetics of the 1940s and 1950s, and CA extend them toward mainstream system science of the 1960s, which is based on spatially distributed systems of reactive elements and culminates in Prigogine's dissipative systems (Prigogine, 1967). As we discussed in Chapter 3, these views originate in engineering, physics, and chemistry, and not in sociology or geography.

Urban regional and CA models apply and further develop a physics approach to complex systems. Elementary urban components—families, houses, land parcels—are regarded, thus, as physical particles, which evolve following objective and external laws for these particles in nature (Weidlich, 2000). In the CA framework, for example, the transformation of a land parcel from agricultural- to dwelling-use is usually ignorant of the will of the parcel owner.

Conceptually, the agent-based framework is necessary when changes of system element states are, at least partially, the result of individual *decision-making*, even though it is evident that this conceptual twist does not necessarily entail difference in model results. For example, the dynamics of a land-use system, in which landowner agents make market and development decisions, can be implemented both as CA and as an agent-based model, and the outcomes of these models may be the same. Intuitively, the more important the human-like features of model agents are—pursuing agenda, and sensing the future, according to Franklin and Graesser (1996)—the higher the chances for an agent-based model to excel are. Surprisingly, it is hard to materialize intuition of this kind. One can easily invent hypothetical examples where agent-based models surpass CA or other nonagent models, but "empirical" proofs are simply absent in much of the literature.

The state-of-the-art in urban simulation theory is still such that, given phenomena, researchers chose either agent or nonagent models according to their experience and then proceed, ignoring the second approach. Comparisons between agent and nonagent descriptions of the same system are still exceptional and limit themselves to comparison between agent-based and analytical, that is given by equations, models (van Dyke Parunak *et al.* 1998; Wilson, 1998; Morale, 2001; Schnerb *et al.*, 2003; Rosa *et al.*, 2003). The majority of agent-based models consider agent systems "as is" and we are not aware of research comparisons between CA and agent-based descriptions of the same phenomena.

5.3   Interpreting Agency

Agents are automata, first and foremost, and, therefore, we have to think about them in terms of states and transition rules, reformulated, maybe, as automata that pursue their own agenda and sense the future. However, they are a particularly sophisticated class of automata, with a wide range of attributes that lend them more human-like qualities that their CA cousins do not possess.

There is no problem with definition of agent states—age and number of family members can be considered as states of a householder agent; velocity, kind of engine, and age of the driver as states of a car agent; the number of employers, daily number of customers, and monthly profit, as states of a "small shop" agent and so on.

Transition rules complicate matters, however. It is hard to imagine a unified formal framework for automating human-like behaviors, which look so different if we compare the behavior of households, cars, or shops. Householders can decide to resettle following increase in family size or improvement of personal economic conditions, but at the same time, their migration decision can be affected by desire to live in a specific area of the city or to avoid places they simply do not like. Car drivers can make wrong turns; they may choose alternate routes because they endured a puncture in a particular location the day before. A shop owner might open a store in a relatively bad location because she failed to follow information about the market potential of her goods. The more "human" agents are, the more specific their formal representation seems. This uncertainty can be regarded as a conceptual disadvantage of the agent-based approach—how can researchers be sure that the notions and formalization they use have the same meaning?

As in many other situations, scholars simply do not worry too much about the lack of conceptual clarity when representing agents' behavior. Agreement between researchers is perhaps what is more important (Wooldridge and Jennings, 1995); if many people share the same intuitive understanding of what agency means, then terminological details hardly matter. Instead, points of agreement become important.

There is a wealth of literature on the topic of agency, defining generally accepted characteristics common to agents. Franklin and Graesser (1996) single-out a range of broad properties for model agents, which are presented in Table 5.1. They describe a

Table 5.1   *Properties of agents in MAS*

Property	Other names	Meaning
Reactive	Sensing and acting	Responds in a timely fashion to changes in the environment
Autonomous		Exercises control over its own actions
Goal-oriented	Proactive, purposeful	Does not simply act in response to the environment
Temporally continuous		Agent behavior is a continuously running process
Communicative	Socially able	Communicates with other agents, perhaps including people
Mobile		Able to transport itself from one location to another
Flexible		Agent actions are not scripted
Learning	Adaptive	Changes its behavior based on its previous experience
Character		Believable "personality" and emotional state

"minimal" agent, as possessing four primary properties, and they distinguish subclasses of agent, with other (application-specific) attributes. Looking ahead to geosimulation applications, distinctions of sub-classes of *mobile* and/or *adaptive agents* are particularly relevant in the context of this book.

At a very general level, agency can be classified into two main cohorts: weak and strong. These two classifications are a by-product of the division of research threads in agent-based research: between theoretical and experimental work (weak agency) and work on developing active agents (strong agency).

The first approach—theoretical and experimental work—focuses on mechanisms that come from the interactions of autonomous agents in a simulation and system outcome, offering explanation of the qualitative characteristics of systems' dynamics, self-organization, and emergence as a consequence of these interactions. The second approach—development work—concentrates on creating agents and formulating rules of their behavior such that agents are capable of accomplishing tasks that are distributed in space and time (Ferber, 1999). The emphasis, in that approach, is on the rules of information exchange between agents, given by communication protocols, coordination of agent actions, etc. This "strong" interpretation of agency includes the "weak" interpretation, but is more commonly conceptualized using mentalistic notions such as knowledge, belief, intention, obligation, emotion, etc.; concepts normally applied to humans (Terna, 1998). This makes it closer to Artificial Intelligence (AI) and Artificial Life (AL).

Interpretations of agency are, for the most part, a function of the particular examples and applications that a designer had in mind when specifying an agent-based model (Franklin and Graesser, 1996). But we can identify several general qualities characteristic of agents as autonomous automata.

Agents are *heterogeneous*. Traditional modeling methodology often refers to an "average individual," possessing mean attributes of a group it represents. Of course, the average individual concept is counter to common knowledge and invokes ecological fallacies (Wrigley *et al.*, 1996), but persists in modeling due largely to the absence of alternatives. The ability to specify each individual agent opens up opportunities for design of heterogeneous collectives. This is of principal importance if we anticipate essential nonlinearity in individual interactions. As we know from general system theory (Chapter 3), if objects' interactions are nonlinear, then the dynamics of a system essentially depend on variance, skew, and other parameters, reflecting heterogeneity of the object population. We cannot avoid employing heterogeneous agents if we believe in nonlinearity of the system we investigate, and we can even investigate the importance of the nonlinearity in this way. Groups amid such heterogeneous structures are defined from the bottom-up as assemblies of independent objects, not defined in advance.

Agents are often designed to be *proactive* (Terna, 1998), and, as we mentioned above, act to realize a goal or set of goals (Franklin and Graesser, 1996). Economic agents can be designed to satisfy utility goals; political agents might be designed with particular agendas to satisfy; and geographic agents could be created to follow a set of spatial paths. Proactivity is one of the basic features that distinguish human-like agents and physical particles.

Individual agents are commonly specified with *perception*. This can be interpreted as an extension of the neighborhood concept in automata. Agents "sense," or are "aware" of, other agents, and cognitive geographic agents are aware of agents in their surroundings first and foremost. However, agent perception is not constrained to a neighborhood. Agents in agent-based models are often endowed with a *cognitive model* of their "world" and the ability to identify nonelementary entities within it. Cognitive geographic agents ought to recognize emerging spatial ensembles as well as the spatial structure of the system as a whole, and geosimualtion models employ cognitive agents to represent human-like behavior of householders, developers, services, institutions etc.

Interactions between agents can be often considered as *communication*—active and intentional querying of other agents, searching for a particular type of information and choosing to ignore extraneous details. In this sense, agents could be regarded as social creatures, and reflect essentially human behavioral features; *bounded rationality*, first and foremost.

Adaptation is another characteristic that we want to stress regarding the design of geographic agents. By adaptation, we mean the ability for agents to change the rules of their behavior based on experience absorbed during their model life span.

Wooldridge and Jennings (1995) would, likely, classify these characteristics, if applied formally, under "yet weak" agency. At the same time, the broader interpretations bring us to a notion of agency in the very common, human, sense (the last row, "character," in Table 5.1), that is, to "strong" agency. Obviously, the spectrum of properties that can be considered as characteristic of strong agents is indefinitely wide. A typology of strong agency, if ever, is essentially more controversial compared to the characteristics of weak agents, and each property in it demands further explanations before formalization is proposed.

We forward readers that may be interested in application of agency in strong senses to reviews of sociological applications of MAS (Carley, 1996). Geosimulation deals with weak agents.

5.4 Urban Agents, Urban Agency, and Multiagent Cities

The urban MAS applications that we consider in this chapter start with "primitive" agents, which possess only some of the properties associated with weak agency; the discussion of applications develops toward including agents" ability to migrate and adapt.

5.4.1 Urban Agents as Entities in Space and Time

The CA models discussed in Chapter 4 focus, predominantly, on land or infrastructure units—land parcels, land-use, land-cover. The unit, cell, itself is the only elementary entity in those models. Few CA applications move beyond this framework, to account for complex units, such as Deltatrons (Clarke, 1997), and in these examples more complex units are still represented by sets of cells. Sets of cells have their own properties—number of elementary cells, for example, but states and transition rules are nonetheless applied to single cells of these sets. CA models are all very much the same in the sense that they do not consider anything but cell-based automata.

The situation with MAS models is more complicated. Representation of urban processes necessitates a variety of agent types and we cannot assume that their states and rules can be formulated in a similar manner; householder residential behavior, construction development, car traffic, and pedestrian walks are qualitatively different phenomena. Consequently, MAS models always begin with some form of declaration regarding the processes considered and the kind of agents involved. This approach has justification in general system theory (see Chapter 3); model processes may be separated according to their *characteristic time τ*.

The notion of time scale is a useful concept for distinguishing urban agents, based on their rate of change and temporal scale of their actions. There is no need, for example, to include cars as agents in a model of household migration. Car-agents are too "fast" and time-average characteristics of traffic—average speed and density on the road and average time to get to work from a given location—are sufficient to represent cars" influence on householder's residential behavior. Likewise, when modeling traffic jams, there is no need to consider householders' migrations—a residential distribution can be assumed as constant over 24-hour or weekly traffic cycles.

CA models do not address the problem of the temporal scales of automata transition rules. In a somewhat hidden manner, they assume the chosen time step; usually one year is just a characteristic time τ of transition for all cells. This is evidently untenable in MAS, where characterization of agents' space-time behavior is fundamental to characterization of agents of different types.

For abstract agents, τ depends on interpretation, while in urban models that we consider in this chapter, agents operate in the following temporal scales:

- Developers: τ is years from a construction perspective;
- Firms: τ is months in terms of the spatial dynamics of services;
- Households: τ is months, with regard to residential dynamics;
- Cars: τ is tenths of second in the context of driver reaction time;
- Pedestrians: τ is seconds with respect to pedestrian dynamics.

The MAS modeler intuitively tends to simplify the situation and account for as few types of agents as possible. As we discussed in Chapter 3, general system theory supports this tendency. Translating the notion of characteristic time into agents' behavior, we can formulate that as follows: account in the model for the agents that determine the phenomena only, set parameters determined by "faster" agents equal to temporal averages, specify the parameters determined by the agents of "slower" types as constant, according to the specific situation of interest.

Starting from this general statement, let us note that in the majority of real-world processes, agents of *several types* act simultaneously at *close* temporal scales. Urban traffic, for example, on a background of "very slow" transport infrastructure, embraces cars, pedestrians, traffic lights, and road police—too many entities to be incorporated in one model. The proper choice of agents is a measure of the "understanding" of the system and the prerequisites for research to be successful.

In the remainder of this chapter, we distinguish between abstract and real-world MAS models. Agent-based modeling is still in its infancy, and there are more abstract or close to abstract MAS models than there are explicit examples. Consequently, many more abstract models are considered in this chapter in comparison to the CA of Chapter 4. Transportation and residential dynamics are the main fields of research where real-world systems have been simulated—relatively intensively—using MAS, and these are the main areas of real-word applications in the forthcoming discussion.

5.4.2   Cities and Multiagent System Geography

Weak agency is a very plausible rationale for developing agent-based models of cities. The motto of modern system theory is that *nonlinear relationships make systems complex.* There is justification, then, for understanding the role of relationships between agents in urban systems, relationships of as simple as possible form, as a first research goal. This approach leaves other considerations in the sidelines; the intricacies of strong agency can wait. In this fashion, then, contemporary geographic modeling considers urban systems as collectives of weak agents. These agents share a common urban environment in which they interact. Research in this area is largely concerned with coordination of agent behavior at the system level, i.e., with their *collective behavior* in a city.

Is there justification for specifying *urban agents*, a distinct group within the general class of autonomous agents? Of course there is! And that is the contention we make in this book. First, we argue that *mobility*, which, in turn, is based on exact definition of *spatial location and spatial relationships*, is a distinguishing principle for urban agents. Recall that CA model cells are located in space "by definition." This is not the case for agents acting in a city. Second, we suggest the capacity for urban agents to *perceive and adapt to the evolving urban environment* as another distinguishing factor for urban agents. These assertions stem from some fundamental notions about the behavior of agents—householders, vehicles, landords, pedestrians, firms, etc.—in urban environments. We will elaborate further upon these ideas as the chapter progresses; urban agents are discussed based on their mobility and their ability to adapt.

5.5   Agent Behavior in Urban Environments

Agent-based models are, by definition, based on rules of agents" behavior. To animate these models, we should specify agents" knowledge about the city and how they react to perceived situations at each moment in time. These are popular threads of research in the social sciences in their own right, outside of agent-based modeling.

Work in psychology and social science has investigated human behavior, most importantly, *choice behavior*, for over a century, and a number of approaches have been formalized and are widely accepted. These include social comparison and social learning theories (Festinger, 1954; Bandura, 1986), which are based on the assumption that people compare their behavior with the behavior of others. Decision and choice theory (Simon, 1976; Janis and Mann, 1977) and theories of reasoned action (Ajzen, 1988) offer perspectives on intentional choice making. Conceptual models, combining various theories in a single framework, have also been proposed (Sonis, 1991; Jager *et al.*, 2000); and many of these ideas have been used in social science MAS modeling. Sociological, economic, and anthropological MAS models have been built to investigate the consequences of different approaches to the description of human behavior and implications for understanding societal dynamics (Epstein and Axtell, 1996; Luna and Stefansson, 2000; Kohler and Gumerman, 2001).

Implementation of these theories can be performed with either weak or strong agency. Urban geosimulation, however, definitely tends to avoid the complications associated with the latter. Few relatively simple assumptions regarding agents" behavior are sufficient in the vast majority of situations, to interpret realistic views of human choice in urban systems, and keep the system understandable. There is a substantial volume of work on "weak" agent-like behavior in urban systems, particularly as regards location and migration, preferences and choice, and decision-making, which can be directly interpreted in geosimulation models.

5.5.1   Location and Migration Behavior

As we claimed in Section 4.2, the ability to change location is one of the main features of agent activity in urban environments. Urban agents" reaction to internal and environmental stimuli is "two-dimensional" in this context. The first dimension represents changes in agent's nonspatial characteristics and is analogous to changes of state in CA cells. The second dimension represents changes in agents' location in a city.

The ability of agent-automata to migrate is one of the novelties of geographic MAS and, thus, extends our capacity to simulate urban systems beyond CA representations of fixed cells. Formalizing migration behavior for autonomous agents is especially important for description of the information spread in geographic MAS. In a system comprised of spatially fixed elements (land units in an urban CA model, for example), information spreads through the system by diffusion. Agents differ, however. No matter whether agents are fixed or mobile, they can react to both closely- and far-located agents or other model objects. This closely resembles the way humans "process" information in cities; they are often able to "know"—via mass media, friends, or in some other way—the state of the entire system or a part, at a distance. Mobile autonomous agents "carry their properties with them" and, depending on their own will, can do so as they migrate to either close or distant locations. As a result, changes in MAS can result from nonlocal prompts.

Phenomena of distant influence and distant interaction are especially important when considering situations in which the capacity of a system is limited, e.g., when vacancies in an apartment building or available sites for redevelopment in a city are in short supply. Here mobile agents differ from spatially fixed elements: they are able to avoid local circumstances and compete for opportunities located over an entire urban space.

The consequences of agent action-at-a-distance may be dramatic; this is the stuff of popular fiction and films—the stranger that wanders into a small town and alters village life indelibly. The Irish examples, of course, are among the finest; consider Synge's *Playboy of the Western World* (1907). In a more academic sense, proponents of general system theory embarked on a search some time ago, looking for rationale to justify kicking of a spatial system out of equilibrium. Fiction is based on one or two strong agents. MAS models aim at formalizing relocation of many weak agents and investigating their consequences for a city. Formalization of location *choice* is thus crucial for model outcomes. A number of formal approaches may be used here, all including mechanisms for expressing location utility.

5.5.2 Utility Functions and Choice Heuristics

Utility functions provide a formal framework for specifying agent choice behavior. Utility refers to the weight attached to a particular choice among a range of choices in a sense such that an agent A, presented with a set of opportunities $\{C_i\}$, is able to estimate the *utility* $U(A, C_i)$ of each of those choices. Often, a complementary *disutility* function $D(A, C_i)$ can be considered. Individual opportunities C_i available to an agent A can be compared, based on the value of their utility or disutility—$U(A, C_i)$ or $D(A, C_i)$. The utility concept is frequently employed in economic-oriented research, in contexts where all the factors influencing choice are supposed to be known. In more general contexts, when the factors governing choice are only partially known, it is often more convenient to formalize the problem in compulsory terms of disutility; *dissatisfaction* and *dissonance* are two synonyms of disutility, frequently used when human agents are considered.

To illustrate the utility idea, consider a householder-agent searching for a new dwelling to rent. In evaluating potential homes, the householder-agent considers two factors: the price of rent and the distance of the property from her job. Ideally, she would like to find an affordable property that is at a reasonable commuting distance from her place of employment. In formal terms, the agent's utility function should specify difference between rent and her income and travel time to work. One can consider these two components of utility independently and employ a two-dimensional utility function to specify them, while components of utility are usually combined to obtain a one-dimensional measure. Combination can be performed formally or by meaningfully translating the components into a general resource, travel time into money-loss in our example. In this way, the overall money-loss yields one-dimensional disutility of dwelling.

Real-world agents often vary in their attitudes to utility and its components. Depending on sensitivity to travel or budget, householder agents in the example above may be indifferent regarding the travel component of utility when a journey takes less than half an hour. Similarly, some agents may be very sensitive to rent prices, while others may be indifferent to small differences. The capacity to vary the perception of utility and its components between agents is a basic advantage of MAS. This variability can be used to reflect not only quantitative, but also *qualitative* variations in agent behavior, and MAS can simulate populations of agents who behave according to different theories of human choice.

Given opportunities' utility, an agent should employ *choice heuristics* in order to select among them. The standard way to do that is to transform the values of utility (or disutility) into *choice probabilities*. The probabilities can be calculated in several ways; the *logit* statistical model is popular. According to the logit model, the probability $P(A, C_i)$ of choosing opportunity C_i of utility $U(A, C_i)$ is calculated as:

$$P(A, C_i) = e^{\beta U(A, C_i)} / \Sigma_k e^{\beta U(A, C_k)} \tag{5.1}$$

where $e = 2.718\ldots$ is a base of natural logarithm, and β is a parameter. Normalization ensures that the values of $P(A, C_i)$ all fall within the interval of $(0, 1)$.

Equation (5.1) has a number of advantages. First, it can be applied to any real-value utility function U, whether negative or positive, because the value of $e^{\beta U(A, C_i)}$ is always positive. Second, we can change the relative attractiveness of opportunities in a choice set by varying the value of β. For example, consider the case of two opportunities C_1 and C_2, having utilities $U(A, C_1) = U_1 = 1$ and $U(A, C_2) = U_2 = 2$. According to Eq. (5.1), the probabilities of choosing C_1 and C_2 are

$$P(A, C_1) = P_1 = e^{\beta}/(e^{\beta} + e^{2\beta}) = 1/(1 + e^{\beta}) \tag{5.2}$$

and

$$P(A, C_2) = P_2 = e^{2\beta}/(e^{\beta} + e^{2\beta}) = e^{\beta}/(1 + e^{\beta}) \tag{5.3}$$

Consequently,

$$P_2/P_1 = e^{\beta} \tag{5.4}$$

By varying β we can alter the relative importance of the opportunities in the choice set. For $\beta = 0$, for example, $e^{\beta} = e^0 = 1$ and $P_1 = P_2 = 0.5$, i.e., the probabilities for each opportunity are equal despite difference in utility. For a positive value of β, P_2/P_1 increases from unity to $U_2/U_1 = 2$ when β grows from 0 to $\ln(2)$. With further increase in β, for example, for $\beta = 2$, $P_1 = 1/(1 + e^2) \cong 0.12$, $P_2 = e^2/(1 + e^2) \cong 0.88$, and P_2/P_1 reaches 7.38, much above U_2/U_1. A negative β-value alters the superiority of opportunity C_2 over C_1. Varying β, thus, we can stretch, shrink, or turn over the differences in utilities; these are very useful functionalities for building models.

5.5.3   Rational Decision-making and Bounded Rationality

Formalization of choice between opportunities is not a prerogative of urban MAS models; most discrete models dealing with humans account for choice heuristics, explicitly or implicitly. In constrained CA models (see Chapter 4), for example, the next state of a cell is chosen from several possible candidates according to the probability of each of the possible transitions as defined by the utility of a cell for potential use (White and Engelen, 1997). The choice of utility function was not focused on in Chapter 4, for simple practical reasons—in the majority of the CA models presented there, the next state of a cell was chosen from several possible options in the simplest way: the cell with the *highest* potential. By contrast, MAS cannot limit themselves to this scheme alone; humans can behave in a *variety* of ways, not just utility-maximizing.

The algorithm for transformation of utilities into probabilities of choice is only one way of defining choice heuristics. As different from CA cells, human agents make choice decisions explicitly, and we mentioned a volume of psychological literature that suggests various methodologies for expressing that behavior. Regarding geographic agents, migration decisions are especially important. Consider some examples: a householder, uncertain regarding the region she wants to inhabit; a car driver, intending to park close to or farther from a target location; a developer, choosing a land parcel to invest in; or an entrepreneur, intending to open a new store, but being uncertain where. All of these agents have to choose one of several locations or abandon their search. The situation is further complicated; in the majority of situations *other* agents are also interested in the same opportunities. Human agents explicitly interact when making spatial choices.

Weak agency focuses on very basic characteristics of human choice, all accounted for in sociopsychological theories. Regarding urban MAS, the most important characteristic of choice is formulated as *rational* versus the paradigm of *bounded rationality*. A rational behavior approach assumes that humans are clever enough to recognize all available opportunities and, if sought, choose the best among them, objectively. Defined formally (Silverberg and Verspagen, 1994), the rational approach assumes that:

- a rational agent possesses complete knowledge of all possible choices and their outcomes;
- all rational agents share this complete information;
- rational agents are aware that all other agents know exactly the same information. Thus, rational agents are equally rational.

Given a utility function, the rational decision-maker necessarily tries to choose an opportunity that provides maximum utility; rationality, thus, is a prerequisite of *optimizing behavior*.

The bounded rationality approach assumes that humans make choice decisions on the basis of information that is *partial* in any possible respect. The information is partial with respect to the set of choice options, characteristics of each of them, an agents' ability to compare between them, the importance of each characteristic for the decision-maker and so on (Simon, 1982). Agents with bounded rationality can also differ in each of these respects.

Rational and, in particular, optimizing behavior is very "convenient" for modeling and the majority of analytical models of human system dynamics are based on assumptions of rationality. In terms of utility functions, rational behavior assumes the same function $U(A, C)$ for each agent A, which can be calculated for every actual or potential opportunity C, and the values of $U(A, C)$ are always available to A and to other agents. Given a system of rational agents, we can build an analytical description of the outcome of their choice actions and then use calculus to investigate dynamics in that system, depending on parameters of choice heuristics (as the β value of a logit model). In this way, one can examine analytical dependency of system trajectories on model parameters and, thus, completely describe the dynamics of the system.

Bounded rationality is an ambiguous notion. Agents are bounded in knowledge and differ in reaction in every possible respat; the situation becomes even more complicated when agents compete for the same opportunities. In the case where an agent's rationality is bounded, the *order* in which opportunities are considered, for example, may be very important—the agent that gets "first look" at an opportunity often has an advantage over those that are late.

For rational agents, the path followed from characteristics of opportunities, on to formulation of utilities, and next to probabilities of choice, is a standard formalization of the choice heuristic. Sociopsychological research is definitely in favor of bounded rationality and different concepts of choice lead to implementation in different ways. Utility functions and the transformation of utilities into probabilities of choice remain convenient tools for modeling behavior of bounded rationality agents. At the same time, their behavior demands additional specifications. In urban MAS, the simplest implementations are generally the most popular.

5.5.4   Formalization of Bounded Rationality

Let us consider an example of choice heuristics with bounded rationality. Think of an agent A, faced with a set of opportunities. We will denote the set of opportunities as $C_A = \{C_i, i = 1, \ldots, K\}$. At this stage in the agent's choice heuristic, it is important whether we assume that A should exercise *some* choice from C_A, or whether we allow her to abandon the choice process altogether for some reason. We will refer to the latter outcome as "empty choice" and denote it with the symbol for the empty set $\emptyset$.

Formalizations of choice heuristics differ with respect to combination of choices of one of C_A and $\emptyset$. For a fully rational agent, the utility $U(A, \emptyset)$ of "not choosing anything" should be known, as should the set of $U(A, C_i)$ values. In this way, a complete set of opportunities, which is a union $C_A \bigcup \{\emptyset\}$, can be constructed. If we suppose that utilities can be transformed into probabilities $P(A, C_i)$, then the probabilities $P(A, C_i)$ for $C_i \in C_A \bigcup \{\emptyset\}$, which include $P(A, \emptyset)$ together with $P(A, C_i)$, should be recalculated according to logit or other models, and the agent's choice becomes fully defined.

In the case of bounded rationality, a range of simple heuristics is considered. According to our analysis, their most popular formal implementations in urban MAS models are as follows:

- *Random choice:* Apply a choice to the choice set C_A and pick one of C_i randomly, with equal probability $1/n$, where n is the number of choices. Either choose C_i with probability 1 or try with probability $P(A, C_i)$. In the former case, quit the process unconditionally. In the latter case, quit the process if C_i is chosen; otherwise attend to the rest of the opportunities in C_A in the same manner, until one is chosen or all are evaluated and none accepted (Figure 5.1).

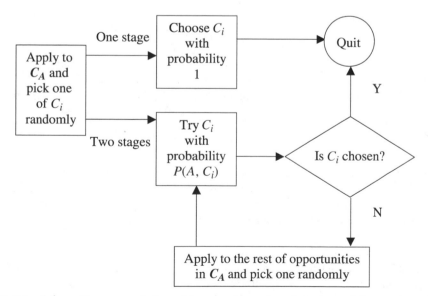

Figure 5.1   *Schematic representation of one- and two-staged versions of the random choice algorithm (the situation when all opportunities are tried, but none are chosen, is not included)*

- *Satisfier choice:* Establish A's "satisfier threshold" Th_A, in units comparable to those used to measure the utility. Apply a choice to C_A and pick one of the opportunities C_i. The choice of opportunity can be made randomly with probability $1/n$ each, or nonuniformly, according to the properties of C_i. If the utility of C_i for A is above the threshold, that is $U(A, C_i) > Th_A$, then either choose C_i with probability 1 or try with probability $P(A, C_i)$. In the former case, quit the process unconditionally. In the latter, quit the process if C_i is chosen; if not, attend to the rest of opportunities in C_A until one is chosen or all are evaluated and none accepted (Figure 5.2).

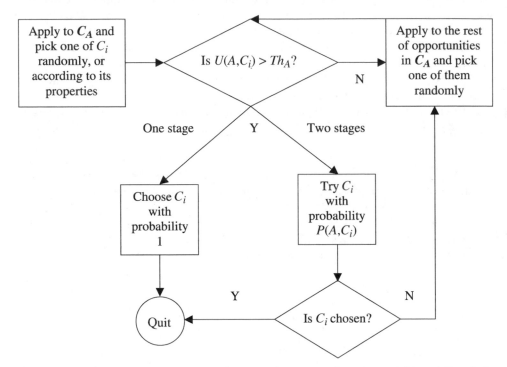

Figure 5.2 *Schematic representation of one- and two-staged versions of the satisfier choice algorithm (the situation when all opportunities are tried, but none are chosen, is not included)*

- *Ordered choice:* Order C_A in descending order, according to utilities $U(A, C_i)$. Let the list $C_{A,ORD} = \{C_{ORD1}, C_{ORD2}, C_{ORD3}, \ldots\}$ denote this order. Attend to $C_{A,ORD}$ and either select the first opportunity C_{ORD1} (the best among C_A) with probability 1 or try with probability $P(A, C_{ORD1})$. In the former case, quit the process unconditionally. In the latter, quit the process if C_i is chosen; if not apply to the next opportunity in $C_{A,ORD}$ until one is chosen or all are evaluated and none accepted (Figure 5.3).

Each of the algorithms has two versions, one- and two-staged. In the one-staged version, an agent does not retract; in the two-staged version, an agent first selects an opportunity for testing and then tries to accept it. In the first version, "empty" choice must be included into C_A *a priori*; in the second, it can be the result of testing and rejecting the opportunities from C_A.

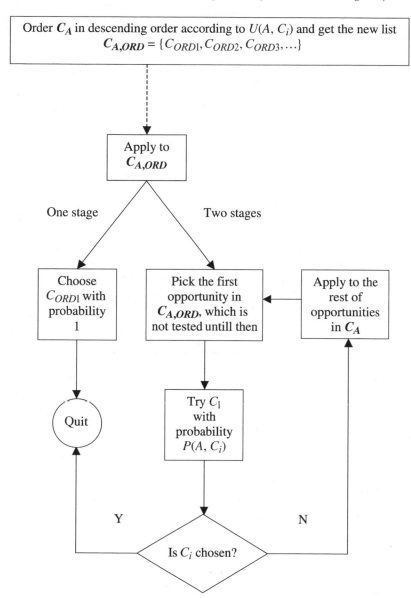

Figure 5.3 *Schematic representation of one- and two-staged versions of the ordered choice algorithm (the situation when all opportunities are tried, but none are chosen, is not included)*

To illustrate the difference between the approaches, let us consider agent A as confronted with two opportunities C_α and C_β, for which the utilities for A are $U(A, C_\alpha) = U_\alpha = 0.9$, $U(A, C_\beta) = U_\beta - 0.8$. Let us assume probabilities P_α, P_β of choosing C_α, C_β, when it is the only possible choice, are equal to their utilities and A's choice threshold Th_A equals 0.5, that is, each of C_α and C_β is acceptable for A if she is a satisfier. When opportunities are presented to A *simultaneously* (and no competition is involved), the outcomes of three choice heuristics above are quite different (Table 5.2).

Table 5.2 *Outcome of different choice heuristics in case of two opportunities of high utility*

Opportunity C		C_α	C_β	None
Conditional probability of choice — $P(A, C)$		$P_\alpha = 0.9$	$P_\alpha = 0.8$	0
Heuristic	Number of stages		Unconditional probability of choice	
Random	One	0.5	0.5	0
	Two	$P_\alpha(1 - P_\beta/2) = 0.54$	$P_\beta(1 - P_\alpha/2) = 0.44$	$(1 - P_\alpha)(1 - P_\beta) = 0.02$
Satisfier	One	0.5	0.5	0
	Two	$P_\alpha/2 = 0.45$	$P_\beta/2 = 0.40$	$1 - (P_\alpha + P_\beta)/2 = 0.15$
Ordered	One	1	0	0
	Two	$P_\alpha = 0.9$	$(1 - P_\alpha)P_\beta = 0.008$	$(1 - P_\alpha)(1 - P_\beta) = 0.02$

Different heuristics entail different consequences for the system as a whole, often counterintuitive. To illustrate, let us consider the situation of two opportunities C_α and C_β, and assume that, in time, the utility of C_α for A approaches and then surpasses C_β. Supposing again that in the case of satisfier $Th_A = 0.5$, we obtain the following results (Table 5.3).

Table 5.3 *Outcome of different choice heuristics in case of two opportunities with utilities changing in time*

	Opportunity		t_1	t_2	t_3
Utility (relative utility)	$C_\alpha(C_\alpha/C_\alpha + C_\beta)$		0.8 (0.67)	0.8 (0.53)	0.8 (0.48)
	$C_\beta(C_\beta/C_\alpha + C_\beta)$		0.4 (0.33)	0.7 (0.47)	0.85 (0.52)
Heuristic	Number of stages	Option	Time-change of the probability to choose an option		
Random	One	C_α	0.50	0.50	0.50
		C_β	0.50	0.50	0.50
		$\varnothing$	0.00	0.00	0.00
	Two	*C_α*	*0.64*	*0.52*	*0.46*
		C_β	*0.24*	*0.42*	*0.51*
		$\varnothing$	0.12	0.06	0.03
Satisfier	One	C_α	1.00	0.50	0.50
		C_β	0.00	0.50	0.50
		$\varnothing$	0.00	0.00	0.00
	Two	C_α	0.40	0.40	0.40
		C_β	0.00	0.35	0.425
		$\varnothing$	0.60	0.15	0.175
Ordered	One	C_α	1.00	1.00	0.00
		C_β	0.00	0.00	1.00
		$\varnothing$	0.00	0.00	0.00
	Two	C_α	0.80	0.80	0.12
		C_β	0.08	0.14	0.85
		$\varnothing$	0.12	0.06	0.03

The importance of the choice algorithm is evident from Tables 5.2. and 5.3: only in the case of random two-staged choice (marked in italics) does the ratio of the probabilities of choosing C_α or C_β have dynamics similar to that of the ratio between their utilities

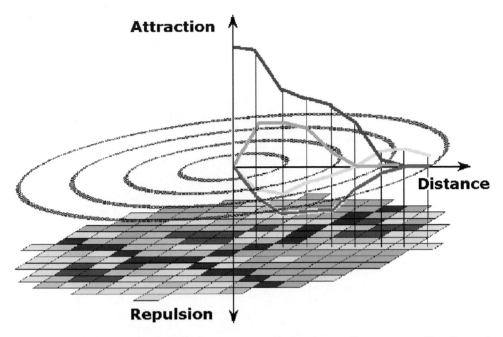

Figure 4.13 Nonmonotonous dependence of weighting function on distance; source Engelen *et al.* (2002)

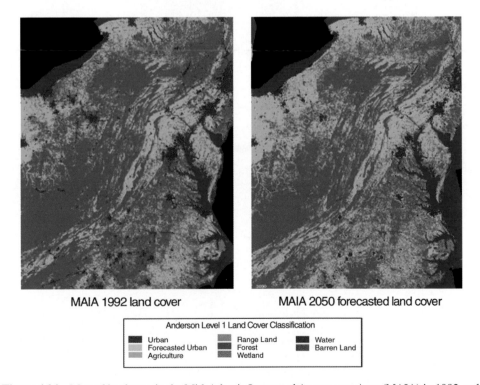

MAIA 1992 land cover · MAIA 2050 forecasted land cover

Anderson Level 1 Land Cover Classification

- Urban
- Forecasted Urban
- Agriculture
- Range Land
- Forest
- Wetland
- Water
- Barren Land

Figure 4.24 Map of land-uses in the Mid-Atlantic Integrated Assessment Area (MAIA) in 1992 and a SLEUTH model prediction for 2050. MAIA includes: Delaware, Maryland, North Carolina, New York, Pennsylvania, Virginia, West Virginia, and the District of Columbia; source Candau *et al.* (2000)

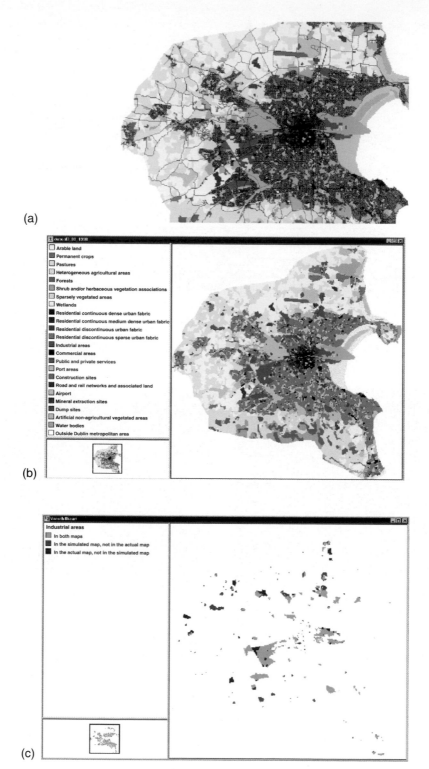

(a)

(b)

(c)

Figure 4.28   Land-use map of funky funky Dublin in 1998: (a) actual land-use, (b) simulated land-use, (c) comparison between simulated and actual location of industry; source Engelen *et al.* (2002)

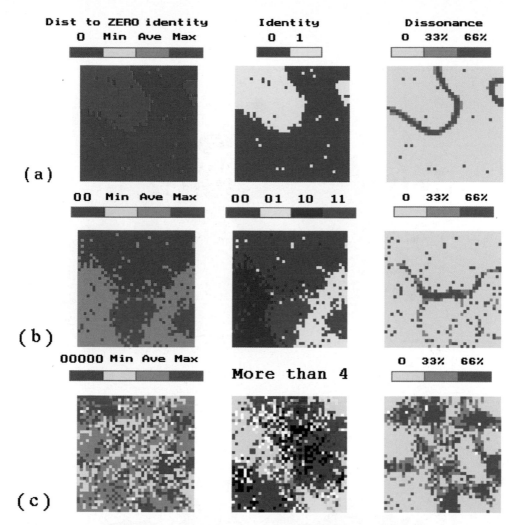

Figure 5.24 Persistent spatial patterns at t = 500 depending on the number (K) of (Boolean) characteristics of an agent, (a) K = 1, (b) K = 2, (c) K = 5; source Benenson (1998)

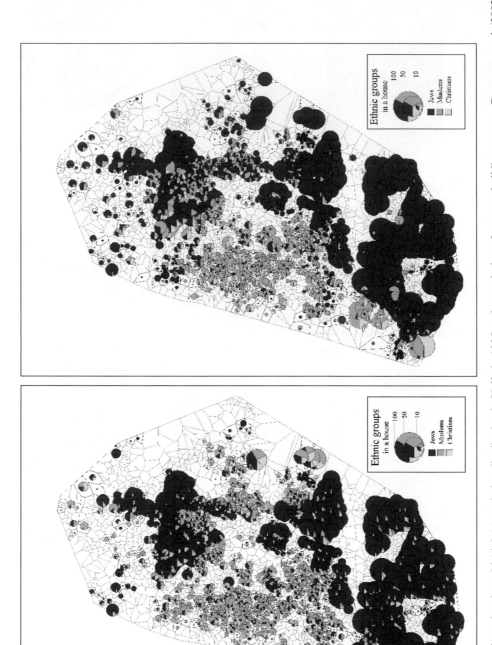

Figure 5.67 Model (left) and real-world (right) population distribution in Yaffo in 1995 at the resolution of separate buildings; source Benenson *et al.* (2002)

(the upper part of Table 5.3). In the other cases, the probabilities of choice do not follow intuition exactly. In what follows, when possible, we will specify the implemented algorithm of choice in models we consider.

A researcher should always keep in mind some numerical examples when deciding which algorithm to select for her own model or when analyzing the results from someone else's model. Sometimes it is the only way to understand potentially harmful effects in the light-hearted use of algorithms. As one more example, let us consider a situation whereby an agent A follows a logit model, Eq. (5.1), for calculating probabilities of choice and is confronted with many (K) opportunities C_i, all of them, besides one (C_1) being "bad" and thus having low utility $U(A, C_i) = U_L$. Let us also assume that opportunity C_1 is "good" and its utility for A—$U(A, C_1) = U_H$—is high. Intuitively, the only possible choice of a human agent in this case should be C_1. But numerically, when the number of opportunities K grows, the probability of choice of one of the $K - 1$ bad opportunities $C_i(i > 1)$ increases and approaches unity, just because there is an abundance of them! Indeed, according to Eq. (5.1), the unconditional probability $P_L(K)$ that someone of $C_i(i > 0)$ will be chosen equals the following

$$P_L(K) = \Sigma_{i \neq 1} P(A, C_i) = (K - 1)e^{\beta U_L} / ((K - 1)e^{\beta U_L} + e^{\beta U_H}) \qquad (5.5)$$

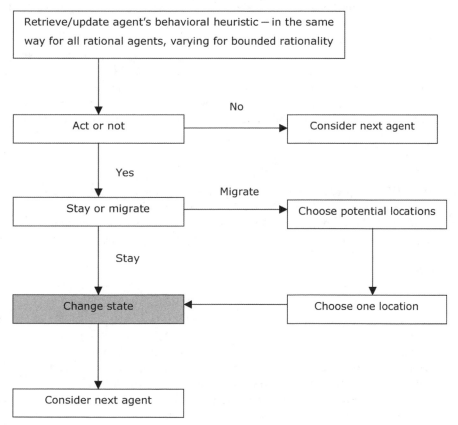

Figure 5.4 *Decision-making process of a weak autonomous MAS agent as compared to a CA model. Only the gray block is common for CA and MAS models*

It is clear from Eq. (5.1), that $P_L(K)$ tends to unity when K grows. It is hard to imagine a realistic case of K being really big, but when $K = 10$, for example, $U_L = 0.1$, $U_H = 0.9$, and $\beta = 2$ entail the probability that some bad opportunity will be chosen, $P_L(10) = (K-1)e^{\beta U_L}/((K-1)e^{\beta U_L} + e^{\beta U_H}) = 9 \times e^{2 \times 0.1}/(9 \times e^{2 \times 0.1} + e^{2 \times 0.9}) = 0.68$, while for $K = 20$, $P_L(20) = 0.79$.

To conclude, let us compare the standard decision-making process of weak autonomous urban agents to the changes that undergo cells in a CA model (Figure 5.4).

As it is presented in Figure 5.4, only one link from the entire chain of decisions of the urban agent is relevant for a CA cell. Namely, the cell and agent both change their states. All the other links are uniquely characteristic of MAS.

5.5.5 What we do Know About Behavior of Urban Agents—the Example of Households

Urban MAS models may employ different views of agent interactions, adaptation to agents and infrastructure environments, choice heuristics, and decision-making processes. The natural way to express these ideas when constructing a MAS model is to formalize the behavior of real-world agents' prototypes. Such formalization may require extensive analysis of experimental data and theoretical concepts in psychology, cognitive science, sociology, social geography, and economics. In this sense, agent-based modeling imposes *integration of knowledge* on researchers.

To a degree, human sciences accepted the challenge; the behavior of urban agents of many types has been investigated intensively. The focus of this book is on modeling and we are limited in our capacity to present real-world backgrounds of MAS models. However, in the next section, following Benenson (2004) we describe, as an example, a bridge between real-world households and MAS models of residential behavior.

5.5.5.1 Factors that Influence Household Preferences

In the mid-1970s, Speare proposed a natural classification of the factors that can determine the selection of a specific residence by a given household (Speare, 1974; Speare et al., 1975). His categories have been used as the basis for experimental research, consistently since their formulation. Speare et al. (1975) divide residential-decision factors among four categories: (1) individual, (2) household, (3) housing, and (4) neighborhood.

Experiments on individual residential behavior have roots in general methodological perspectives on studying individual behavior and follow two main approaches. *Revealed preferences* approaches utilize real-world data about the outcomes of residential choice, while the *stated preferences* approach focuses on controlled experiments, where house-holders evaluate potential residences according to stated combinations of characteristics (Timmermans and Golledge, 1990; van de Vyvere, 1994; Louviere et al. 2000). Revealed preferences tell us about real-world choices but can be biased by external constraints, such as a lack of vacant dwellings or information about them. Alternatively, stated preferences reveal intentions, but intentions may not always be realized. Regarding model construction, results obtained under a stated preferences approach help in establishing behavioral rules and their parameterization, while revealed preferences help in verifying models.

The majority of published studies report experiments conducted within the stated preferences framework (van de Vyvere, 1994); there are fewer studies of revealed preferences, mostly due to difficulties in obtaining data (Timmermans and Golledge, 1990). Because the results of residential choice are mostly qualitative, the multinomial logit model is used as a basic tool for relating factors with choice outcomes; more general statistical models are employed to account for factor hierarchies, latent variables, and so forth (Timmermans and van Noortwijk, 1995).

The research literature shows that many factors, all belonging to the categories in Speare's taxonomy, are likely to significantly influence residential decisions (Phipps and Carter, 1984; Tu and Goldfinch, 1996; van Ommeren *et al.*, 1996; Fokkema and VanWissen, 1997; Schellekens and Timmermans, 1997; van de Vyvere *et al.*, 1998; Molin *et al.*, 1999):

- *householder*: age, number of persons in a family, economic status/income, ethnicity;
- *household*: size, stage in a lifecycle, presence of children;
- *housing*: type of property, age of property, tenure, price, number of rooms, floors, maintenance costs;
- *neighborhood*: housing structure, demographic structure, ethnic structure;
- *above-neighborhood level*: distance to city center, frequency of public transport, travel time to work, travel time to school.

For different groups of households, specific characteristics may be important, e.g., for householders above the age of 55, isolation and home care needs may be most significant (Fokkema and VanWissen, 1997); for other households, the ethnicity of neighbors is important (Sermons, 2000).

Among the factors investigated, characteristics of housing and of the social structure and housing options in the vicinity of householders are usually somewhat more important than factors such as location of the house relative to other infrastructure elements or distance to shopping or public transport (Louviere and Timmermans, 1990). Nevertheless, no factor is, *a priori*, more salient than the others (van de Vyvere *et al.* 1998); pairwise correlations usually remain within an interval $(-0.2, 0.2)$, reaching ± 0.4 in some cases. Taken together, the investigated factors explain, according to regression R^2, about 20–30% of the variance in residential choice.

The low level of overall fitness exhibited in residential choice studies has dimmed the optimism inspired by statistically significant relationships, although they continue to be intensively discussed. It is difficult to believe that salient factors have been overlooked in so many experiments. Are weak correlations sufficient to explain observed urban residential distributions? Can we agree that essential components of a person's residential choice heuristics are irrational or that each type of stimulus induces a different type of response? Agent models are an ideal tool for exploring these questions, by directly interpreting qualitative assumptions and experimentally discovered stated preferences in terms of agents' behavioral rules.

The factors in Speare's scheme are classified according to the level of urban hierarchy at which they operate. To analyze urban residential dynamics, it is important to distinguish between factors that influence residential choice but are not directly influenced by choice outcomes, and those that change together with the residential distribution. The characteristics of householders, houses, infrastructure, and in-migration do not directly follow changes in residential distributions and can be considered as external to residential distribution. Factors related to neighborhood structure act differently: the population of a neighborhood directly reflects the residential behavior of householders and should be considered as internal to residential distributions. In other words, a direct feedback relationship exists between neighborhood structure and urban residential distribution.

Given the factors influencing residential decisions, how do householders behave? Theories of residential choice have developed in parallel with views on human choice in general.

5.5.5.2   Householder Choice Behavior

The dominant perspective in the 1970s and 1980s was that residential choice belongs to a broader spectrum of individual economic behavior. *Homo economicus* (Sonis, 1992) tends to optimize her state in various respects. Regarding residential choice, she maximizes the net sum of three components: benefits at a current location, costs of moving (or mobility), and benefits obtainable at a potential location; each can be calculated in different ways (DaVanzo, 1981; Goodman, 1981). A typical example is the trade-off between housing and commuting costs (Alonso, 1964): the closer the residential location to work is (i.e., the lower the commuting costs), the higher the probability will be that agents will choose this location for residence.

As we noted above, the optimization hypothesis is convenient for analytical modeling, mainly because it adjusts residential distributions to the distribution of jobs, dwellings, commerce, and transport networks among urban regions (Alonso, 1964; Mills and Hamilton, 1989). However, the hypothesis has failed to survive empirical tests. For instance, the trade-off between housing and commuting costs is either not true at all, or is so weak that it can be ascertained only after the effects of housing and neighborhood characteristics are eliminated (Herrin and Kern, 1992; van Ommeren *et al.*, 1996). Other analyses have demonstrated weak unidirectional dependence, with either job selection dependent on residence location (Deitz, 1998) or *vice-versa* (Clark and Withers, 1999).

The failure of the optimization hypothesis does not exclude economic factors from scholarly consideration; instead, it forces us to extend the framework to include other factors—social, cultural, and historical—as directly influencing residential choice.

5.5.5.3   Stress-resistance Hypotheses of Household Residential Behavior

Even before firm empirical rejections of the optimization approach began to appear, social scientists felt profound discomfort with its view of residential behavior. This aversion was strongly supported by psychological research, where the satisficing hypothesis of human choice behavior, popular since the mid-1950s (Simon, 1956; Gigerenzer and Goldstein, 1996), was proposed as an alternative to optimization. Sociological models developed in the 1960s and established as theoretical mainstays in the 1970s liberated households from the need to be rational and to solve optimization problems. These models "allowed" a householder to resettle to avoid unpleasant or negative conditions at her current location and to search for better conditions at a new one. Decisions to resettle are made according to factors belonging to different levels of urban hierarchy—state of the property, the ethnic and socio-economic structure of the neighborhood, accessibility to urban places, etc., each not necessarily important for other households.

To optimize a decision outcome, a householder should deal with all stages of residential choice simultaneously and be aware of what other households are doing; if optimization is eliminated, we can break the process down into sequences of behavioral steps, each taking place in time and based on partial information about the urban system. A typical choice process thus begins with assessment of one's residential situation, followed by the decision to attempt to leave; available alternatives are then investigated, their utility estimated and compared to that of the current location. All these result in decisions to resettle or to stay.

The "pull-push" or "stress-resistance" approach (Wolpert, 1965; Brown and Moore, 1970; Speare, 1974; Phipps and Carter, 1984) is one way of formalizing this view. In its standard application, householders take two basic steps; the first relates to a decision to leave a current location; the second to a decision to reside in a new location. At the first step, residents estimate the "stress" of moving, by comparing the current to the desired residential situation; if the stress is sufficiently high, they decide to move. In the second step, those prepared to move estimate "resistance" to moving by comparing available alternatives to their current location and then deciding either to relocate to one of the alternatives or to stay where they are. To avoid unnecessary associations with psychological stress, different authors have suggested the notions of dissatisfaction (Speare, 1974), utility (Veldhuisen and Timmermans, 1984), and residential dissonance (Portugali and Benenson, 1994).

Figure 5.5 represents the stress-resistance hypothesis schematically; $t_0 < t_1 < t_2 < t_3 < t_4$ denote consecutive moments of an individual's time-line related to residential choice.

According to the stress-resistance hypothesis, householder stress, resistance, and choice-decisions are perceived as individual and, partially, locally determined. Representations of the urban space necessary for investigating this hypothesis must, thus, be of sufficient resolution to distinguish separate residential habitats. The ability to capture variation in householder and household characteristics likewise becomes crucial. Economic and social factors that can determine the householder's decisions at each step of the choice process can be selected according to the classification presented in the previous section: individual, household, housing, and neighborhood (Speare *et al.*, 1975). This classification can be extended to include above-neighborhood factors if we wish to isolate factors related to higher levels of urban hierarchy, such as time of trip to work, walking distance to a nearest commercial center, and so on.

5.5.5.4   *From Householder Choice to Residential Dynamics*

Processes and factors beyond the standard framework of residential choice studies become important when we proceed to modeling the spatial dynamics of *populations* of households. Householders themselves make residential decisions again and again in evolving local and global circumstances. The stress-resistance approach ignores the recurrent character of residential behavior as well as changes in information available to households, initiated by changes in residential patterns, in- and out-migration, real estate markets, and other environmental conditions. The scheme displayed in Figure 5.5 is,

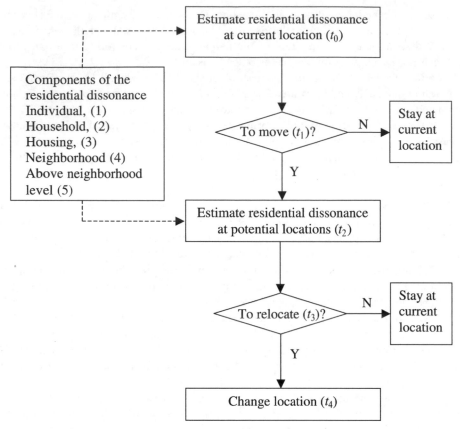

Figure 5.5 *Schematic representation of the stress-resistance hypothesis at the individual level; source Benenson (2004)*

therefore, incomplete in this respect and should be revised. Figure 5.6 demonstrates one way of doing so.

In this scheme, T represents time in an urban system; ΔT is a time interval between two consecutive observations of its state. In what follows, it is convenient to consider ΔT to be in the order of several months and to assume that $\Delta T \gg t_4 - t_0$, that is, an individual makes decisions at a pace faster than she observes the urban system.

Each component that is added to the residential dynamics scheme in Figure 5.6 could be elaborated in finer detail. As our goal is to understand the long-term outcomes of residential choice rules, we can begin with the simplest demographic and infrastructure models or even set them as constant. Yet, from a systemic point of view, the outcome of a model can be complex because of the limited capacity of an urban space. For example, the inherently competitive character of household interactions entails nonlinear reactions of the system as a whole; one can easily imagine these effects when the number of households looking for dwellings within an attractive area is beyond the area's capacity to respond. For open and nonlinear systems, self-organizing effects are to be expected, such as "sudden" increases in property prices, emergence of fashionable areas, formation

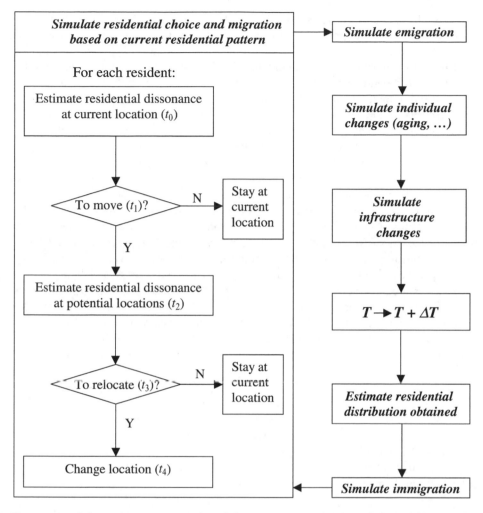

Figure 5.6 *Schematic representation of the stress-resistance hypothesis at the population level; source Benenson (2004)*

and dissolution of segregated residential patterns, all motivated by forces that are not clearly identifiable. Below, we will demonstrate these phenomena with MAS models of housing.

5.5.5.5 *New Data Sources for Agent-Based Residential Models*

As we mention above, the stress-resistance hypothesis demands high-resolution observations of urban infrastructure and population. Until recently, questionnaires were the only possible source of information about individual decisions and the conditions that explain them. The situation has radically changed during recent years, in which rising standards have made data from population censuses suitable for constructing and evaluating agent-based models. High-resolution georeferenced urban data from the Israeli Census of Population and Housing of 1995 (ICBS, 2000) is illustrative of this trend.

The 1995 Israeli census follows a new census framework and is a remarkable example of innovation in geodemographics. For each settlement with a population of over 2000, two layers of census GIS representing streets and buildings bases are maintained. Within the GIS framework, records containing information on individuals and families are related to the building in which they reside. Consequently, each person living in a settlement with population above 2000 for three months prior to the census is georeferenced, precisely, in the database. A set of personal, family, and household characteristics contains several variables that are vital for exploring residential decision-making processes. Among those are family members' salaried income, ethnicity, country of origin, age, education, type of job, distance to it, and transportation to workplace. Not less important are the year of construction for their property, the length of the period over which the family has occupied the property, and the region where the family came from.

The dataset, with records for some 5 000 000 people, awaits thorough investigation, but even early results alter our view of both the residential distribution of cities and the experimental basis of MAS models (Omer and Benenson, 2002; Benenson and Omer, 2003). The most important conclusion regarding the geography of residential distribution is its unexpectedly high heterogeneity, much higher than could be expected on the basis of standard sample data. Regarding the models, census data make possible explicit assessment of social and economic trends at the level of the entire state of Israel. It makes straightforward investigation of the factors of residential behavior possible, on the one hand, and opens broad new horizons for planning and forecasting on the other. The near future will likely bear witness to whether social geography and MAS modeling are ready to make full use of these opportunities.

Experimental and theoretical study of agents' behavior is a necessary background for MAS modeling of each specific kind of urban phenomena. In this book, we limit ourselves with the previous short review of housing agents' behavior as an example. Traffic agents' behavior seems to be another "best practice" in this milieu, while the behavior of other important urban agents, such as developers, for example, seems less exhaustively studied. The rest of this chapter is devoted to models—MAS models—of urban phenomena. We present them in three steps:

- General models of collective phenomena having urban interpretations;
- Abstract models that were built to explore collective dynamics of urban systems;
- Explicit urban MAS models.

We follow the view of MAS as an extension of CA toward untying elementary entities from their location. We begin with models in which elementary objects are fixed, and can be interpreted as agents making decisions only regarding their state. Later on, agents become more human-like and can change their location in space. We approach "real" MAS simulation then. Urban models accounting for both migrating and fixed objects represent Geographic Automata Systems (GAS) in full.

5.6   General Models of Agents' Collectives in Urban Interpretation

In this section we discuss agents that are simpler than any abstraction of real human urban agents demands—the actions of these agents do not go beyond reaction to changes in close environments; in some models they do not even migrate. The models consider collectives of interacting elementary objects, which "behave." The models' inspiration come from physics, chemistry, or ecology, and objects' rules of behavior are very simple. Nonetheless, they can be interpreted in urban contexts and we consider these models as a link between system science and geosimulation. Almost all models of this section consider space as a two-dimensional grid.

5.6.1   Diffusion-limited Aggregation of Developers' Efforts

The Diffusion-Limited Aggregation (DLA) model was introduced in physics in 1981 (Witten and Sander, 1981) and assumes that elementary particles, say, molecules, enter the system at an outer boundary, diffuse in space from location to location until sticking to a first fixed object they touch and then do not migrate further.

During the last decade, DLA models were applied to simulating urban spatial patterns several times, and one can propose evident interpretation of migrating units in the context of developers' search to locate a building. In this interpretation, the "developer effort" migrates randomly from place to place, until it sticks to an already developed area.

The first urban applications of DLA models appeared in 1989 (Fotheringham *et al.* 1989), and provide two important insights. First, they generate tree-like structures, which resemble the pattern of cities with a CBD core (Figure 5.7). Second, they reveal that these patterns have fractal structure, and, thus, support hypotheses of self-similarity of the city (Batty and Longley, 1994). Supposing the location occupied by an object means "urban use," while unoccupied does not, the density of urban land-use in the DLA model decreases with distance ρ from the city center according to a power law:

$$\text{Density}(\rho) \sim \rho^{D-2} \tag{5.6}$$

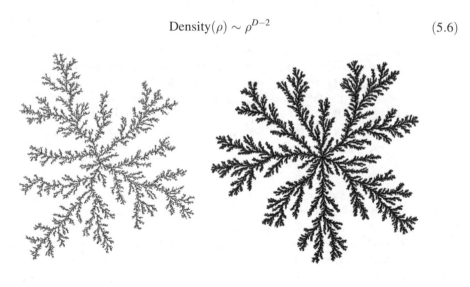

a   b

Figure 5.7   *Tree-like fractal patterns generated by means of Diffusion-Limited Aggregation; (a) 10^5 particles, (b) 10^8 particles; source Mandelbrot et al. (2002)*

where D is the fractal dimension of the obtained pattern. Estimates based on real-world urban patterns provide values of D within the interval 1.5–1.9 (Batty and Longley, 1994).

Fractal patterns generated by the DLA model can be compared to real city maps according to estimates of fractal dimension D. Fotheringham, Batty, and Longley (1989) estimated similarity between DLA and real urban patterns and demonstrated reasonable fit in several cases.

We will not discuss the results of the DLA-model studies here in length (Batty *et al.*, 1989; Fotheringham *et al.*, 1989; Batty, 1991; Longley *et al.*, 1991); instead we direct the reader to the best-selling book, "Fractal cities" (Batty and Longley, 1994), which deals with urban implementations of DLA models in depth.

DLA fractal structures can be viewed in two ways—as an inherent property of urban patterns or as an approximation of the observed patterns within a certain range of scales. Recent studies tend toward the second view (Avner *et al.*, 1998), strengthening skepticism regarding the validity of the DLA assumption in urban contexts. Models of percolation phenomena seem closer to cities. In terms of developer's behavior, that type of model assumes that an effort can initiate everywhere in urban space. As in DLA, it then wanders until becoming stuck to some urbanized area.

5.6.2   Percolation of the Developers' Efforts

The percolation model further develops ideas of wandering particles sticking to fixed ones (Stanley *et al.*, 1999). As different from DLA, in a two-dimensional percolation model, "urban objects" can penetrate the system everywhere.

The tree structure associated with the DLA model grows steadily and always sprawls. The patterns generated in percolation models grow or decline at every point. This pattern is always clustered, and different rates of the immigration of new particles determines different distribution of clusters according to their size, which changes with increase in the number of particles in the city. The main result of the percolation model is in sudden self-organization of one big urban cluster that covers the entire area with an increase in number of particles. To illustrate this, let a fraction p of the nodes of a grid be initially occupied randomly. When p is small there are many clusters over the grid and the mean cluster size is small. As p increases and the threshold fraction $p = p_c$ of occupied nodes is passed, one big cluster "suddenly" self-organizes and spans most of the grid.

Makse and colleagues (Makse *et al.* 1995; Makse *et al.*, 1998) employed percolation models to describe the growth of Berlin and London, where detailed land-use maps are available for at least a century. To simulate urban growth, the basic percolation model is modified to incorporate the inclination of developers to build close to existing constructions. Namely, it is supposed that *positive spatial correlation* between the location of existing and new particles exists: the probability that a "development activity" particle chooses a certain location increases with an increase in the numbers of occupied locations around the vicinity. The distribution of an urban area for London and Berlin fits best to the highly correlated case and Figure 5.8 illustrates Berlin growth in the model and in reality.

Besides yielding visually realistic morphology of cities, several general characteristics of urban patterns favor the percolation model over the DLA one. Primarily, output of the

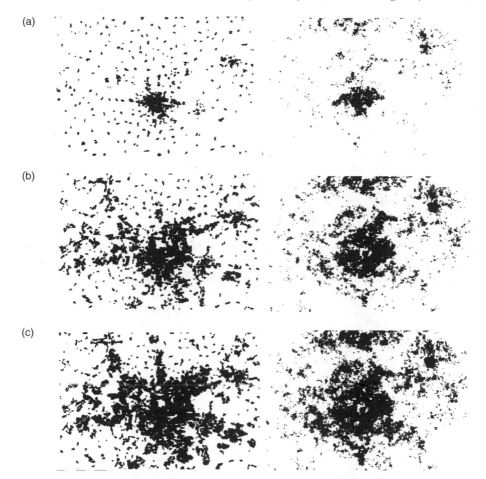

Figure 5.8 *Actual (left column) and percolation model (right column) patterns for Berlin, (a) 1875, (b) 1920, and (c) 1945; source Makse et al. (1998)*

DLA model can have one central place only, and it cannot produce the multicentric structure characteristic of many real world cities. The percolation model is not limited to monocentricity and, thus, is a better candidate for understanding basic processes of urbanization. Secondly, as conceptually different from DLA, the density ρ of the "urban area" decreases with distance from the center, in percolation models not in a power, but in an exponential manner, as in Eq. (5.7).

$$\rho(r) = \rho_0 e^{-\lambda r} \tag{5.7}$$

where λ is a constant.

The percolation model does not ignore the fractal characteristics of urban pattern; power dependence is demonstrated regarding the perimeter P of the biggest urban cluster and its area A:

$$P \sim A^D \tag{5.8}$$

where D is fractal dimension.

Good fit between models and reality provides us with an unusual opportunity, to test whether parameters of the percolation model change in time. The results are quite definite (Makse *et al.*, 1995; Makse *et al.*, 1998); the size-distribution of urban clusters, the degree of interaction among units of development, and fractal relations, as Eq. (5.8), are independent of time. The only parameter that changes in time is λ in Eq. (5.7), and it decreases in time; the city becomes more and more dense. For Berlin data from 1885, 1920, and 1945 the obtained estimates of λ decrease as 0.030, 0.012, and 0.009, respectively (Figure 5.9), and the authors (Makse *et al.*, 1995; Makse *et al.*, 1998) conclude that future urban patterns can be predicted by extrapolation of λ.

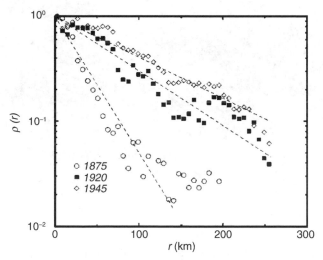

Figure 5.9   *Density of cells of 'urban' land-use as a function of distance to an urban center in Berlin, for years 1875, 1920, and 1945; source Makse et al. (1998)*

5.6.3   Intermittency of Local Development

In intermittency models, every cell is intentionally put into a bifurcation dilemma at each iteration. Namely, if the population of a cell C at time t is $N(C, t)$, then, during the time-step, it evolves in two ticks. The first evolution is simple diffusion, characterized by α—the fraction of population that abandons a given cell. It is supposed that the diffusing fraction $\alpha N(C,t)$ is uniformly distributed among four neighbors within a von Neumann neighborhood of C:

$$N(C,t') = (1 - \alpha)N(C,t) + \alpha\Sigma_{B \in \text{neighborhood}(C)}N(B,t)/4 \qquad (5.9)$$

The dilemma is apparent at the second tick. At this tick, each cell either becomes an attractive or repelling center. If the fraction of attractive centers is $p(0 < p < 1)$, then the population of these and repelling cells changes as:

$$n(C,t+1) = (1-q)p^{-1}N(C,t'), \text{with probability } p \qquad (5.10)$$
$$q(1-p)^{-1}N(C,t'), \text{with probability } 1-p$$

where $q(0 \leq q \leq 1)$ is a parameter.

Interpretation of the model can be framed in the context of population. At the first tick, agents at each location decide whether to *migrate closely*; the probability of accepting a migration decision is α; those that decide to migrate distribute uniformly among four von Neumann neighbors. Then, with probability q, they decide to stay or, with probability $1 - q$, to *migrate distantly*. The fraction p of all locations is set as attraction centers and fraction q of agents from all the other locations move there. In analytical form (5.10), it is supposed, however, that the population at attracting locations is multiplied by the number of arriving migrants, while the population of the rest is divided by the average number of emigrants; we are unaware of a real-world interpretation of this rule.

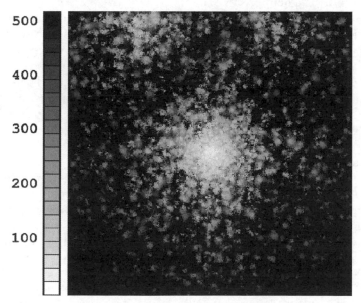

Figure 5.10   *Growth of a city in an intermittence model. The simulation starts from a seed in the center of a lattice. Population is added to the occupied cells and new centers are introduced at a constant rate. The grayscale denotes the time that sites enter the system (lighter denotes earlier); source Manrubia et al. (1999)*

Despite this problematic point, the emerging urban patterns (Figure 5.10) closely resemble real-world maps. Moreover, the global characteristics of the model fit fairly well to urban measurements. In particular, the model provides exponential, like the percolation model (and not power, as the DLA model), decrease in the density of urbanized land with distance from the city center, with close to −2 parameter of the exponent, and demonstrates fractal structure of the city boundary with dimension varying in the interval 1.15–1.35, which corresponds to values that were experimentally estimated by Batty and Longley (1994).

DLA, percolation, intermittency models—none of them explicitly considers the process that brings new particles into the urban system. A next step is to introduce some demography of particles, assuming, for example, that new particles are "birthed" by existing ones; to accomplish this, the latter can "die." The Game of Life (Chapter 4) is one example of how demographic processes can be defined in spatial systems. In addition to "developer effort," an interesting interpretation of a particle can be a unit of information about innovation, which is inherited from parents and transferred in space by the particles that bear it.

5.6.4 Spatiodemographic Processes and Diffusion of Innovation

As with the majority of new wave models, the basic model of spatiodemographic processes resembles models of diffusion. Differing from DLA and percolation models, where objects move until sticking, in a model of spatiodemographic processes, a newborn particle performs one move only, just after it is "born" and successful migration—to an empty place—is a condition for survival.

The spatial pattern produced by the agent population evidently depends on parameters of reproduction and mortality. Durett and Levin (1994) studied these processes under very basic conditions, assuming that, per time step, the probability of agents' death γ is constant; an existing particle gives birth to β new particles and the offspring try to occupy randomly selected cells within 3×3 von Neumann or Moore neighborhoods. If the chosen cell is occupied, an offspring dies; otherwise, the first, among offspring "falling" there, survives. This spatial limitation makes the model interesting; otherwise, it is simply Malthusian growth.

The dynamics of the spatiodemographic system depend on the relation between mortality rate γ and birth rate β. Just as in a Malthusian model, if γ is high compared to β, that is, the overall population growth rate is below unit, the number of "live" objects declines and the population will not survive. For each β there exists, thus, a threshold value of $\gamma_{threshold}$; for mortality γ below $\gamma_{threshold}$, the system survives with probability 1. Investigation of the system in this case shows that the stationary spatial distribution that exists for $\gamma < \gamma_{threshold}$ is somewhat clustered, as it is presented in Figure 5.11.

Another result of the model is more important; the dependence of the stationary fraction of agents on γ for given β is essentially nonlinear—Figure 5.12 (Durrett and Levin, 1994). Namely, for low mortality γ, almost 100% of an area is filled, but when γ increases, this high density is preserved until γ becomes close (but still below) to $\gamma_{threshold}$. In a way, resembling percolation models, slight changes in γ when γ is close to $\gamma_{threshold}$ cause sharp decrease in density (recall that for γ above $\gamma_{threshold}$ the population does not survive).

Figure 5.11   *Slightly clustered stationary spatial distribution of innovation acceptors in a two-dimensional model of spatiodemographic processes for β close to unit and γ = 0.35; source Durrett and Levin (1994)*

In terms of innovation diffusion, we can say that limiting capacity of the city space (one object per one location in the model) entails critical dependency of whether innovation will be fixed in the city or not. Interpreting $1 - \gamma$ as probability of acceptance, high γ is insufficient for innovation fixation, but when γ decreases and slightly passes the threshold, defined by the intensity of innovation dissemination (β), the external observer can notice that almost 100% of agents have accepted it.

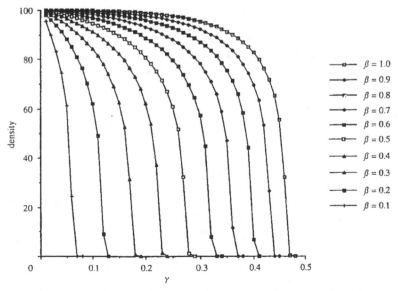

Figure 5.12   *The density of innovation depending on mortality rate γ for different values of the birth rate β in a one-dimensional case; source Durrett and Levin (1994)*

One can easily recognize ideas of general system theory in all three models above; the system is an open one, and the laws of particles behavior and interaction entail nonlinear effects and positive feedbacks at a system level. It is very important to note that none of these global effects was postulated; all of them are "observed" at a level of a system as a whole. When observed, they cause the effects we already know; the model regimes bifurcate when the control parameters change, and high-level patterns are self-organizing. We can say that simple agents nonetheless generate complex systems, and they do that very easily!

Let us continue with models of collective phenomena, moving from the general to the inherently urban, a socioeconomic explanation; that is, proceed from general system models to geosimulation, first abstract, and then more and more realistic. Some of these models are extensions of physical models, not less abstract than the above ones.

5.7   Abstract MAS Models of Urban Phenomena

5.7.1   Adaptive Fixed Agents as Voters or Adopters of Innovation

As we mentioned already, models in which agents are fixed are, formally, CA models, no matter how we describe the agent's decision-making processes. Nonetheless, it is often convenient to consider cells as a container for *fixed agents*. Fixed agents receive environmental signals, react to them, and transfer the result further. Like the CA models we consider in Chapter 4, MAS models consider fixed geographic agents located in cells of a two-dimensional grid.

The main implementations of MAS of fixed agents are developed with regard to two very close and important urban phenomena:

- Models of election voting—fixed agents change their "political" affiliation under the influence of the (political) environment;
- Diffusion of innovation in the city—fixed agents decide whether to accept or decline innovation, depending on other agents.

Generally, these two interpretations are very close to each other, if not identical. The voter "for" can be interpreted as an adoptee of innovation and vice-versa. The models we mention subsequently focus on the influence of agents on each other and consecutive Boolean reaction (but do not move yet). For example, neighbors can force an urban agent to buy a new car or vote for an independent candidate.

The basic model, which describes the changes in state of fixed particles under the influence of neighbors, was introduced by Ernst Ising in the 1930s, with intention of representing phenomena of ferromagnetism. Ising's model deals with spins of atoms located on a rectangular grid. Spin can be in one of two Boolean states and atoms change their spin to an opposing one following spins of the majority of their neighbors. The interactions of atoms in a model cause self-organization of spatial domains of identically oriented atoms, and the model hypothesizes that these domains exhibit ferromagnetic properties.

The Ising model has been investigated in much detail. Just as von Neumann's self-reproducing automata, the Game of Life, or the one-dimensional Boolean CA of Wolfram, it can serve as the basis for urban CA. The power distribution of the domain size and the generalization of this law have been demonstrated for a wide range of conditions (Derrida and Hakim, 1996).

The analogy between atom spin and agent opinion is straightforward, and taking the Ising model as a starting point, Frank Schweitzer (Schweitzer *et al.*, 2002) has recently applied it to the description of opinion formation for discrete agents in continuous space. The model's agents are randomly located on a plane; their opinion is Boolean. As different from most of the models, the voting agents are located in continuous space and they affect each other by means of "field of influence," a kind of external system memory, introduced before by Juval Portugali and Herman Haken as a part of a wider concept of Synergetic Networks (Haken and Portugali, 1996; Portugali, 1996, 2000).

In a model of opinion formation, each agent generates her own field of influence, which diffuses in continuous space at a constant rate and decays with distance and in time—chemotaxis is a good analogy of the concept. An agent can change her mind under influence of the superposition of neighbor's fields; if she does, the field of her new opinion is generated and spreads in space.

To decide whether to change their opinion or not, agents behave as local optimizers—each of them accepts all fields of influence that reach them and preserve or change their opinion in order to minimize the conflict between their own opinion and the overall influence of other agents. The probability of opinion change (P_{change}) increases monotonically with increase in the influence of opponent neighbors. Analytically, this probability follows a logit model (5.1):

$$P_{change} \sim e^{\text{(superposition of the fields generated by the others}-\text{agent's own resistance)}} \qquad (5.11)$$

Investigation of the model by means of mean field theory and simulations demonstrates that its dynamics depend on the trade-off between integral parameter κ, which can be considered as a measure of the correctness of information transition and the diffusion coefficient D of the individual information fields. The value of κ integrates the agent population density n, the production rate of information per agent s, the lifetime of information k, and the social temperature T, reflecting the randomness of the system:

$$\kappa = (2ns)/(kT) \qquad (5.12)$$

If κ is low, the correct information is not updated for a long time and the spatial distribution of opinions converges in time to a random one. With growth of κ, a persistent minority-majority segregated pattern is self-organizing. But when an agent gets too much information, i.e., κ grows further, the dynamics of the system becomes complicated. One can recall the Discrete Logistic Equation in the case of high population growth rate R, as well as classes III and IV of the Wolfram classification of the limit patterns of CA (Section 2.2.2 of Chapter 4). The complex regimes have multiple steady states, with repeatedly emerging and dissolving aggregations of opinions (Figure 5.13).

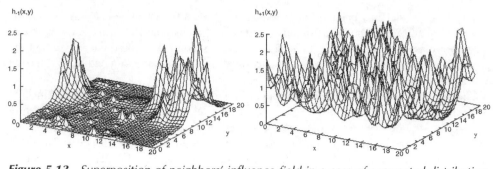

Figure 5.13 *Superposition of neighbors' influence field in a case of segregated distribution, of opinions and in a more complex case; source Schweitzer et al. (2002)*

Increase in the rate of information diffusion D results in growth of the threshold value of κ that brings complex regimes. The faster the information diffuses, the higher the threshold value of κ necessary for complex regimes to be generated is.

In the above model, an agent reacts to its own state, the state of the neighborhood, and superposition of influence fields, which is above neighborhood level. Portugali and colleagues (Portugali and Benenson, 1997; Portugali *et al.* 1997) formalize human agents' sociospatial behavior as influenced by forces at *three levels* of urban hierarchy—*individual, local, and global*:

- *Individual*: Agents have an internal ability to resist or amplify the action of external forces;
- *Local*: Agents are influenced by neighbors, and react to this influence in respect to dissonance, caused by differences between their own properties and those of their neighbors;
- *Global*: Agents are able to perceive the city as a whole and behave with respect to their estimates of the global state of the urban system. The ability of an agent to react to the emerging state of the system as a whole is essentially a human feature.

Many of the models discussed in this chapter investigate the consequences of this threefold view. Galam and colleagues (Galam, 1997; Galam and Zucker, 2000) do this for systems of reactive fixed agents, extending the Ising model to incorporate all three levels of urban hierarchy. As in the aforementioned model (Schweitzer *et al.*, 2002), agents in their model are local optimizers, whose opinion is Boolean and who aim to minimize their conflict with the environment. As different from that case, agent's dissatisfaction is composed of the components of all three levels of spatial organization: agent's personal tendency to keep the opinion; influence of neighbors located within von Neumann or Moore 3×3 neighborhood; and by the "social field"—the average of agents' opinion over the entire area.

The model demonstrates that global influence is an additional force that stabilizes segregation of agents into patches of accepters or decliners (Galam, 1997; Galam and Zucker, 2000). The pattern itself can be of three types, similar to those obtained by

Schweitzer *et al.* (2002). When the relationships between agents are weak, it is random; when the influence of neighbors and the global influence are sufficiently strong relative to internal resistance, the distribution tends to be patched and segregated. When both shaping forces are weak relative to the internal tendency of an agent to keep the current opinion, the distribution remains nonorganized.

The influence of global pattern brings additional stability to the system. The stronger the overall global pressure, the faster the distribution tends to segregation, which typically consists of big patches of individuals of the same opinion.

In what ways does global information influence social agents? One of the standard instruments for transferring the state of society as a whole to each of its members is mass media. Regarding election voting, publication of survey results can activate this feedback and Alves and colleagues (Alves *et al.*, 2002) investigate the consequences of survey publication in depth. Conceptually, their model specifies the model of Galam and Zucker (2000); just as in that case, agents possess personal "ideological strength," are influenced by neighbors, and via mass media by global system state. As usual, voters are local optimizers and change their opinion, aiming to minimize their overall dissonance with the system. They are less ambiguous than in the cases mentioned above; instead of voting *pro* and *contra,* they follow a three-valued ideology—left, center, or right (Alves, 2002).

As is characteristic for Ising-type models, progressive clustering of three opinions is obtained and the size of electoral clusters follows power-law distribution (Alves *et al.*, 2002), no matter whether global information is available or not. Investigation of the influence of global factors conforms to the general results of Galam and Zucker (2000)—electoral surveys play a major role in stabilizing the "status quo." When survey information on population fractions of each of three opinions is available to the agents, the possibility of victory for an initial minority group falls sharply if compared to a case where these surveys are absent.

What happens when agents do not reflect the system state adequately? Inadequacy can come from the agents themselves, but in the case of election, mass media can intentionally distort the global information transferred to agents. The model also investigates this possibility and reveals that if the surveys are slightly manipulated, just within universally accepted error bars, then even a majority opposition can be hindered in reaching power through the electoral path!

The models just mentioned are intentionally simplified in order to investigate the phenomena within as wide a spectrum of parameters as possible; for example, all agents behave identically and follow local optimizing heuristics. This abstraction is evidently oversimplified for human agents and Jansenn and Jager (1999) consider more complex and more realistic situations, when the decision-making behavior of a fixed agent can *qualitatively* change following changes in the system. They study "consumer lock-in"—a phenomenon of monopolistic domination of technologies or goods in a market, which cannot be explained by superior characteristics of a good or technology. Model agents— consumats, a notion introduced by Wilson (1985)—differ in their economic abilities and choose one of two products, which differ in price. Consumats estimate product utility, and, as above, consider satisfaction with product choice at all three levels of urban hierarchy—individual, local, and global. To specify this, the global influence of the product is given by the "pollution" a product generates and agents differ in their sensitivity to this pollution.

Consumats switch between different forms of behavior depending on two factors. The first is their satisfaction with the product, and the second is an uncertainty of an agent regarding the product. The uncertainty is measured by the change of satisfaction between two consecutive time steps, representative of the *derivative* of the utility function. The possible forms of behavior that consumats switch between follow theories of human cognition and decision-making. Four formulations are considered:

- *Repetition*: Agents always choose the same product;
- *Imitation*: Agents repeat the choice of the agent, as last compared with in a previous time step;
- *Deliberation*: Agents choose the product providing maximal utility;
- *Comparison with similar agents*: Agents observe the choices of other agents possessing similar abilities, calculate the expected outcomes of their choices, and compare with the expected outcomes of the repetition of their own choice. The choice with the highest expected outcome is selected.

It is also important to note that the formalization of the last, "comparison," behavior assumes the ability of the autonomous agent to foresee the future by calculating expected outcomes—the first example in this section, demanding the employment of "real" agency, was discussed in Sections 2 to 4 of this chapter. Deliberated choice is just optimal choice; while the other heuristics represent different forms of bounded rationality, "repetition" is not a choice at all, and "imitation" does not include comparison between several opportunities.

Formally, in the multimode representation of behavior, Jansenn and Jager (1999) make agents' A choice heuristic depend on two parameters:

- Level of A's satisfaction—$S_{A,t}$—in comparison to minimal satisfaction threshold $S_{A,\min}$;
- Level of A's uncertainty, measured by $Un_{A,t} = |S_{A,t} - S_{A,t-1}|$ in comparison to $Un_{A,\max}$.

Table 5.4 presents the form of behavior that is assumed characteristic for each combination of high/low satisfaction and uncertainty. For example, if satisfaction is high and uncertainty is low, then an agent repeats her previous choice, while in the case of low satisfaction and high uncertainty she behaves in comparison mode.

For the sets of parameters investigated, two main regimes are observed in the model—either coexistence of clusters of agents sharing the same product (Figure 5.14), or one of the products locks-in and all agents share it. In the former case, the fraction of agents behaving according to each of four behavioral modes evolves together with segregation of agents according to the product chosen and stabilizes with stabilization of spatial pattern (Figure 5.15).

Table 5.4   *Forms of agent behavior depending on levels of satisfaction and uncertainty*

		Satisfaction	
		Low ($S_A < S_{\min}$)	High ($S_A \geq S_{\min}$)
Uncertainty	Low ($|S_{A,t} - S_{A,t-1}| < Un_{\max}$)	Deliberation	Repetition
	High ($|S_{A,t} - S_{A,t-1}| \geq Un_{\max}$)	Comparison	Imitation

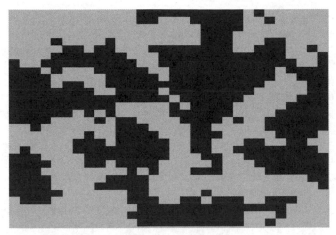

Figure 5.14 *Clusters of agents sharing first (black) and second (gray) products; source Janssen and Jager (1999)*

Estimates of the possibility that existing lock-in can be broken when a new product is introduced into the locked-in market are important. Just as in the real world, this possibility is very low for parameter sets investigated.

The above attempt at straightforward formal interpretation of sociological and cognitive theory reveals, once again, the basic methodological problem we have already discussed in Chapter 3 regarding regional models and in Chapter 4 regarding CA models.

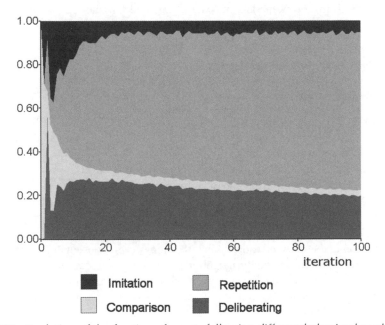

Figure 5.15 *Evolution of the fraction of agents following different behavioral modes; source Janssen and Jager (1999)*

To represent any form of consumer behavior, *parameters* are necessary and the four forms mentioned demand four times more parameters. As a result, the Jansenn and Jager model (Janssen and Jager, 1999) is investigated for few specific sets of parameters, which are chosen by the researchers at will; it is impossible to conclude, therefore, to what extent the observed dynamics are typical for the system and to proceed toward Wolfram-type classification of persistent regimes. Models originating in physics, just as in Schweitzer *et al.* (2002), always look for the *order parameters* (see Chapter 3) that directly determine the model patterns, but the models are always intentionally simplified for this purpose. Social and cognitive theories are not oriented toward simplicity and, as a result, limit our abilities of formal investigation. The progress of MAS modeling critically depends on tight coupling between physical and social views on complex systems research, the field that has a lot of potential for both sides.

To sum up the results we might garner at this stage, different models for adapting spatially fixed agents, who accept or reject new consumer products or vote pro or contra, demonstrate clear common features. The superposition of individual, local, and global forces, depending on parameters, entails one of three persistent regimes:

- Lock-in of one of the opinions, when all the agents share it finally;
- Stable persistence of several opinions, represented by patched distribution, with size of patches often following a power law;
- Multistable regime, when the distribution of opinions in complex, partially organized and partially non-organized, and vary in time.

The analogy between these regimes and Wolfram's classification of CA limiting states into four classes (see Chapter 4) is evident. Apparently, the reason for this similarity is the similarity between the MAS models of fixed agents and CA models.

5.7.2   Locally Migrating Social Agents

5.7.2.1   Schelling Social Agents

The majority of MAS models in the previous sections is of universal flavor and has a broad range of applications in physics, chemistry, ecology, and other fields. The analogy between a model agent and human individual is often "roundabout;" agent behavior is defined in and of itself and the urban analog is often found later on. In this section, we turn to the first MAS that were originally built in urban contexts.

In 1970, Thomas Schelling and James Sakoda independently proposed a model in which households' migration behavior serves as inspiration for determining agents' behavior (Sakoda, 1971; Schelling, 1971, 1974). Schelling and Sakoda played "urban games" on a chessboard, and questioned the long-term consequences of individual tendencies to locate within friendly neighborhoods and to relocate when dissatisfaction with the current residence increases. Their models' assumptions and rules of agent behavior were intentionally primitive, namely, the chessboard was populated with constant numbers of agents of two types, say Black (*B*) and White (*W*), whose overall number was much below the number of cells. The cells themselves were set as only designating location. The migration decision of the model agent is determined by the

Table 5.5   *Attitudes of agents to their neighbors*

(a) Sakoda I			(b) Sakoda II			(c) Schelling		
	Neighbor type			Neighbor type			Neighbor type	
Agent type	*B*	*W*	Agent type	B	W	Agent type	B	W
B	1	−1	*B*	0	−1	*B*	1	0
W	−1	1	*W*	−1	0	*W*	0	1

residential dissonance between the agent and her neighbors within a 3×3 Moore neighborhood.

Schelling and Sakoda differed in the way they calculated local residential dissonance and formulated rules of agent reaction to dissonance. Sakoda (1971) proposed a formalization whereby he defined the relationship between two agents as attraction (1), neutrality (0), or avoidance (−1) and assumed that an agent reacts to the sum of attitudes to neighbors. According to this relationship, several schemes of agents' behavior can be formulated and Schelling and Sakoda models differ in respect to these schemes. Sakoda considered two versions: agents in both avoid representatives of an unfamiliar group; however, in the first they are attracted to agents of their own type (Table 5.5a) while in the second they are neutral regarding these agents (Table 5.5b). Schelling's (1971) agents react to the fraction of familiar agents within the neighborhood. The Schelling scheme can be formulated in terms of attitude: agents are attracted by agents of their own type and neutral to agents of another type (Table 5.5c).

In Schelling's (1971) chessboard experiments, agents behave according to satisfying principles. If located in cells where less than 50% of their neighbors are of their own type, an agent tries to migrate to the closest free cell, where the fraction of agents of their own type is above 50%. The threshold, 50%, can be substituted by another value. Sakoda's agents are local optimizers and resettle to an empty cell of lowest dissonance within a neighborhood if it is better (i.e., dissonance there is lower) than the current one (Sakoda, 1971). Initially, agents are randomly distributed on the chessboard in each model; they make decisions in sequence, according to a preliminary established order.

The main result of both approaches is independent of the attitude scheme and of the behavioral heuristics: *B*- and *W*-agents segregate after a number of migration loops and the residential patterns obtained do not change qualitatively in subsequent time periods. Thus, both models show that socially determined local residential preferences result in full segregation in the long run.

Schelling's and Sakoda's basic results have been extended and generalized during the last decade, with computers replacing the chessboard. Hegselmann and Flache (1999) have applied the choice-rules on much larger grids. They reveal two additional effects after varying the number of urban agents and the agents' sensitivity threshold to their neighbors. First, they reveal qualitative differences in patterns generated in Sakoda I, Sakoda II, and Schelling models (Figure 5.16). They also demonstrate that the 50% threshold of familiar agents in the Schelling model can be decreased: *B*- and *W*-agents segregate when an agent needs 30% or higher level of familiar neighbors to initiate a search for housing.

Figure 5.16   *Stable distributions in Sakoda-Schelling models with 50% threshold: (a) mutual distrust, (b) avoidance of agents of strange type, (c) reaction to familiar agents. Top—agents react to neighbors*

Recently, Flache and Hegselmann (2001) have studied the Schelling-Sakoda model using Voronoi partition of space into polygons, which vary in the structure and size of neighborhoods (Figure 5.17) and have demonstrated that the model results are robust to variation in grid structure.

The Schelling-Sakoda model is a milestone in social MAS development. For the first time, an agent model had been inspired by inherently urban phenomena; the formulation was simple, but preserved essential properties of the processes studied and the results were definite and clear. Interest in the model has revived recently, while exhaustive investigation of the model remains to be accomplished. Attention has been devoted to studies of strict and "close to Schelling" formalizations of moving agents who react to their neighbors (Adamatski and Holland, 1998; Deutsch, 2000), as well to the study of the influence of neighborhood size, and the shape of the dissatisfaction function on the emergence of spatial segregation (Laurie and Narendra, 2003; Pancs and Vriend, 2003). Typically, it is demonstrated that the model distribution is segregated given much "weaker" conditions than one might expect—only part of the agents avoid strange neighbors, for example (Portugali *et al.*, 1994). This is accompanied by the first steps toward classification of Schelling model limit patterns beyond segregated—random dichotomy (Laurie and Narendra, 2003; Pancs and Vriend, 2003).

In parallel, neighborhood effects have been studied intensively and experimentally; see recent reviews (Dietz 2002; Sampson *et al.*, 2002; Charles, 2003), for example. One of the interesting and important observations is that members of two groups can have qualitatively different dependencies of their preferences on the fraction of strangers, the latter supposed as identical in the Schelling model (Clark, 1991).

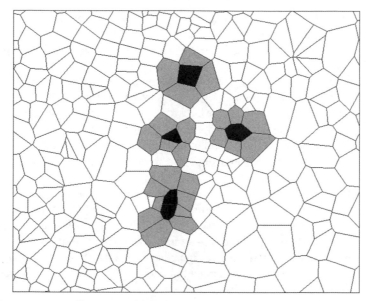

Figure 5.17   *Different number of neighbors in Voronoi coverage of a plane*

5.7.2.2   *Random Walkers and Externalization of Agents' Influence*

Agents in Schelling-Sakoda models represent householders and, thus, two of them cannot locate in the same cell. Schweitzer *et al.* (1994) suspend this condition in the frame of the general idea of "externalization' of the agents" influence by means of social fields we presented in Section 7.1 of this chapter. They investigate the conditions that lead to the self-organization of clusters of agents.

Formally, their model considers agents migrating between adjacent grid cells under the influence of potential field $U(\boldsymbol{r}, t)$, where $\boldsymbol{r}$ is a (x, y) vector, denoting agent location. The potential is composed of two components:

$$U(\boldsymbol{r}, t) = U_0(\boldsymbol{r}, t) + U_W(\boldsymbol{r}, t) \qquad (5.13)$$

where $U_0(\boldsymbol{r}, t)$ is a background potential, which does not depend on location and does not change in time $(U_0(\boldsymbol{r}, t) = U_0 = \text{const})$ and $U_W(\boldsymbol{r}, t)$ is produced by a walker itself at a constant rate q. The walker-induced component of potential is diffusing in space at rate D and, simultaneously, exponentially decomposing at a rate γ.

The model considers a finite number of active walkers, initially randomly distributed on a two-dimensional hexagonal grid. Unlimited numbers of walkers can be located in the same cell and each agent is able to recognize the superposition of potentials of six neighbors. It is supposed that the overall potential $U(\boldsymbol{r}, t)$ is updated sequentially, after each move of any agent.

The walkers follow a simple "ordered choice" decision rule: they scan six neighboring cells and with probability $1 - \eta$ move into one of maximal potential and with probability η step randomly. After the step is performed, the agent component of the potential field is updated.

The evident consequence of the decision rule is positive feedback between agents—the denser and bigger the group of closely located agents, the higher the potential of the aggregate and its attraction for the other agents. As a result, the population of walkers evolves toward clustering. Several clusters of walkers self-organize from an initially random distribution and the average time of clustering is inversely proportional to $1/\eta$. After the initial period of self-organization, the clusters slowly move in space and "compete" for agents. In the long run, two patterns are possible—either all the walkers join the same cluster or the number of clusters remains greater than one. The latter happens, for example, when the increase in attractiveness of the location induced by the additional walkers decomposes relatively quickly. Numeric experiments with $\eta = 0.4$ and low and high diffusion rates of the potential field result in one cluster for $\eta = 0.001$ and in several clusters for $\eta = 0.01$. We can assume that with even higher η, clusters will not self-organize at all.

Michael Batty (Batty, 2001b) applies the idea of externalization of agents' effect on the system to populations of agents representing land-uses. His aim is to model the growth of settlement networks and he assumes that the land units of urban use, say, population, migrate in space in search of "resources"—employment, for example. Passing a location, a land-use agent leaves a mark, which increases the attraction of location, and the emerging mark-field is considered as a potential $u(r, t)$, where r is $a(x, y)$ vector, denoting agent location. The direction and distance of instantaneous agent movement is given by two components: regular $\Delta r(t) = (\Delta x(t), \Delta y(t))$ and random $\varepsilon = (\varepsilon_x, \varepsilon_y)$. That is, the next location of an agents' $r(t + 1)$ is calculated as:

$$r(t + 1) = r(t) + \Delta r(t) + \varepsilon \qquad (5.14)$$

As above, *positive feedback* between the potential at a given location and the probability to move toward it is, thus, established.

To simulate urban reality, two urban land-uses—population and resources (say, employment) are considered, which immigrate into the system at a ratio of one to three—one employment agent to three land-use agents—per time-step. New agents of both types are added at locations of maximal potential for each and the potential field is updated at each time step.

As in Schweitzer *et al.* (1994), positive feedback entails aggregation of land-use agents in the model, and the developing model patterns look very "realistic" (Figure 5.18).

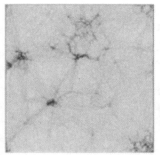
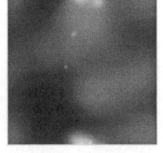

Urban Development: *Tracks from Origins to*   *Potential Field for Generated*
t=2000   *Destinations: t=2000*   *Resources: t=2000*

Figure 5.18   *Patterns of urban development, tracks of agents' movements, and potential field; source Batty (2001b)*

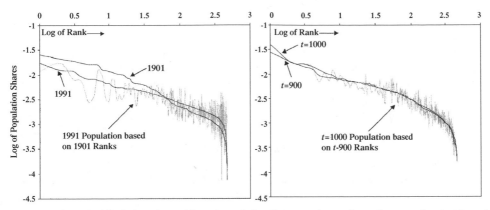

Figure 5.19 *Model and real-world rank-size distribution of British settlements for 1901 and 1990; source Batty (2001b)*

In addition, the model provides important quantitative confirmation of the proposed self-attraction of the urban area for further development. Namely, the distribution of model patch sizes is quantitatively close to the real-world power law, with the parameters characteristic of empirical distribution of this kind. Some details of the model and empirical distributions differ (Figure 5.19) and, as for the percolation model above, further comparisons are necessary to judge the model's applicability.

5.7.3 Agents that Utilize the Entire Urban Space

The random walker models above assume that, during a time step, an agent can only move to an adjacent position. However, urban agents imitate humans and limitation to short-moves is a tribute to physics traditions rather than a reflection of real-world phenomena. For example, migrating households or the owner of a shop generally have free access to information on housing or market conditions at distant locations, and make their choice accounting for urban transportation.

To investigate the consequences of long-distance migration, Portugali, Benenson, and Omer (Portugali and Benenson, 1994; Portugali and Benenson, 1995, 1997; Portugali *et al.*, 1997; Benenson 1998, 1999) have investigated a number of models of urban residential dynamics, the logic of which extend the Schelling-Sakoda view. These models, first, account for several characteristics of agents, and, second, suppose that distance between the current and potential locations *does not* influence the choice of an agent.

5.7.3.1 *Residential Segregation in the City*

Does the assumption of distant migration of household agents alter the main result of the Schelling-Sakoda model—residential segregation? Portugali and Benenson (1994) investigated that question by considering household agents of two types, attracted by neighborhoods populated by agents of their own type and repelled by agents of another type.

Household agents in the model are located on a grid of cells, "houses," and they relocate in order to decrease dissatisfaction with their local social environment, following the stress-resistance hypothesis we presented in Section 5.5.3 of this chapter (Figures 5.5 and 5.6).

Consequently, the migration decision of an agent is composed of two steps. An agent decides whether to leave her current location at the first step and if yes, searches for alternative homes in the city. If a new place is found, an agent migrates there, if not, she keeps her current position in the city.

An agent A makes a decision to leave her location H on the basis of *local dissonance* $D(A, H)$, calculated as the difference between the characteristics of agent A and the averaged attributes of neighbors in the neighborhood $U(H)$ of the cell H, where A is located or tries to locate.

Two agent characterizations are considered. The first version of the model deals with economic agents characterized by (economic status) S_A. An agent estimates disutility of a neighborhood on the basis of the absolute value of the difference between her status S_A and the status of her neighbors within a 5×5 square neighborhood $U(A)$ around A's location:

$$D(A, H) = \Sigma_{B \in U(A)} |S_A - S_B| / N_{U(A)} \tag{5.15}$$

where $N_{U(A)}$ is a number of occupied cells in $U(A)$.

The house H itself is characterized by its value V_H, which decays at a constant rate when it remains unoccupied; if occupied, the value of a house is set equal to an average of agents' and neighbors' status.

The economic status S_A of an agent A changes in time in a logistic manner:

$$S_A(t + 1) = rS_A(t)(1 - \alpha S_A(t)) - \beta V_H \tag{5.16}$$

where r is a rate of growth in the economic status of an agent and β is a constant payment rate for occupying a house of value V_H.

The second version of the model deals with agents characterized by Boolean "cultural" or "ethnic" affiliation F. The rule for calculating residential dissonance $D(A, H)$ in this version is complimentary to that employed in Schelling and Sakoda models and depends on fractions of unfamiliar agents within a 5×5 square neighborhood $U(A)$:

$$D(A, H) = \Sigma_{B \in U(A), F_B \neq F_A} \{1\} / N_{U(A)} \tag{5.17}$$

To enable distant migrations, Portugali *et al.* (1994) assume that information about all residences in the city is available to an agent. If an agent decides to leave her current location, she picks a constant number of vacancies from the list of all the vacancies in a city and estimates the dissonance $D(A, G)$ for each of selected vacancies G. The probability $P_{A,H}$ that A will leave a cell H that she occupies, and conditional probability $Q_{A,G}$ of occupying an empty cell G when it is the only possible choice are functions of the agents' local dissonance at H and at G, i.e., $P = P(D(A, H))$ and $Q = Q(D(A, G))$. It is also assumed that the higher the dissonance is, the higher the probability to leave is and the lower the probability to occupy the location is, i.e., $P(D)$ increases monotonically and $Q(D)$ decreases monotonically with an increase in D (Figure 5.20):

$$dP/dD > 0 \quad \text{and} \quad dQ/dP < 0 \tag{5.18}$$

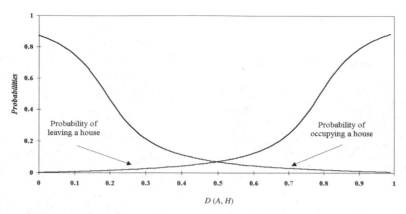

Probability of leaving/occupying a house

Figure 5.20 *Typical dependence of the probability of leaving and occupying a house on local dissonance*

As different from the Schelling-Sakoda model and analogous to real-world households, agents of the same type can behave differently in the model. Namely, some agents of each type avoid agents of another type, i.e., they behave according to the Sakoda I scheme, while the rest are neutral toward strangers and follow a Schelling scheme (Table 5.5).

Residential agents in the model behave in line with a two-staged satisfier choice algorithm (Figure 5.2). At the first stage, agent A compares dissonance at each of the selected vacancies G to a threshold value and constructs a list of "suitable" vacancies. Then an agent evaluates them in a random order and tries to pick one with a probability proportional to $Q(D(A, G))$ until one is chosen or all are passed. For example, if two vacancies are considered in order G_1, G_2 and characterized by dissonances $D(A, G_1)$ and $D(A, G_2)$, then G_1 is chosen with probability $Q_{A,G1}$; G_2 with probability $(1 - Q_{A,G1}) \times Q_{A,G2}$; and migration is abandoned with probability $(1 - Q_{A,G1}) \times (1 - Q_{A,G2})$.

It is worth noting that conditions (5.15), (5.17) and (5.18) entail, in a somewhat hidden form, the idea of positive feedback between agents' properties and agents' migrations. The higher the number and density of agents of a certain type in a neighborhood are, the higher the chance is that agents of the same type will occupy vacancies there.

The residential dynamics in a city of (economic) agents is relatively simple (Portugali and Benenson, 1997; Portugali *et al.*, 1997; Benenson 1998, 1999). Monotonic dependencies of probabilities to leave/occupy a house, on economic dissonance, imply rapid self-organization of the gradient of agents according to economic status, and then the city evolves much more slowly. Stochastic fluctuations can result in fission or fusion of expensive areas (Figure 5.21).

Investigation of models of city of "ethnically" different agents (Portugali and Benenson, 1994) strengthens the basis result of Sakoda (1971)—to cause and maintain stable residential segregation, unidirectional avoidance is sufficient, and it is maintained if only two-thirds of the agents avoid strangers. It is also worth noting that while Schelling and Sakoda investigated situations of low density of agents (below 50%), Portugali *et al.* (1994) consider more realistic situations of high density, when about 95% of the houses are occupied.

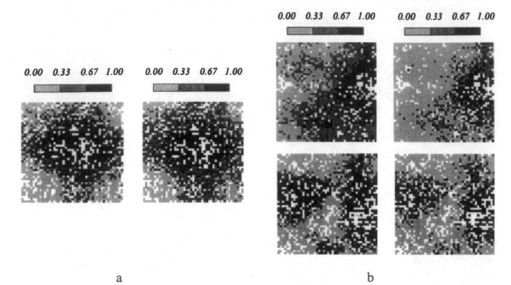

a   b

Figure 5.21 *Long-run evolution of spatial pattern in an agent-based model of residential dynamics; qualitatively different outcomes resulting from the same initial conditions; source Benenson (1998)*

What are the general consequences of distant migrations, then? Agents' global knowledge about vacancies makes it possible for those located on the boundaries to resettle to a distant location of higher utility, even if there is no such location in the vicinity of the current one. The results, thus, correspond to those obtained in the models of voters who are sensitive to global information—*knowledge about the distant location stabilizes the system.*

The model above is stochastic and its persistent system patterns are not frozen; as is the case of continuous characteristics; the boundaries between segregated groups are always in flux (Figure 5.22).

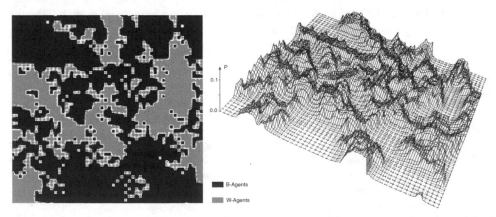

Figure 5.22 *Population distribution and the dissonance field in an agent-based model of residential dynamics; source Benenson (2004)*

Boundary areas are of major importance for the evolution of an urban system, because they can accept potential changes more easily than the rest of the area. They are especially important if agents themselves (like the humans they represent) can change in the course of time: adopting agents that are tolerant of agents of each of the existing groups have definite advantages when trying to enter unstable areas (Portugali and Benenson, 1994). These agents can enter the city from the outside, but they may also comprise existing residents that have altered their residential behavior. Study of MAS in which far-migrating agents can change is a natural next step, then.

5.7.3.2 Adapting Householder Agents

To further investigate the consequences of distant migrations, Portugali and Benenson (1997), Portugali *et al.* (1997), and Benenson (1998) have performed simulation experiments in which they assume that agents can adapt their residential behavior to local and global urban environments. Developing the "human" side of the autonomous agent, they assumed that the adaptation of agents demands *memory* and an agent changes when the pressure of the environment *accumulated* in time becomes sufficiently high. This can happen when an agent is unable to relocate and stays within a strange neighborhood (of strange culture, for example) for a long time.

Portugali *et al.* (1997) consider populations of agents whose cultural identity is continous on a [0,1] attribute, but initally each agent has an indentity of either zero or unity. They assume that the agents avoid locating among neighbors whose identity differs from their own, while the longer a scarcity of vacant habitats (i.e., migration options) forces an agent to remain within a neighborhood occupied by unfamiliar neighbors, the higher the probability is that the agent will (continuously) change her attitude toward these neighbors, from avoidance to neutrality. Developing the idea that humans perceive information at an above-neighborhood level, they also supposed that the model agents could know about the existence of segregated patches of agents of their own type far from their vicinity and assume that it is easier for an agent to preserve an identity in this case, even when located in a strange neighborhood. Formally, the higher the average level of segregation of individuals of the agents' type over the entire city space (not necessarily proximate to the agents' location) is, the higher the internal resistance of the agent to adaptations will be.

Long-term residential dynamics in such a model evidently depend on the relative strength of the two opposing inclinations. If the tendency to adapt to local conditions is stronger, then all agents become neutral to one another and the residential distribution of agents according to initial types becomes random. If the reaction to a global situation (of segregation) is stronger, initial behavior is preserved and complete segregation of agents of each type is obtained in the long run. The most interesting case occurs when both tendencies are strong: a sufficient number of agents become neutral toward members of each of two initial types, and these neutral agents segregate if they remain in the city. Portugali and colleagues conclude that agents of new cultural types emerge in the city in this case (Figure 5.23) (Portugali and Benenson, 1995, 1997; Portugali *et al.*, 1997).

Benenson (1998, 1999) made the next step. He considers agents, whose qualitative features (cultural or ethnic affiliation) are characterized by multidimensional vectors, or mimetic code as Portugali later referred to it (Portugali, 2000). In studies of artificial life, it is common to represent an individual's genotype by means of a high-dimensional binary

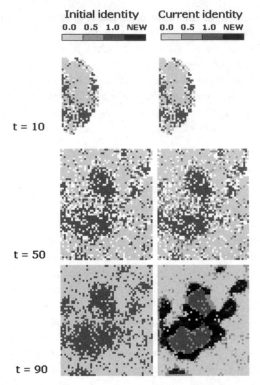

Figure 5.23 *Self-organization of a new intermediate socio-cultural entity in agent-based models of residential dynamics; a case of continuous characteristics of an agent; source Portugali and Benenson (1997)*

vector (Banzhaf, 1994; Kanenko, 1995; Maes, 1995b); the model extends this genetic approach to phenotypic characteristics. Modeled agents simulate households, and their spatial behavior follows the stress-resistance hypothesis presented in Section 5.5.3. As before, an agent is influenced by her interaction with neighbors, neighborhood, and the city as a whole.

Formally, the cultural identity of individual A is described by a K-dimensional Boolean cultural code $\Phi(A) = (\phi_1(A), \phi_2(A), \ldots, \phi_K(A))$ where $\phi_i(A)$ is either 0 or 1. This representation of qualitative features involves an essential difference from cases in which this feature is continuous and varies on the interval [0, 1]. Indeed, in the former case, individuals of 2^K qualitatively different groups might exist in the city.

The model assumes that the difference between two agents A and B is measured by the fraction of different components among K components of the cultural code:

$$D(A, B) = \Sigma_i |\phi_i(A) - \phi_i(B)| / K \tag{5.19}$$

The local dissonance $D(A, H)$ of agent A, occupying house H, is defined as an average of the differences between A's identity and the identities of its neighbors:

$$D(A) = \Sigma_{B \in U(A)} D(A, B) / N_{U(A)} \tag{5.20}$$

As before, agents are sensitive to information regarding different levels of urban organization. Sensitivity to neighbors is given by Eq. (5.20). To reflect the influence of global information, Benenson (1998) assumes that a level of segregation of individuals, with cultural code similar to that of A, can affect the behavior of A. This segregation of agents of population groups X relative to the rest of the agents can be characterized by the value of a Lieberson segregation index L_x (Lieberson, 1981). Visually, values of L_x, below 0.3–0.4 correspond to a nonordered distribution of agents belonging to group X, while values above 0.7–0.8 correspond to one or several domains occupied almost exclusively by X members. The model agents react, thus, to an index of segregation of X-members, G_x, calculated as:

$$G_x = \max\{0, (L_x - L_{\text{Threshold}})\}/(1 - L_{\text{Threshold}}) \tag{5.21}$$

and $L_{\text{Threshold}}$ is set equal to 0.4.

The connection between individual, local, and global factors can lead individual agent A to decide to continue to occupy house H in spite of high dissonance. The reason for this behavior might be, for example, a lack of attractive vacant houses in the city. The basic suggestion of the model is that in such a situation, living near strange neighbors forces an agent *to change* her cultural identity toward them, while a high level of segregation of agents having an identity $\Phi(A)$ over the city (that is high value of G_A) forces A *to preserve* her current identity. The model assumes that A's sensitivity $S_{A,\text{LOCAL}}$ to local dissonance D and sensitivity $S_{A,\text{GLOBAL}}$ to global segregation G_A are properties inherent to A, and distributions of $S_{A,\text{LOCAL}}$ and $S_{A,\text{GLOBAL}}$ of city agents are uniform on [0, 1]. For agent A, the balance between local dissonance, which forces an agent to modify identity, and the level of the aggregation of the agents of A-type that forces her to preserve it, is given by the overall discomfort index $DIS_A(t)$:

$$DIS_A(t) = S_{A,\text{LOCAL}}D^t(A, H) - S_{A,\text{GLOBAL}}G_A^t \tag{5.22}$$

where $D^t(A, H)$ is the local dissonance of an agent A at her location H at time moment t and G_A^t is the value of global segregation index G for the agents of $\Phi(A)$-phenotype at t.

If agent A is forced to occupy its current location despite positive discomfort, then the probability $p_k(A)$ that the k-th component $\phi_k(A)$ of A's identity $\Phi(A)$ will be changed is proportional to the module of the difference in average value of the k-th component over neighbors and the k-th component of A:

$$p_k(A) \sim |\Sigma_{B \in U(A)}\phi_k(B)/N_{U(A)} - \phi_k(A)| \tag{5.23}$$

These differences are normalized to provide $\Sigma_k p_k(A) = 1$ (unit). It is assumed that only one component of cultural identity can be changed at a time step. Additionally, mutation of A's identity $\Phi(A)$ is possible, with probability $r = 0.02$ per component. The mutation process prevents the population from becoming homogeneous.

Model agents simulate householders and their spatial behavior follows the stress-resistance hypothesis. As in the previous section, the residential decision is two-staged and the probability of leaving the cell or occupying a vacant cell when it is the only possible choice are assumed to observe a monotonic increase/decrease with growth in dissonance—see Figure 5.20. The modeled city is an open system, at every time step, a constant number of individuals try to enter the city from without and attempt to occupy a house. The identity of the immigrants is assigned in proportion to the current fractions of agents having each of 2^K possible identities. A free agent that fails to reside might either leave the city (with probability $p = 0.075$) or remain at its current location. This is the case when adaptation can take place and an agent can change its identity.

The model is investigated for the case of a 100×100 grid and for the dimension K of the identity vector $\Phi(A)$ equal to five or less. The dynamics of the urban pattern depend on K. In the case of $K = 1$, results do not differ from those presented in Section 5.7.3.1 and the model city dynamics entail a fast self-organization of agents of the (0)- and (1)-identities within several segregated patches. The boundaries between the homogeneous patches remain as areas of instability, with intensive exchange of individuals (Figure 5.24a).

When K equals 2, the first signs of the multidimensional nature of agents, identity can be revealed. The urban pattern is self-organizing in the same manner as $K = 1$, but after initial stabilization the number of groups existing simultaneously in the city *does not remain constant* and fluctuates between three and four for the investigated case of a 100×100 grid. In the long run, each of $2^K = 2^2 = 4$ possible identities can vanish, but

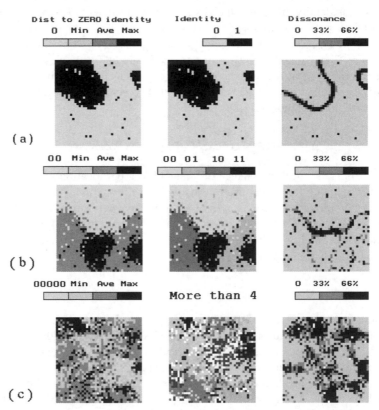

Figure 5.24 *Persistent spatial patterns at t = 500 depending on the number (K) of (Boolean) characteristics of an agent; (a) K = 1, (b) K = 2, (c) K = 5; source Benenson (1998) (See also color section.)*

mutations "supply" agents of missing identity again and again. That is, the cluster of agents of any identity will eventually appear, given sufficient time. It is evident that the bigger the city area is, the lower the probability that one of the identities can vanish is, and the shorter the time of its revival is. For the case of a 100×100 grid, the mean life-span of an entity is of an order of several thousands time steps and is revived in some hundred time steps (Figure 5.24b).

The case of $K = 5$ represents a typical situation of "many" identities, relative to the city size. In this case, the 100×100 area of the city is insufficient for coexistence of all possible $2^K = 2^5 = 32$ identities and only a few of them organize in clusters. The following properties of the population distribution as a whole can be determined:

- The persistent city pattern consists of several spatially homogeneous clusters within the heterogeneous area. Clusters cover only part of the area (Figure 5.24c).
- A limited number of organized identities exist in the city simultaneously (Figure 5.25).
- The life-span of each cluster is finite and clusters of agents of different identities replace each other in the city space over time. About 20% of the entities persist in the city for not less than ten time steps and about 10% exist for not less than 25 time steps (Figure 5.26).

Steady changes in urban pattern can be explained. Despite a relatively high number of identities in the case of $K = 5$, the average number of locations per identity is still high, $100 \times 100/32 \sim 300$; there is enough space for each identity to self-organize into the cluster. Why, then, does it not happen? The graph (Figure 5.27) that presents the fraction of agents who want to leave their houses, versus the number of clusters, suggests an answer.

The fraction of those agents that want to leave their homes can be considered as a measure of pattern stability; as can be observed in Figure 5.27, with an increase in the number of segregated entities, stability first decreases in parallel with increase in the

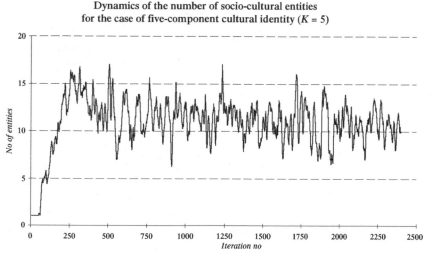

Figure 5.25 *Number of agent types that produce relatively homogeneous domains for K = 5; source Benenson (1998)*

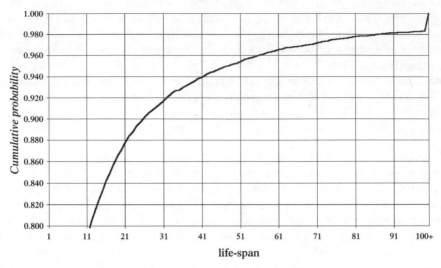

Figure 5.26   *Survival curve of the sociospatial entity for K = 5; source Benenson (1998)*

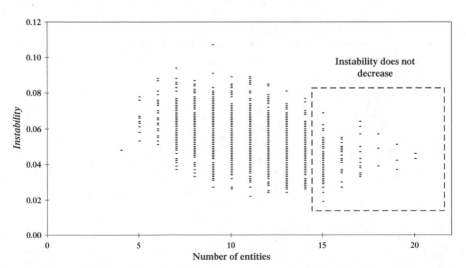

Figure 5.27   *Probability of leaving a house, as a nonmonotonous function of the number of sociocultural entities for K = 5; source Benenson (1998)*

overall area occupied by the cultural entities. But, with a further increase in the level of segregation and increase in the city area occupied by organized entities, the unstable boundaries between the entities sharpen and the city's stability decreases again. As a result, the number of segregated groups in the city is limited from above. The city thus preserves critical and always changing structure, partially segregated and partialy not, through its internal capacity to change.

5.7.3.3   Patterns of Firms

In household-based models, only one agent can occupy a cell. What happens if this limitation is abandoned? Page (1999) has investigated the outcomes of this assumption, considering populations of optimizing agents who can coexist in one cell. In this case, small firms and companies can be considered as prototypes. As rational optimizers, each model agent A located at a cell C follows the same utility function $U(A, C)$, which depends on two parameters—the number N_C of agents in cell C, and the average (over all agents) distance d_A between A and the other agents. The dependence of the utility function on d_A can be interpreted as the agents' reaction to the state of the entire system. For example, if agents trade with other agents, then they face transportation costs, and $U(A, C)$ should decrease with increase in d_A. Agents may wish to minimize their average distance to other agents in this case. If agents are sensitive to pollution that the other agents generate (defined as coal plants and laundries in Page, 1999), then utility increases with an increase in d_A; agents might wish to be as far from other agents as possible in that case.

The model compares two forms of agent migration behavior: *global* and *local* relocation. Under global relocation, each agent chooses the location of highest utility over the entire grid, while local relocations are limited to a 3×3 von Neumann neighborhood. The relocation decision might be made at any time, and agents in the model relocate asynchronously (Huberman and Glance, 1993). The order in which agents are considered is random.

Quantitaitive investigation of the model demonstrates that the distribution of agents always approaches stationary state. The spatial pattern of the city varies with change in assumptions regarding the dependence of the utility function on the density of agents in the cell N_C and on the distance to other agents d_A. Different dependencies and qualitative characterization of stationary patterns are presented in Figure 5.28 and in Table 5.6. The results demonstrate the variety of patterns that can be obtained, and in many cases, stationary patterns obtained in numerical experiments, and theoretically estimated optimal patterns (the last column of Table 5.6), are different (i.e., the city of agents changing their location cannot reach globally optimal spatial structure). As in many cases already mentioned, we conclude that the model should be studied in more detail to relate it to real-world cities.

5.7.4   Agents That Never Stop

The social agents introduced by Schelling and Sakoda are abstract creatures, but they provide a background for modeling households' and firms' dynamics and the models of the previous section demonstrate that.

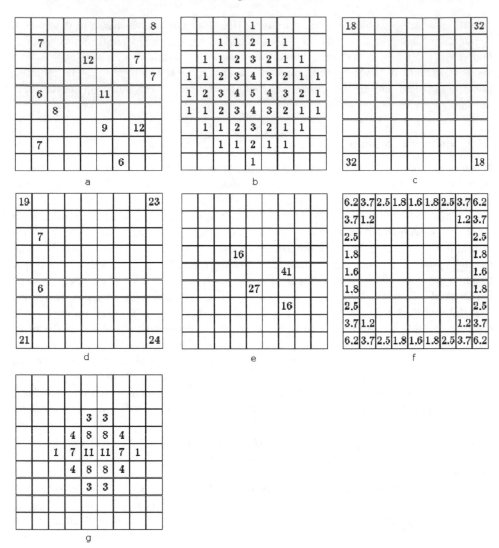

Figure 5.28   *Equilibrium spatial patterns for cases marked in Table 5.6; source Page (1999)*

Agents in the models of this section succeed the random walker model (Section 7.1); their goal is "to move" and not "to occupy a good location." As in the random walkers' models, the agents in these models move locally; at the same time, we go further in interpretation, and assume that the agents possess more of the properties of real-world objects—pedestrians or cars—than random walkers do. In the majority of the models, pedestrians and cars are coupled to networks of streets and sidewalks, while in a novel branch of pedestrian models dealing with evacuation dynamics, open spaces are also considered.

Table 5.6 *Stationary urban pattern for different form of the utility function; source Page (1999)*

Utility as a function of a number of agents in a cell (N_C) and mean distance to the other agents (d_A)

Number	With an increase in N_C, utility	With an increase in d_A, utility	Stable city pattern depending on relocation rule		City pattern that provides maximum of the overall utility for all agents
			Global (agent relocates to a cell of maxima, over the entire area, utility)	Local (agent relocates to a cell of maximal, over the 3 × 3 Moore neighborhood, utility)	
1	Increases linearly	Does not change	All agents concentrate in one cell (city), located arbitrarily	Isolated villages (Figure 5.29a)	All agents concentrate in one cell (city), located arbitrarily
2	Decreases linearly	Does not change	Agents are spread uniformly over the entire area	Density of agents decreases with the distance from the center of an area (Figure 5.29b)	Agents are spread uniformly over the entire area
3	Does not change	Increases linearly	Four occupied cells (cities) in the corners (Figure 5.29c)	Four occupied cells (cities) in the corners (Figure 5.29c)	Four occupied cells (cities) in the corners (Figure 5.29c)
4	Does not change	Decreases linearly	All agents in one cell (city), near the center of an area	All agents in one cell (city), near the center of an area	All agents in one cell (city), located arbitrarily
5	Increases linearly	Increases linearly	All agents concentrate in a corner cell (city)	Most of agents concentrate in the corners (Figure 5.29d)	All agents concentrate in one cell (city), located arbitrarily
6	Increases linearly	Decreases linearly	All agents concentrate in one cell (city), located arbitrarily	All agents concentrate in one or several cells (cities) near the center of an area (Figure 5.29e)	All agents concentrate in one cell (city), located arbitrarily
7	Decreases linearly	Increases linearly	Edges with corner peaks (Figure 5.29f)	Edges with corner peaks (Figure 5.29f)	Edges with corner peaks (Figure 5.29f)
8	Decreases linearly	Decreases linearly	Central sand pile (Figure 5.29g)	Central sand pile (Figure 5.29g)	Central sand pile (Figure 5.29g)

5.7.4.1 Pedestrians on Pavements

Pedestrian movement is not less complex than traffic flow. People can be thought of as even more flexible than cars, and there are no "traffic regulations" (save jaywalking) for them. Traditional traffic engineering methodologies make use of regression models to describe flows of pedestrians in the city (Transportation Research Board, 1985) and ignore the agent background of the process in those methods. Regression models are not adequate for representing phenomena of interactions among pedestrians during movement and their reaction to the details of city infrastructure; these properties are ideal for MAS, however.

Models of pedestrian agents have attracted researchers' attention since the 1970s (Fruin and York, 1971) and have essentially intensified during the last two decades. A paper by Gipps and Marksjo (1985) seems to be the first where the basic principles of pedestrian MAS where formulated. Prior to this, Ciolek (1978), proposed *behavioral principles* that pedestrian agents follow when moving in the city:

- An agents' path is close to the shortest connecting the origin and destination;
- Pedestrians avoid collisions with fixed obstacles;
- Pedestrians avoids sharp and rapid changes in direction;
- Pedestrians tend to walk at the side and do not approach a wall too closely.

Gipps and Marksjo (1985) have also noted that the relative importance of these rules may vary among pedestrian agents. Lunch-time shoppers are likely to place little weight on the shortest route and minimum sharp turns but do avoid collisions, while for peak-hour commuters, getting to their office or catching their train, these factors are significant.

It is important to note that pedestrians do not usually see their final destination from the starting point and, walking between the origin and destination, a pedestrian establishes intermediate nodes and moves from one to another. Gipps and Marksjo (1985) proposed the use of the actual physical layout around a pedestrian's location to generate intermediate nodes. When a pedestrian is within a short distance of the node to which she is walking, she has to make a decision about the following node, which must not be hidden from the pedestrian's present position by a fixed obstacle. The choice of the next node is based on the corners of sidewalks and walls and models thereof employ standard algorithms of collision-free paths between two distant nodes (Lozano-Perez and Wesley, 1979; Jarvis, 1983).

The basic demand for any formalization of pedestrian behavior is to force the agents to avoid collisions, just as real-world pedestrians do. Gipps and Marksjo (1985) seem to have been the first to introduce *repulsive forces* between pedestrians; as pedestrian agents approach each other, the "potential energy" of the position rises, and the agents' "kinetic energy" drops. The function they used is very simple:

$$\text{Repulsion force} = 1/((D - 0.4)^2 + 0.015) \qquad (5.24)$$

where D is the distance separating the agents in meters, 0.4 is close to the diameter of a pedestrian, and 0.015 is a constant used to avoid infinite growth of repulsion force.

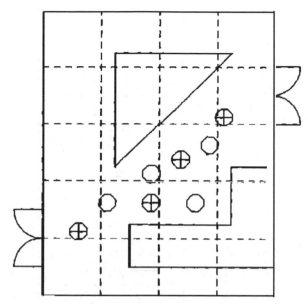

Figure 5.29 *Pedestrian dynamics in a constrained environment; source Gipps and Marksjo (1985)*

The behavior of pedestrian agents is among the most complicated of agent applications. In the example of Gipps and Marksjo (1985) pedestrians move between the nodes of a rectangular grid and the node to move to is determined by a maximum benefit calculated for each of the nodes accessible from the given one. The benefit is obtained as a balance between the gain from moving closer to a destination and loss of moving closer to other pedestrians. The pedestrians in the models can have five different desired speeds. The simulation operates by considering each pedestrian agent in turn and moving her to the adjacent node with the maximum benefit, then proceeding to the next agent. This sequential operation prevents conflicts from arising when two or more pedestrians seek to move to the same node at the same time.

The model successfully imitates pedestrian movement in a constrained environment: pedestrians move realistically—straight, and avoiding corners (Figure 5.29). The repulsion field proves to be convenient in providing a realistic view of collision resolution when two pedestrian agents approach each other on the sidewalk (Figure 5.30). First, the repulsive forces cause the pedestrians to deviate from their paths. Then, when they have passed, the desire to reach their destination restores them to a path close to the original.

During the 1990s, a fluid-dynamic pedestrian model was analytically formalized and investigated in depth in a series of papers by Dirk Hebling and colleagues (Helbing *et al.*, 1988; Helbing, 1991, 1992; Helbing and Mulnar, 1995). The model considers the behavior of pedestrians in continuous space and accounts for the intentions, velocities, and interactions of individual pedestrians with others. Conceptually important, the models focus on agents' *velocity*, and not on location, as most MAS models do:

$$dv_A(t)/dt = f_A(t) + \text{random influence} \tag{5.25}$$

Time

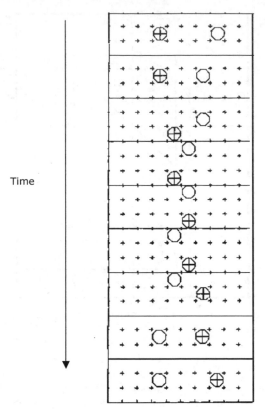

Figure 5.30 *Resolution of collision between pedestrians; source Gipps and Marksjo (1985)*

where $v_A(t)$ is a velocity vector of a pedestrian agent A, and the location of pedestrians is derived from the velocity-based model; $f_A(t)$ is a superposition of four factors:

- Pedestrian agent A wants to walk in desired direction e_A (toward the next destination) with a desired speed v_0;
- Pedestrian agent A moves under the pressure of time. A deviation of the actual velocity $v_A(t)$ from the desired one v_0 leads to acceleration of movement;
- Buildings, walls, streets, and other borders repel pedestrians and repulsion of the borders decreases monotonically with distance from them;
- Other pedestrians influence the motion of a pedestrian by repulsive potential created by each pedestrian; this repulsion also decreases with distance.

The model (5.25) imitates important characteristics of the emerging collective behavior of pedestrians:

- When the density of pedestrians is above a critical level, lanes of pedestrians walking in the same direction self-organize (Figure 5.31).

Figure 5.31 *Formation of lanes when the density of pedestrians is above a critical level; source Helbing and Mulnar (1995)*

- At narrow passages (for example, doors) the walking direction of pedestrians oscillates between those trying to enter from one and another side (Figure 5.32).
- At street intersections, roundabout pedestrian traffic is formed spontaneously (Figure 5.33).

The abilities of analytical models, as Eq. (5.25) above, are always limited and engineering simulations of pedestrian dynamics vary in implementation of the conceptual principles

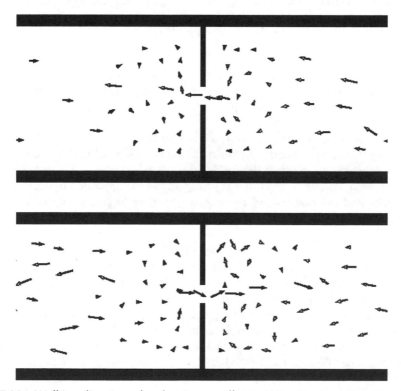

Figure 5.32 *Walking direction of pedestrians oscillates at narrow passages; source Helbing and Mulnar (1995)*

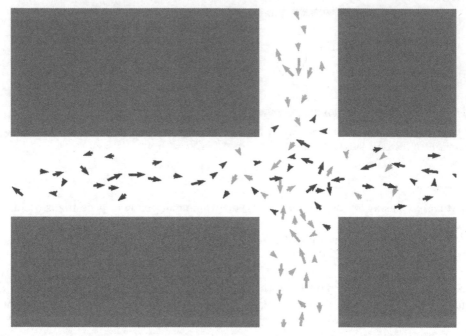

Figure 5.33   *Roundabout traffic at an intersection, with rotation direction changing from time to time; source Helbing and Mulnar (1995)*

above in discrete cellular space and discrete time. Blue and Adler (2001), for example, simplify behavioral rules and their model still remains capable of effectively capturing the behaviors of human-like (and polite) pedestrians:

- Pedestrians move ahead when there is a free cell before them;
- Two pedestrians do not occupy the same cell and in a case of collision one of them is randomly selected to step into the cell;
- Pedestrians do not step in front of agents moving in the opposite direction, while they can step behind an agent moving in the same direction to give the opposite pedestrian the opportunity to pass ahead.

As similar to the basic dynamic regimes of CA and a number of other models we consider in this book, three modes of persistent bidirectional pedestrian flow are obtained in this model. The first, regular, mode is the directional behavior described above: if a walkway is randomly filled with pedestrians, the flows will tend to migrate to one side over time (e.g., toward the right), forming two sets of contiguous directional lanes. Two less organized modes of pedestrian behavior are also obtained. One is short-lived "interspersed flow," where pedestrians pick their way through a crowd without forming distinct directional flow lanes. The other mode, called "dynamic multilane flow," is characterized by dynamic groupings in directional lanes by avoiding people coming from the opposite direction and by following a person going in the same direction. These modes are wonderfully illustrated Online at www.ulster.net/~vjblue.

Another "polite" resolution of the conflict between pedestrians trying to step into the same cell is proposed by Weifeng and colleagues (Weifeng *et al.*, 2002). Namely, if two pedestrians moving in opposing directions try to step into the same cell, one of them makes a step back. The model successfully generates phenomena of directed movement and, in addition, demonstrates that back-stepping essentially increases the density of pedestrian jams.

5.7.4.2 Depopulating Rooms

Evacuation from rooms is an important application of MAS. In evacuation models, pedestrian agents are not glued to the street network. The most important parameter of evacation research is the time over which an area empties and models study its dependence on various parameters. Hebling and colleagues (Helbing *et al.*, 2000) extend the model of pedestrians' behavior in continuous space to investigate the mechanisms of panic and jamming caused by uncoordinated motion of agents trying to escape danger. They consider panic behavior to be a result of uncoordinated motion in circumstances when people move, or try to move, considerably faster than normal. In these situations, people show a tendency toward mass behavior, that is, to do what other people do, and interactions among individuals become physical in nature. Among many consequences of mass behavior, Helbing and colleagues (Helbing *et al.*, 2000) specify the following:

- Passage through a bottleneck becomes uncoordinated and arching and clogging are observed at exits; escape is further slowed by fallen or injured people acting as "obstacles;"
- Jams build up and physical interactions in the jammed crowd add up and cause dangerous pressures, which can bend steel barriers or push down brick walls;
- Alternative exits are often overlooked or not efficiently used in escape situations.

The basic version of the model considers a room with one exit, filled with individuals, which embody parameters that reflect soccer fans who want to leave a room as quickly as possible (Figure 5.34).

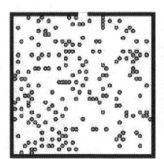  

Figure 5.34 *Typical stages of room depopulation dynamics; source Burstedde, et al. (2001)*

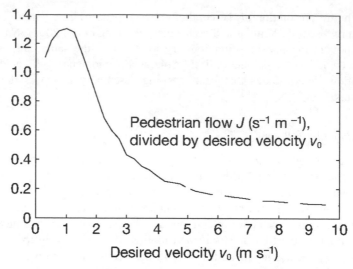

Figure 5.35   *The decrease in efficiency of evacuation with increase in desired velocity; source Helbing et al. (2000)*

Realistic parameters determine the critical value of desired velocity, which is somewhat above 1.5 m/s ($\sim$6 km/h). Until this value is reached, the faster that agents wish to leave the room, the faster they do so. If the desired velocity is higher, the result is opposite. Agents physically push each other, their exit from the room becomes clogged, and the mean exit time and, especially, the mean time divided by desired velocity, i.e., the "emotional time," grow (Figure 5.35).

The tendency to escape as fast as possible can have counterintuitive effects. For example, the wider parts of escape routes can increase the evacuation time by inducing clogging at narrow cross-sections (Figure 5.36).

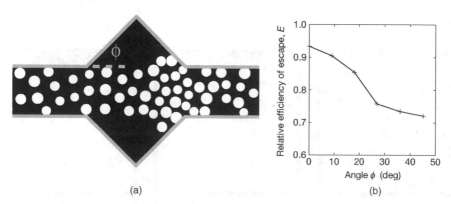

Figure 5.36   *(a) Jams in wider parts of an escape channel, and (b) the efficiency of escape as a function of the form of local widening; source Helbing et al. (2000)*

The most crucial consequences of panic behavior are exhibited in situations in which pedestrians have to find an *invisible exit*. The model demonstrates that neither individualistic nor herding behavior perform well in this context. Pure individualistic behavior means that each pedestrian finds an exit only accidentally, while pure herding behavior implies that the entire crowd will eventually move into the same and probably blocked direction, so that available exits are not efficiently used. The escaping agent can behave differently in this respect: she can set the direction of neighbor movement by herself, can follow the average direction of the neighbors within a neighborhood of a given radius, or react to both individual and neighbors' stimuli. In the latter case, the model's direction of movement is considered as a linear combination of the averaged direction of neighbors with weight p and an agents' own direction with weight $(1 - p)$. The value of p can be naturally considered as a "panic parameter"—low values of p correspond to deterministic behavior, high values to a situation of panic, when "herding" behavior is exhibited. The model demonstrates likely effects of panic; some level is necessary for intensification of the escape process, but beyond a threshold level (about 0.4 in the model) the effectiveness of the evacuation decreases (Figure 5.37).

The highest chances of survival are observed when some people behave individually and detect exits, while herding guarantees that the others imitate their success. Burstedde *et al.* (2001) and Kirchner and Schadschneider (2002) confirm these results with models of slightly different agent behavioral form from that used by Helbing *et al.* (2000) and also demonstrate that variation of model parameters allows description of different types of behavior, from regular to panic.

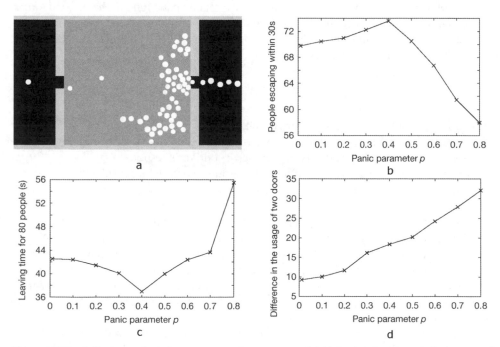

Figure 5.37   *Influence of panic on evacuation process. (a) Use of only one exit in a case of high-level panic, (b, c, d) dependencies of evacuation parameters on panic level; source Helbing et al. (2000)*

Pedestrians are in constant motion, and, differing from the models in previous sections, the MAS is used to explore modes of movement, and not persistent stationary patterns. Parameters of real-world pedestrian movement are determined by specific morphologies of streets, malls, squares, and after discussion we will consider these real-world models of pedestrian dynamics in more detail.

Before we get to that, we would like to pay attention to another group of constantly moving agents, cars in urban traffic.

5.7.4.3   Cars on Roads

Traditionally, traffic engineering focuses on analytical descriptions, based on "average" traffic—flows of vehicles between origins and destinations (Transportation Research Board, 1985). Thinking of vehicles in terms of agents, MAS enable the simulation of the individual cars that comprise that flow: individual cars, trucks, and their drivers who, actually, make a decision regarding road to take, speed to keep, and place to park.

The agent-based view of cars has a two-decade history and now is one of the mainstreams of transportation modeling and applications. In this section, we limit ourselves to general presentation of the topic and several examples of basic MAS models of car traffic. In later discussion we shortly consider the real world application of MAS models car traffic.

Long ago, the topic was given some theoretical consideration in the physical sciences; the MAS view was applied, and so-called car-following theories were formulated in the mid-1950s (Gipps, 1981). These theories considered cars on a one-lane road, and the motion of each car A was described by a differential equation of the second order. The equation included reaction to location and velocity of the car ahead of a given car, as well as driver reaction time as a delay (more details are available in recent reviews; Wolf, 1999; Chowdhury *et al.*, 2000; Schadschneider, 2000).

$$d^2 x_A(t + \tau)/dt^2 = \kappa_A(dx_B(t)/dt - dx_A(t)/dt) \tag{5.26}$$

Where $x_A(t)$ denotes location of a car A (and, thus $dx_A(t)/dt$ is A's velocity), B is a car ahead of A, the sensitivity κ_A of a car A is a function of the A's velocity and the distance between the cars B and A, and τ is the timing of a driver's response.

Collective phenomena, such as traffic jams, detours, etc., can be considered as emergent and self-organizing, and this view is supported in multiple documented examples (Figure 5.38).

In parallel, the macroscopic, top-down, view of car traffic as a flow was being developed. Theoretically, this view is based on equations analogous to the Navier-Stokes equations for fluids. Practically, it specifies a series of relationships between parameters of car flow such as density and speed, and the characteristics of the road network (Chowdhury *et al.*, 2000). The top-down view became dominant in real-world engineering applications, and this is reflected in most traffic engineering manuals (Transportation Research Board, 1985).

Despite a clear trend in favor of top-down description, the agent-based approach did not lose its luster; perhaps because all of us are, ultimately, driver agents (particularly in the United States!). Differing from most other urban CA and MAS applications, road traffic MAS were inspired by engineers, and this seems to be one main reason why they were

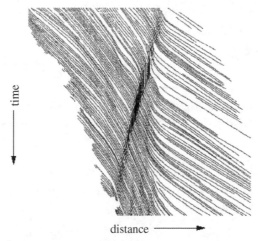

time

distance ————▶

Figure 5.38   *Trajectories of individual vehicles on a single lane of a multilane highway, which show spontaneous formation of a traffic jam; presented in Treiterer (1965); referred to in Wahle, et al. (2001)*

among the first such urban applications actually developed in practice as "down to earth" simulations. Another reason is data supply: the quality and extent of urban traffic data was always exceptional compared to other fields of urban studies. As a result, one of the first MAS simulation systems—MULTSIM, developed in the early-1980s (Gipps, 1986b)— was very well developed compared to other MAS models of the time. This system was based on the car-following equations we mentioned; it considers cars as separate agents that enter and leave a system context. Agents' choice algorithms were applied to lane-changing and way-finding; agents' cognition is activated to determine how far an agent-driver might see ahead or behind in the same lane and how far to look ahead in adjacent lanes before switching position (Rickert *et al.*, 1996). MULTSIM included characteristics of intersections, stop lines, signal timings, public transport stops, passenger flows, bus or transit lanes, etc.; these attributes later became part of standard modern transportation GIS. Vehicle agents are generated by means of a user subroutine, which specifies the type of vehicle (motorcycle, private cars of various classes, trucks, buses, etc.), its length, the desired speed of the driver, maximum acceleration that the driver will employ, and the parameters responsible for changing lanes. MULTSIM users could generate simulations to study volumes of traffic and their characteristics and the model provided the framework for pioneering studies of jams and lane-changing phenomena during the 1980s (Gipps, 1986a).

The description of car-agent behavior in MULTSIM was based on multiple differential equations, as in Eq. (5.27). From the point of view of the MAS modeler, it is too complicated. A boom in interest in traffic MAS models began in the early-1990s, stimulated by a paper by Kai Nagel and Michael Schreckenberg (Nagel and Schrecken-berg, 1992), who provided simple, but sufficient description of car movement dynamics (Figure 5.39).

In their model, the road is divided into cells of 7.5 m length. Each cell can either be empty or occupied by at most one car. The speed v_A of each vehicle-agent A can take

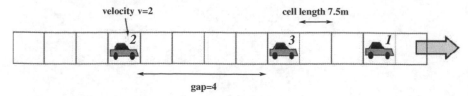

gap=4

Figure 5.39 *Representation of a road in a model of Nagel-Schreckenberg: each lane is subdivided into cells of a length of 7.5 m; source Wahle et al. (2002)*

one of the $v^{MAX} + 1$ integer values from 0 to v^{MAX}. The state of a vehicle at time $t + 1$ can be obtained from that at time t by applying the following rules to all cars simultaneously (the model is based on synchronous update):

1. Acceleration: $v_A(t + 1) = \min(v_A(t) + 1, v^{MAX})$;
2. Braking: $v_A(t + 1) = \min(v_A(t), d_A(t) - 1)$;
3. Random deceleration with probability p: $v_A(t + 1) = \max(v_A(t) - 1, 0)$;
4. Driving: $x_A(t + 1) = x_A(t) + v_A(t)$;

(5.27)

where $x_A(t)$ denotes the position of car-agent A on a road, and $d_A(t) = x_B(t) - x_A(t)$ denotes the gap between A and the car B ahead. The typical value of v^{MAX} equals five and in this case the model can be considered in time-units of a second, just as the timing of drivers' reaction (Wagner *et al.*, 1997). In a theoretical sense, the model demonstrates that no jam occurs at sufficiently low densities, but spontaneous fluctuations give rise to traffic jams at higher densities and provide likely traffic patterns. In applied contexts, the model generates good match with real traffic data (Figure 5.40).

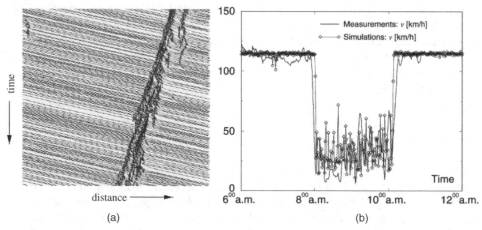

(a)   (b)

Figure 5.40 *Some results of the Nagel-Schreckenberg model. (a) Traffic pattern obtained with $v^{max} = 5$ and $p = 0.5$; source Wahle, et al. (2001); (b) change in the average velocity on a road during the day, compared with model results; empirical time-series by Neubert et al. (1999); referred to by Schadschneider (2000)*

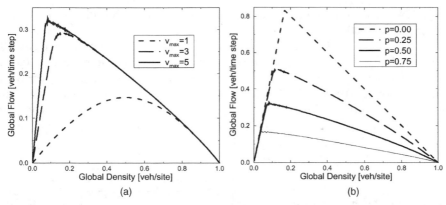

Figure 5.41 *Traffic engineering density-flow diagram, obtained by the Nagel-Schreckenberg model; (a) for different values of the number of velocity grades v^{max}, (b) for different probability of random deceleration p and $v^{max} = 5$; source Wahl, et al. (2001)*

The model developed by Nagel and Schreckenberg successfully reproduces funda-mental characteristics of traffic engineering density-flow diagrams (Figure 5.41) and served as a basis for many others, where car agents directly perceive location and the velocity of neighboring cars (Nagatani, 1993; Emmerich and Rank, 1995; Gu *et al.*, 1995; Nagatani, 1996; Rickert *et al.*, 1996; Wahle *et al.*, 2000; Barlovic, 2001).

In MAS models of car traffic, car/driver agents accelerate and brake following the states of vehicles around them (type, gap to the next car, speed), local conditions (type of road, perceived traffic conditions ahead), local road structure (ramps, roundabouts) and global parameters (distance to goal) (Wylie *et al.*, 1993; Wagner *et al.*, 1997; Wang and Ruskin, 2002). Agents also accelerate and decelerate randomly, to mimic erratic movement (Rickert *et al.*, 1996). Neighbors-outliers are often accounted for, to simulate collision-detection and collision-avoidance, as well as other nuances of driver behavior aimed at simulating signal stopping behavior (Barrett *et al.*, 1999) and movement at junctions, with "gap acceptance" functions that determine how long drivers must wait at a junction before they can proceed (Wylie, 1993). In models where collections of vehicles are simulated as traffic queues (Rickert *et al.*, 1996; Barrett *et al.*, 1999; Cetin *et al.*, 2001), entrance and departure from vehicle queues is also simulated, with vehicles leaving a queue freeing up space on a link, allowing another driver to join the queue.

To explore theoretical and real-world scenarios, MAS traffic geosimulations facilitate the introduction of car/driver behaviors. Perception of traffic conditions by driver agents—relatively simple behavioral rules to describe cars around them, buffers between adjacent cars in the same lane, gap opportunities for merging at junctions (Wylie *et al.*, 1993)—can often be sufficient for representing traffic congestion (Nagel and Schreken-berg, 1995).

The main result of the theoretical studies is in imitation of qualitative characteristics of traffic flow (Figure 5.41). We have mentioned already that for low densities of cars on a road, the models demonstrate laminar flow, while they demonstrate jams for high densities. Measurements of real traffic demonstrate that there is an interval of densities between low and high, where the state of the traffic depends on the history of the system. If the density of the cars increases, there is high chance of preserving laminar motion, while jams can persevere despite a decrease in density (Chowdhury *et al.*, 2000; Schadschneider, 2000).

Developing toward adequate description of real-world car-agents, much effort has been invested in adequate description of lane-changing behavior, and its representation in MAS vehicle models is generally quite sophisticated (Chowdhury *et al.*, 1997; Wagner *et al.*, 1997; Hidas, 1998; Fukui *et al.*, 2002; Hidas, 2002). MAS models are usually formulated with "strips" of one-dimensional CA, thereby facilitating true lane-changing (Figure 5.42).

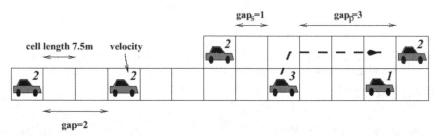

Figure 5.42   *Schematic representation of lane-changing on a two-lane road; source Wahle et al. (2001)*

The movement of vehicles between lanes can be represented, in this regard, as dependent on a number of factors: the number of empty sites in a vehicle's neighborhood ahead in the same lane, ahead in adjacent lanes, and behind in adjacent lanes; velocity; and hindrance in the current lane (Rickert *et al.*, 1996; Wagner *et al.*, 1997). From the agents' point of view, change of lanes is also described as a decision-making process, as in Figure 5.43 (Hidas, 2002) and the model provides good correspondence to real-world speed-flow dependencies (Figures 5.44).

We have only touched upon a vast literature in this section; for more details, we refer the reader to recent reviews (Wolf, 1999; Chowdhury *et al.*, 2000; Schadschneider, 2000; Torrens, 2004a).

5.7.5   Multi-type MAS—Firms and Customers

The models we have considered thus far deal with agents of one specific type, agents that react to urban infrastructure and their own distribution in urban space. This view is sufficient for investigating many basic urban processes—diffusion of innovation, migrations of householders and firms, movement of pedestrians along sidewalks, and behavior of drivers on roads. Each of these groups can be extracted from the general urban evolution process according to many parameters. One of them is the characteristic time of the processes the agents are involved in; other groups of urban agents can be supposed to be either "still" or so fast that only their average characteristics are important for the subsystem being studied. For example, in householder models, infrastructure can be "frozen," because it changes much more slowly than residential distribution. Nonetheless, there definitely exist situations where one-type MAS are inadequate and system dynamics

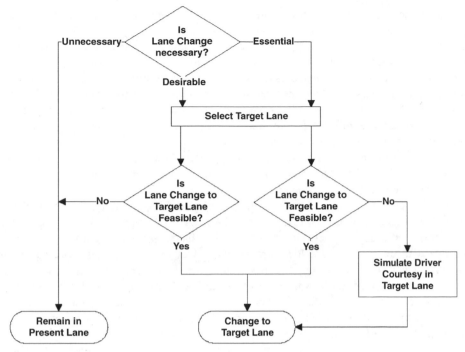

Figure 5.43 *Typical representation of the sequence of lane-change decisions; source Hidas (2002)*

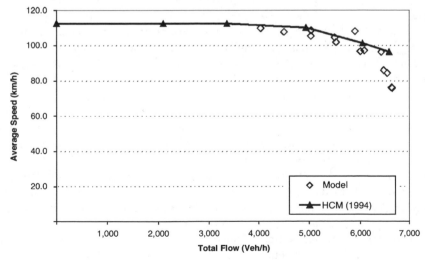

Figure 5.44 *Lane-change models provide likely description of the dependency of average speed on traffic flow; source Hidas (2002)*

depend on the behavior of agents of more than one type. The immediate examples here are economic systems, for example, spatial behavior of service firms depends on the distribution of customers, who are decision-making agents themselves, and the rates of relocation of customers and services can be considered as similar (Otter *et al.*, 2001). There are other pairs of urban agents that can hardly exist separately in quite a number of situations: developers and householders should be considered together in newly established developing settlements, drivers and pedestrians in tourist areas of the city, etc.

In ecological modeling, MAS consisting of agents of more than one type, as predators and prey, are standard. It is not so in urban contents, where most potential situations seem unrevealed and those that are recognized are definitely underinvestigated. In this section, we consider existing MAS models that account for agents of more that one type; they all consider service firms and customers.

Urban MAS with agents of several types have a very short history. In the pioneering models of Benati (1997), each cell of a small 15×15 grid represents a fixed customer, and firm-agents migrate between neighboring cells to serve as many customers as possible. Customers, who are local optimizers, select the firm that is closest to them. The firm-agents follow satisfier heuristics and possess the ability to foresee the consequences of their relocation. A firm moves to one of neighboring cells within a 3×3 von Neumann neighborhood if the number of customers in a neighborhood is below a threshold and can be increased after the move. The system is studied for a very limited situation of nine firms, which are initially located randomly on a plane. After some twenty iterations, firms form three clusters of two, three, and four firms; the location of firms within clusters fluctuate, but each cluster serves the same ring of customers.

Nagel and co-workers investigate a system of service firms and customers in depth (Nagel *et al.*, 2000) and combine the principles of spatial competition with the mechanism of price formation, aimed at understanding the formation of low prices of Western retail networks. Their idea is that this is the consequence of the ability of consumers to patronize another shop. As in Benati's models (Benati, 1997), they assume that fixed customers are located on grid nodes, and choose one of the closest firms according to the information they get from their neighbors. Firms compete for customers by establishing prices and do not interact directly; they can go out of business for two reasons: losing too much money, or losing too many customers. The first situation corresponds to a price that is too low; the second corresponds to a price that is too high. If the firm goes out of business, its customers are taken over by the remaining companies. Reduction in the number of companies is balanced by the injection of start-up firms.

A model time-step consists of five substeps, each intentionally simplified: Trade $\rightarrow$ Bankruptcies, go out of business $\rightarrow$ Surviving firms change prices $\rightarrow$ New firms are introduced $\rightarrow$ Consumers change their patronage. Let us present these steps in some detail.

Trade: All customers have an initial amount M of money, which is completely exhausted in each time-step and replenished in the next. Every customer agent A knows the firm F she buys from and, thus, orders $Q_A = M/P_F$ goods from it, where P_F is F's price. A company that has n_F customers and a product price of P_F will thus produce and sell $Q_A = n_F M/P_F$ units of product and collect $n_F M$ units of money.

Company exit: The model assumes a function $S(Q)$, which gives the cost for production of amount Q of goods, is the same for all firms. The firm profit is given as

$$\Pi_F = n_F M - S(n_F M/P_F) \tag{5.28}$$

If Π_F is less than zero, then the company immediately goes out of business. The cost function employed in the study is $S(Q) = Q$; with this choice, companies with prices $P_F < 1$ exit as soon as they attract at least one customer.

Price changes: Prices are changed randomly; the firm to change the price is randomly selected and its price is randomly increased or decreased by δ.

Company introduction: It is assumed that each cell contains a "sleeping" company, which can awake when conditions permit. The firms to be activated are picked randomly with constant probability p_{inj} and one consumer is attached to such a company. The price of the introduced company is set to the price that the attached customer has paid before, randomly increased or decreased by δ.

Consumer adaptation: All customers whose prices are increased, either via company exit or via price changes, search for a new shop among those serving its nearest neighbors. If a neighbor is also searching, nothing happens; if not, and if the neighbor is paying a lower price, the consumer will accept the neighbor's shop. If a consumer does not improve its situation, it will remain with her old shop, and she will cease her search. This is repeated until consumers cease searching.

The prices that firms establish converge towards unity, just because the firm that (randomly) decreases its prices to this level still survives, while attracting customers of other more expensive firms. This rule includes comparison between the prices of neighbors; "if the price goes up, try another firm," and this is still sufficient to provide convergence to minimally possible prices, while in less regular manner (Figure 5.45).

Some likelihood has also been demonstrated between the model distribution of the size of the cluster of firm's customers and real data; similar time-lag between cost change and price adjustment in the model and in the US economy has also been noted.

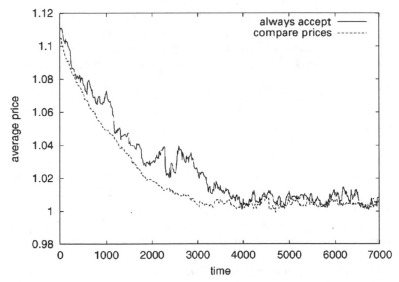

Figure 5.45 *Evolution of prices. Lower curve denotes agents searching for another firm, comparing prices of neighboring ones; upper curve denotes agents that select a new firm randomly among neighbors; source Nagel et al. (2000)*

In general, the model by Nagel (Nagel *et al.*, 2000) demonstrates that sharing between economic and spatial behavior can be done in a relatively simple way. We see a lot of potential in MAS models including agents of more than one type; this is an area that is almost unexplored at the moment.

To conclude the section, we see the abstract models we have considered thus far in this chapter as part of the background of geosimulation. They are all based on weak agents, maintain strong connection with general system theory, and demonstrate the advantages of urban systems as a source of inspiration, both regarding novel theoretical outcomes and the ability to explain phenomena and reveal important characteristics.

Of course, there is a second, not less important, component to geosimulation: modeling of real-world systems. Let us move on to that topic.

5.8   Real-world Agent-based Simulations of Urban Phenomena

There are still very few applications of MAS models to real-world urban situations; as we claimed several times, much less than one could expect on the basis of existing high-resolution GIS, for example. As you will hopefully have gleaned from reading this far into the book, geosimulation is in better shape as an endeavor; we anticipate an explosion of applied papers in the near future. In what follows, we present several encouraging examples that we are aware of.

5.8.1   Developers and Their Work in the City

In earler sections of this chapter, we considered several physical models, which were based on very abstract agents, which we, nonetheless, interpreted as urban developers. To represent expansion of real-world cities, Benguigui and colleagues extend the idea of the "percolating agent" toward behavior of real-world developer agents and applied it to the Israeli city of Petakh-Tikva (Figure 5.46) (Benguigui, 1995; 1998; Benguigi *et al.*, 2001).

Benguigui and colleagues, model (Benguigi *et al.*, 2001) accounts for one additional feature of real-world developers that the percolation model ignores; namely, that the human developer visits a site *several times* before making a decision to develop it from "unused" to "urban" use. First, a developer randomly visits a neighbor within a 3 × 3 von Neumann neighborhood. This opens up an area of investigation that is then extended and includes von Neumann neighborhoods of both initial and visited sites. An agent stays at the visited site and at a second step visits one of its von Neumann neighbors. The process is repeated; the visited site is registered and the investigated area is either extended or remains the same if the site for the visit is one of those visited before. During the process of visiting, a developer agent thus extends her knowledge about the utility of the sites in the vicinity.

The visiting process repeats until one of the "recognized" sites is visited V times. When this happens, the site is marked as developed and "added" to the city. One might interpret V in terms of investment in a development permit, for example, and the value of V evidently controls the rate of city growth. The smaller the V, the higher the rate of city

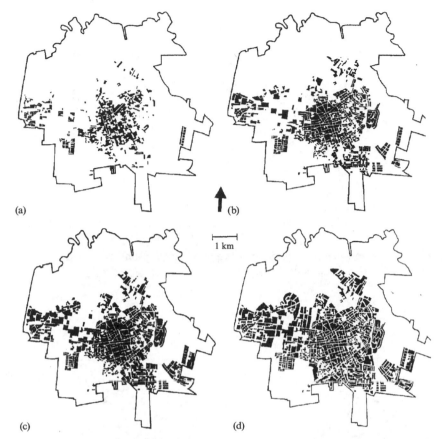

Figure 5.46 *Dynamics of the built-up area of Petakh-Tikva, 1949–1996; (a) 1949, (b) 1961, (c) 1971, (d) 1996; source Benguigui et al. (2001)*

expansion, and the less compact its structure becomes; theoretical investigation of this model is documented by Benguigui *et al.*, (1995, 1998).

Constant value of V is convenient for theoretical study, but does not work well in the case of Petakh-Tikva. To simulate the growth of the city, V is assumed to change in time, with the growth of the city and non-monotonously (Figure 5.47). These changes can be considered as reflecting the general tendencies of Israel construction policy—growth of Petakh-Tikva slowed toward 1955, but accelerated thereafter, and the reader can compare this dependence with that introduced in the SLEUTH model we considered in Section 3.7.2 of Chapter 4.

Simulated urban patterns, generated with a value of V that changes in time, visually fit available maps for Petakh-Tikva (Figure 5.48). Quantitative comparison made on the basis of fractal dimension of the real and modeled Petakh-Tikva, confirms the match (Figure 5.49).

The models of Benguigui *et al.* (Benguigui, 1995; 1998; Benguigi *et al.*, 2001) takes us back to the general physical models considered in this chapter. There is an important point

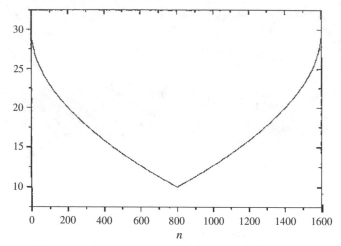

Figure 5.47   *Dependence of the number of attempts V on the number of particles in the urban aggregate; source Benguigui et al. (2001)*

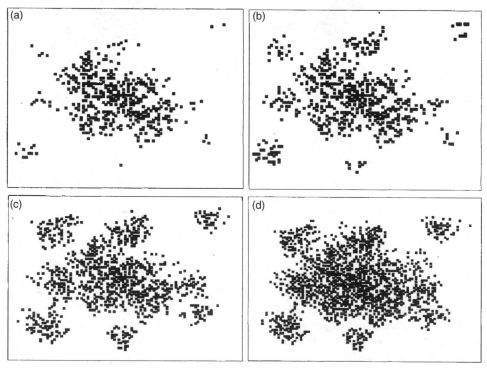

Figure 5.48   *Model snap-shots, mapping built-up area (Figure 5.47). (a) 1949, (b) 1961, (c) 1971, (d) 1996; source Benguigui et al. (2001)*

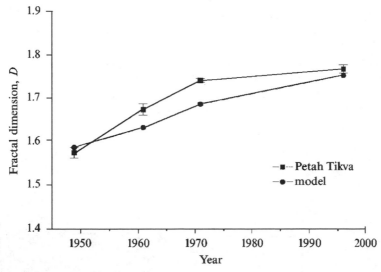

Figure 5.49 *Fractal dimension of the Petakh-Tikva urban pattern in the model and reality; source Benguigui et al. (2001)*

here, a step beyond the boundaries of abstract constructions and an attempt to estimate parameters on the basis of real-world data. The results are encouraging in two respects: first, likely simulation of Petakh-Tikva is provided; imitation of the observed situation required change to the basic physical model in only one respect. The results of the Petakh-Tikva simulations demonstrate that when the concept is simple and clear, the conceptual models can be adopted with limited alteration.

To turn a non-urban land unit into an urban one, Petach-Tikva developer agents follow an empirical decision rule, the parameters of which are tuned to provide good model fit. A decision rule of this kind can be justified based on the factors that are not included into the Petach-Tikva model, economic ones first and foremost. This direction is investigated in depth by Irwin and colleagues (Irwin and Geoghegan, 2001; Irwin and Bockstael, 2002), who assume that a user of an undeveloped land parcel (say, forestry) bases the decision on expected profit when deciding whether to turn it into residential. They demonstrate good fit of model patterns to data on urban sprawl in central Maryland and based on this correspondence explain fragmented patterns of development in the rural-urban fringe by influence of negative externalities that create a repelling effect among residential land parcels.

5.8.2 Pedestrians Take a Walk

The earlier discussion of pedestrian dynamics constitutes a basic overview of the important processes; to extend them toward simulation of pedestrian movement in real-world cities, we have to specify several parameters and processes. A number of factors are important in applying pedestrian models to real-world cities: classes of real-world pedestrian agents, the ways they establish targets, and the elements of the environment a pedestrian agent perceives.

Most of these factors remain largely unexplored in model specification, although the first attempts are beginning to be made (Dijkstra and Timmermans, 2002; Turner and Penn, 2002). The way pedestrian agents "see" and perceive the environment is in the heart of the transition from abstract models to human pedestrians and Turner and Penn (2002) employ Gibson's theory of perception for this purpose. They assume that elementary movements by agents are determined by their "fields of view," which account for the real geometry of space an agent utilizes (Figure 5.50).

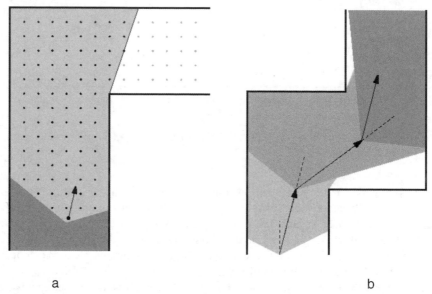

a

b

Figure 5.50 *(a) Visibility field in constrained urban space; (b) consecutive positions of pedestrians with corresponding visibility fields; source Turner and Penn (2002)*

In varying parameters that describe agents' visual perception, Turner and Penn (2002) succeeded in generating movement patterns that are very similar to records obtained in actual building contexts (London's Tate Britain gallery, for example) (Figure 5.51).

MAS geosimulation accounting for real, i.e., nonuniform and complex, geometry of the environment always runs into performance problems: for each agent, it is generally necessary to recalculate geometry-based characteristics anew, at each location. Turner and Penn (2002) propose a solution to this problem that is worth noting. Namely, for each point of a grid (and the grid serves as an approximation of the space) over which pedestrians move, information on visibility is prestored in a lookup table (Figure 5.52). In this case, reestimation of geometry-based characteristics is substituted by retrieving these characteristics from the database. Retrieve operations in modern DBMS are very fast and, as a result, large collectives of pedestrians over large spaces can be considered.

Several software environments have been developed to represent real-world pedestrian behavior. Torsten Schelhorn and colleagues developed STREETS, using Swarm and GIS (Haklay *et al.*, 2001); PEDFLOW is another model, implemented in the parallel

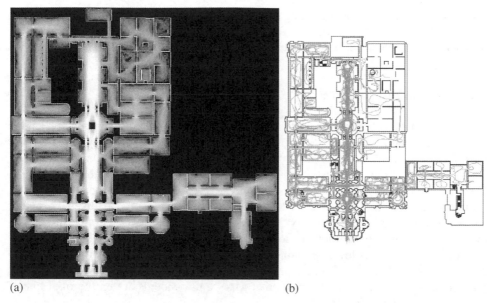

(a)   (b)

Figure 5.51 Trails left by agents through Tate Britain; (a) model simulation, (b) actual movement traces of 19 people; source Turner and Penn (2002)

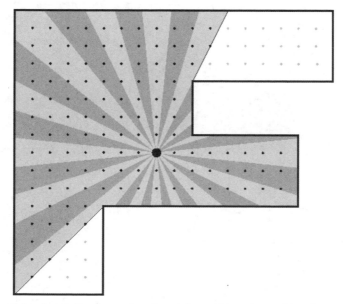

Figure 5.52 Field of visibility for each point of the grid is estimated in advance, according to predefined precision (32 angular directions in the figure) and stored in the database; source Turner and Penn (2002)

processing language Occam (Kerridge *et al.*, 2001); and Dijkstra and Timmermans (2002) have developed a 3D CAD interface for animating pedestrians in a city. The theory and techniques for simulation of pedestrian traffic in streets are relatively well developed.

5.8.3   Cars in Urban Traffic

Reliable estimation of urban road network capacity is one of the most significant planning and management issues for cities. Estimation of capacity is necessary for a variety of reasons: the installation of new infrastructure, evaluation of air emissions, safety studies, etc. The estimation task seems perfectly suited to evaluation with MAS models. However, data needs are at an exceptional level when urban car traffic is considered. Fortunately, this is an area of research that is relatively data-rich. The quality and extent of car traffic databases is extraordinary compared to the databases on the other aspects of urban studies, and, in addition, a flow of real-time traffic data at many locations— junctions and dangerous locations along the roads mostly—is available for many Western cities (Wahle *et al.*, 2001) (Figure 5.53). Consequently, real-time tuning of the models is possible. It is not surprising, therefore, that MAS traffic models are on the front line of real-world geosimulation applications.

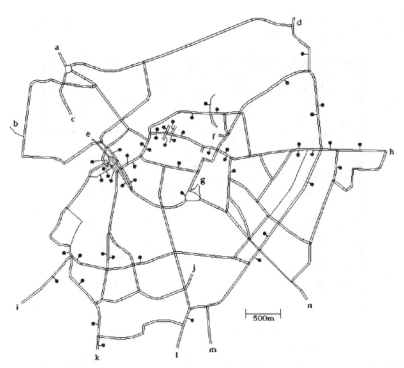

Figure 5.53 *Part of Duisburg, Germany, which served as a study area in real-world simulation of car traffic; at check points, data on the all lanes of the road are available; source Wahle et al. (2001)*

According to the paradigm of geosimulation a MAS approach to real-world vehicular traffic facilitates simulation of individual vehicles, and ensembles of vehicles, represented in a very realistic fashion with a rich set of state descriptors (Wagner *et al.*, 1997). The PARAMICS model is an example. In that model, individual agent-vehicles are described with state variables representing vehicle length, maximum acceleration and deceleration, cornering speed, desired destination, preferred traveling speed, current position, and current direction (Wylie *et al.*, 1993), and heterogeneity of the vehicles according to these characteristics is accounted for. The other leading (operational) model—TRANSIMS—operates with similar detail, adding further characteristics describing vehicle drivers: a record of the household to which the vehicle belongs, pollution, and others from a 23 type scheme (Barrett *et al.*, 1999).

From the very beginning of their development, MAS models of vehicular traffic were considered as a step toward simulation of real-world transport situations in cities. To translate theoretical models into reality, two background components are necessary: first, the behavior of cars at junctions and other nonsmooth elements of the road should be simulated; second, information on the real urban network—lights, crossings, lanes, turn permissions, etc. should be included. Recent progress in both of these aspects is significant. First of all, GIS of urban road networks are simply "perfect." To satisfy the demands of the automotive industry standard, such as car navigation systems and Intelligent Transport Systems, road and highway GIS are often available, containing data that one could only have dreamed about a decade ago, at precision of meters and usable for locating individual vehicles by Global Positioning Systems and satellites (Brown and Affum, 2002). Similarly, modeling theory is sophisticated. Discretization of the road network has been modeled for specific road situations—junctions, roundabouts, ramps, interchanges (Figure 5.54)—and modifications of the basic models for each of them have been investigated, tested, and demonstrate satisfactory fit to reality.

However, outstanding issues remain to be addressed with as much satisfaction: the flow of cars into the system, the destinations of each of them, the way-finding algorithms that drivers apply, modal split between public and private transport; these are only a few. Rickert and Nagel (2001) point out, for example, a problem of consistency; if a person expects congestion, he or she may make different plans than they would under expectation of no congestion. Yet, congestion occurs only when plans interact during their simultaneous execution. In short, plans depend on congestion, but congestion depends on plans; chickens and eggs.

Consideration of these problems represents a move from engineering and into much more vague domains: human habits; cognition; choice, location and capacity of dwellings, shops, and businesses (Fox, 1995; Kitamura *et al.*, 1999) Figure 5.55. Regarding planning interventions, one can use synthetic data, generated according to stated or revealed behavioral information, and most of the models apply this approach, including TRANSIMS (Nagel *et al.*, 2000; Rickert and Nagel, 2001; Claramunt *et al.*, 2002). In this kind of modeling, synthetic data are applied on the basis of real-world description of the road network, and the model is verified based on comparison to integral characteristics of car traffic in the simulated city (Figure 5.56). Often, traffic externalities such as noise, atmospheric, and water pollution are accounted for (see Adler and Blue, 2002 for a review).

Another branch of traffic simulations deals with real-time prediction of traffic. One can even argue that weak agents are not sufficient here; on this basis there is an argument for

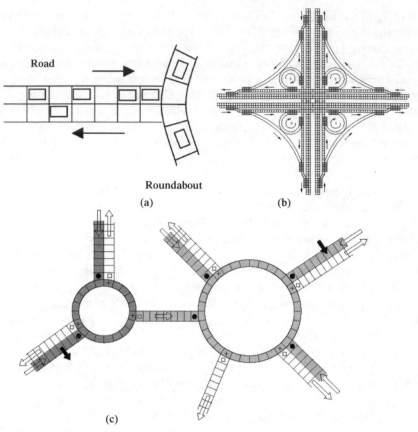

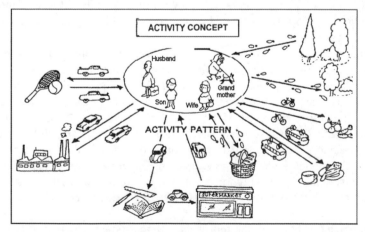

Figure 5.54 *Discretization of the road network for representation of roundabouts, junctions, and interchanges in cellular mode (a) as proposed by Wang and Ruskin (2002), (b) as proposed in Wahle et al. (2001) (c) as proposed in Dupuis and Chopard (1998)*

Figure 5.55 *Schematic representation of household activity patterns in an urban area; source Brog and Fri (1983); referred to in Fox (1995)*

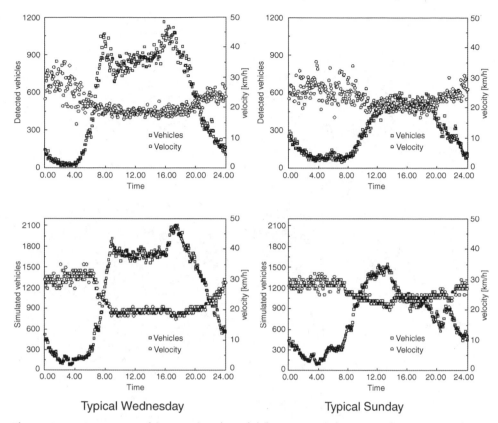

Typical Wednesday   Typical Sunday

Figure 5.56 *Experimental (top row) and model (bottom row) dynamics of average car density and velocity in different days of the week in real-world simulation of car traffic in Duisburg, Germany; source Wahle et al. (2001)*

extending such systems to include strong agents having Belief, Desires and Intention (BDI-agents); this approach is popular in AI research (Rossetti *et al.*, 2002), but seem far from practical at the current stage of traffic modeling.

Another approach focuses on real-time tracking of conditions of car traffic, based on data fed via multiple checkpoints in the city (Wahle *et al.*, 2001; Logi and Ritchie, 2002). Again, this is popular and popularly investigated, based on the existence of control points for continuous transmission and reception of traffic information; the goal is to assist in regulating traffic, on-the-fly (Adler and Blue, 2002; Wahle *et al.*, 2002). Recent developments in parallel computing have bolstered the capabilities of such systems; the state-of-the-art is now at the level of handling ten million cars; zoning and varying resolution could expand that capability further (Dupuis and Chopard, 1998; Cetin *et al.*, 2001) (Figure 5.57).

5.8.4   Citizens Vote for Land-use Change

Cellular Automata are a standard tool in conventional land-use modeling; we have mentioned several times that it is quite common, in those implementations, to ignore

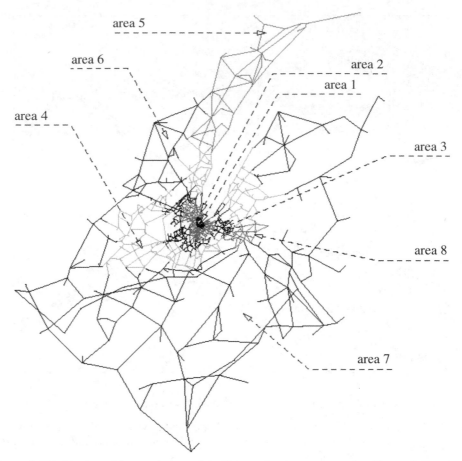

Figure 5.57 *Zoning of the road network in Geneva for the purpose of traffic modeling. In the central zones, all roads are taken into account, in the peripheral zones only important roads are communicated; source Dupuis and Chopard (1998)*

the fact that decisions regarding changes in land-use are made by *humans*. Ligtenberg *et al.* (2001) do actually pay attention to the human basis for land-use change in their models, considering agents that *vote* for land-use change. Agents exist in cellular space and possess global and local knowledge about the use of land units. None of the agents migrate and, formally, the model is an extension of the voting models mentioned in earlier sections. Voting among urban agents depends on their location and their land-use goals. The model is applied to simulating development of the municipality of Nijmegen in the Netherlands (Figure 5.58).

As in the constrained CA model (see Chapter 4), an agent wants to turn a cell into the use that provides the highest utility. But, she is not alone; there are many other agents, who also want to change the use of the cells. The model assumes that, in the case of conflict, agents in the model vote for the next use. Three "collective agents" participate in

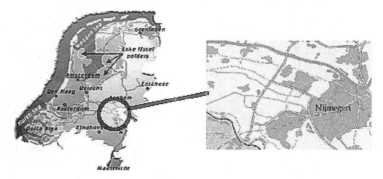

Figure 5.58 *The location of Nijmegen municipality; source Ligtenberg et al. (2001)*

the voting: the "Municipality," "Nature and Environment," and "New Rich" agents. Agents decide *where* to locate a new piece of considered area, and this piece is considered necessary for urban development. Different agents have different goals in mind. To decide which land-use of a given unit they favor, agents estimate the utility of alternatives on the basis of White and Engelen's distance-dependent weighted function (White and Engelen, 1997). This is different for different agents: Nature and Environment agents want to preserve environment, while New Rich agents want to use the areas adjacent to green for dwellings, etc. (Figure 5.59).

The municipality is always a planning agent; two others influence the planning decision. Two model experiments compare scenarios of the superiority of the Nijmegen municipality, considered as an agent with a dominant voice to a scenario, where the rights

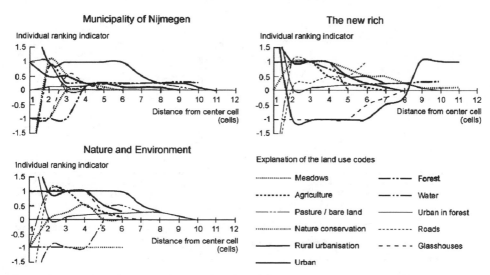

Figure 5.59 *Dependence of the weights that different agents assign to different factors on distance from their location; source Ligtenberg et al. (2001)*

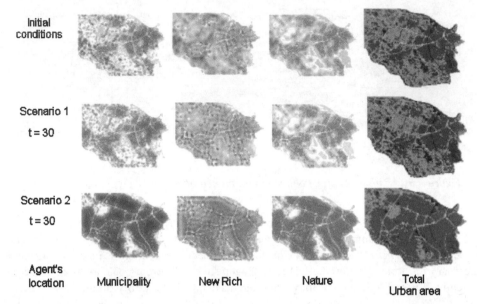

Figure 5.60 *Initial conditions and scenario outcomes in year 30; source Ligtenberg et al. (2001)*

of all three agents are equal. Both are based on spatial data for Nijmegen, but do not involve real planning procedure. The scenarios entail different land-use dynamics (Figure 5.60) and spatial outcomes after 30 annual time-steps (Figure 5.61).

The approach seems valid for real-world modeling assuming the number of land-uses will decrease and the weight functions will be justified experimentally.

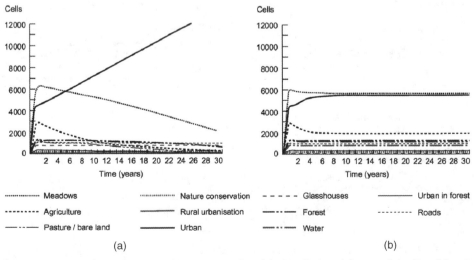

Figure 5.61 *Land-use dynamics in two scenarios; (a) superiority of the municipality, (b) equal rights of all three agents; source Ligtenberg et al. (2001)*

5.8.5   In Search of an Apartment in the City

A number of factors should be considered when formulating models of residential dynamics in real-world cities: components of the utility function, the way householders perceive neighborhoods and neighbors, etc. The review we presented previously introduces a variety of data and views on the topic; as we claimed several times in this book, in simulating real-world residential dynamics, one has to properly guess a *few* major factors that govern the modeled situation. Different situations may be determined by different factors and here we present a recently developed agent-based model of residential migration in the Yaffo area of Tel-Aviv as an illustration (Benenson *et al.*, 2002).

The Yaffo simulation follows the stress-resistance model we presented before. To remind the reader, householders' residential behavior is considered as a two-staged process. At the first stage, a householder agent decides whether to leave her current home. If the decision is yes, she searches over the available vacancies, and can relocate to one of them if it is better than her current situation. If she fails to find better homes, she either remains at the current location or leaves the city altogether.

Yaffo is an area in the southern reaches of Tel-Aviv and is populated by Arab and Jewish householders. The selection of Yaffo for construction of a real-world model of residential dynamics is not random. First, one can assume that ethnicity strongly influences the residential behavior of Yaffo agents, and that the relationships between agents—representing Arab and Jewish households—are similar to the theoretical attraction-avoidance relations. Second, quite a lot is known about Yaffo's infrastructure: during the Israeli Census of Population and Housing of 1995, made by the Israeli Central Bureau of Statistics (ICBS), high-resolution GIS coverage of streets and houses was constructed and released for all Israeli cities, including Tel-Aviv (ICBS, 2000). We introduced the ICBS database as an example of a new generation of urban GIS in Section 4.3 of the Introduction to this book.

Yaffo covers about 7 km^2; its infrastructure was established, mostly, in the early 1960s, when the majority of Yaffo's buildings were constructed; for the most part, they have not changed since that time. GIS layers of houses and streets, constructed in 1995, are used as proxies for the entire period of 1955–1995; further proliferation of infrastructure dynamics is avoided. Figure 5.62 illustrates a GIS view of Yaffo; houses are marked according to their architectural style, which enables use of this characteristic in the model. The architectural style of about 90% of Yaffo's buildings can be characterized as either "oriental" or "block" in nature, with the remaining 10% approaching one of these two styles. Put formally, in the model this is represented as follows: architectural style of a building (S) is defined as a continuous variable, whose values range from 0 (oriental) to 1 (block). Only residential buildings are taken into account; the dwelling capacity of a building is estimated by the number of floors and the foundation area, assuming that average apartment area in Tel-Aviv covers 100 m^2.

Yaffo's demography: According to ICBS data, Yaffo's population was about 40 000 in 1995, composed of a Jewish majority (about 70%) and an Arab minority (the remaining 30%). After Israel's War of Independence (1948), only 3000 of the original Arab inhabitants remained in Yaffo, almost all of them concentrated within the small neighborhood known as Adjami (Portugali, 1991; Omer, 1996). Beginning in 1948, the Arab population of Yaffo grew in number and spread throughout the area, whereas the

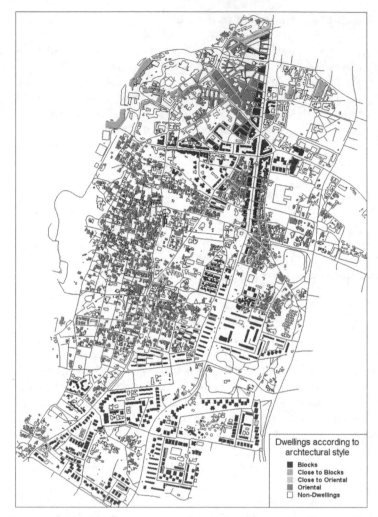

Figure 5.62   *Map of buildings' architectural style in Yaffo; source Benenson et al. (2002)*

Jewish majority declined by gradual out-migration. Available statistics for the Arab population in Yaffo are presented in Figure 5.63.

The fraction of ethnically mixed families in Yaffo is below 1% (ICBS, 2000) and family ethnicity is defined according to the ethnicity of the family head. The Arab population of Yaffo is divided into two major cultural groups—Muslims and Christians; the differences in their residential behavior are not relevant to this particular discussion.

Factors determining residential choice in Yaffo: Direct data on the residential preferences of Yaffo inhabitants are not available. According to indirect evidence (Omer, 1996; Omer and Benenson, 2002), however, two factors can be considered as influencing the residential decisions made by Jewish and Arab agents in Yaffo: the Jewish/ Arab ratio within the neighborhood and the architectural style of buildings. 1995 distributions of salaried income for Yaffo's Jews and Arabs are similar (Benenson

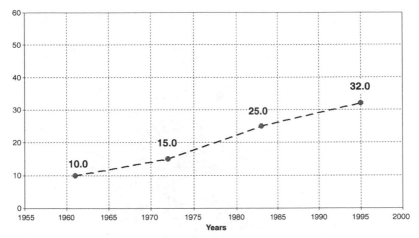

Figure 5.63 *Dynamics of the fraction of the Arab ethnic group in Yaffo; source Benenson et al. (2002)*

et al., 2002); hence, two potentially important factors of householder income and housing price are dispensed with here.

As with any choice model, a stress-resistance model calls for a utility function to be defined. Experiments demonstrate that the utility of a habitat strongly depends on the properties of the neighborhood and neighbors; consequently, the "neighborhood" should be defined in a real-world context for Yaffo. The neighborhood for Yaffo's residential buildings is constructed according to a Voronoi tessellation of the built area, tessellated on the basis of house centroids (Halls *et al.*, 2001; Benenson *et al.*, 2002). We considered this approach in Chapter 2: two buildings are considered as neighbors if their Voronoi polygons have a common boundary. To apply it in reality, and account for roads as frontiers between proximal houses, an additional condition is implemented—houses should be on the same side of a main road (Figure 5.64).

Yaffo households differ qualitatively in their ethnicity. Consequently, the disutility or "residential dissonance" of Yaffo householder agents is also considered as a qualitative-valued function. Based on results from studies by Itzhak Omer (Omer, 1996), six different levels of dissonance D_i are defined (Table 5.7). To avoid by-side numeric effects, they are quantified as stochastic variables. The standard deviations of a given level D_i of dissonance is set equal to $STD_i = 0.05 \times \sqrt{(D_i \times (1 - D_i))}$; the last column of the table below presents levels that correspond to 95% confidence intervals.

According to Table 5.7 if we assume, for example, that the dissonance between an Arab agent, located within a purely Jewish neighborhood, and her neighbors is "high," then her decision to leave will be based on a dissonance value calculated according to a normal distribution with mean 0.8 and standard deviation 0.02.

Discrete levels of dissonance make it possible to qualitatively define relations between Jewish and Arab residential agents in Yaffo. Estimates for extreme cases are shown in Table 5.8; for these "unmixed" situations, it is assumed that Arab agents strongly avoid houses that are block style and prefer houses of oriental architecture; Jewish agents prefer the newly built block-houses, although they also accept oriental houses. Arab and Jewish agents strongly avoid homogeneous neighborhoods populated by agents of the other ethnic type; avoidance of Arab agents by Jewish agents is maximal.

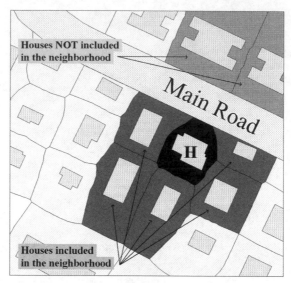

Figure 5.64   *Definition of neighboring buildings based on adjacency of the Voronoi polygons and road network; source Benenson et al. (2002)*

Table 5.7   *Residential dissonance estimates; source Benenson et al. (2002)*

Dissonance level	Zero	Very low	Low	Intermediate	High	Very High
Assumed average	0.00	0.05	0.20	0.50	0.80	0.95
Standard deviation	0.000	0.011	0.020	0.025	0.020	0.011
95% confidence interval	(0.000, 0.000)	(0.029, 0.071)	(0.161, 0.239)	(0.451, 0.549)	(0.761, 0.839)	(0.929, 0.971)

Table 5.8   *Initial estimate of the dissonance between an agent and a house (D_h) and between an agent and a homogeneous neighborhood (D_n); source Benenson et al. (2002)*

	$D_h = D_h(A, H)$		$D_n = D_n(A, U(H))$	
	House's architectural style		Neighbors common identity	
Agent's identity	Oriental ($S = 0$)	Block ($S = 1$)	Arab—$U(H)_B$	Jewish—$U(H)_J$
Arab—A_B	Zero	High (*Low*[a])	Zero	High (*Low*[a])
Jewish—A_J	Intermediate	Zero	Very high	Zero

[a]Denotes changes applied in "Arab Assimilation II" scenario below

Residential dissonance for "mixed" situations is linearly interpolated based on values set up for unmixed situations. The dissonance $D_h(A_E, H_s)$ of an agent of identity A_E regarding dwelling in a house H of style S is calculated as

$$D_h(A_E, H_S) = D_h(A_E, H_0)(1 - S) + D_h(A_E, H_1)S \qquad (5.29)$$

where A_E is either A_J or A_B, and H_0 stands for a house of oriental style and H_1 of block style.

For a mixed neighborhood $U(H)_b$ with fraction b of Arab agents and $1 - b$ of Jewish agents, the dissonance $D_n(A_E, U(H)_b)$ regarding neighbors is calculated as

$$D_n(A_E, U(H)_b) = D_n(A_E, U(H)_J)(1 - b) + D_n(A_E, U(H)_B)b \qquad (5.30)$$

where $U(H)_J$, $U(H)_B$ refer to homogeneous Jewish and Arab neighborhoods respectively.

Combination of different factors into a one-dimensional utility function is a standard problem that a modeler of real-world situations is faced with. In the case of Yaffo, disutility is defined by only two factors and is assumed that disutility is high if it is high according to any one of its components.

To employ this condition, the overall dissonance $D(A_E, H_S, U(H)_b)$ between an agent of identity A_E located in house H_S of style S within a neighborhood $U(H)_b$ having a fraction of Arab agents b is formally expressed as

$$D(A_E, H_S, U(H)_b) = 1 - (1 - \alpha_h D_h(A_E, H_S))(1 - \alpha_n D_n(A_E, U(H)_b)) \qquad (5.31)$$

where α_h, $\alpha_n \in [0, 1]$ denote the influence of house style and neighborhood ethnicity.

Stress-resistance hypothesis applied to Yaffo: Yaffo is relatively small in area and it is therefore possible to ignore the distance between residential locations and assume that agents' residential choice and decisions to migrate depend only on residential dissonance. A three-step algorithm specifies a stress-resistance hypothesis of residential choice and migration for Yaffo model agents.

Step 1: Decide to migrate. The probability P that an agent A will decide to move is specified as linearly dependent on overall residential dissonance $D = D(A_E, H_S, U(H)_b)$ given in Eq. 5.31 at the agents' location:

$$P(D) = P_0 + (1 - P_0)D \qquad (5.32)$$

where $P_0 = 0.05$ is the probability of sporadic departure. If the decision is to move, A is marked as a potential migrant; otherwise, agent A remains at her current location (with probability $1 - P(D)$) and is ignored until the next step in time.

Step 2: Scan residence. Each potential migrant A randomly selects ten houses H_v, $v = 1–10$, from the set of houses currently containing vacant residences. The probability $Q_v(D)$—attractiveness—that agent A will decide upon occupation of a vacancy v is calculated for each selected residence H_v, as complementary to the probability of leaving Eq. (5.32):

$$Q_v(D) = 1 - P_v(D) = (1 - P_0)(1 - D_v) \qquad (5.33)$$

where D_v is an estimate of A's dissonance at potential location H_v.

Step 3: Occupy one of the scanned dwellings. An agent A sorts information about all vacancies H_v according to their attractiveness $Q_v(D)$. After the sorted list of opportunities is constructed, A attempts to occupy the most attractive vacancy H_{best} with probability $Q_{best}(D)$; if A fails to occupy this vacancy, she turns to the second-best option, and so on.

In addition to internal migration, Yaffo exchanges households with the external world. Potential Jewish or Arab migrants that fail to resettle either leave Yaffo with probability L_J (for Jewish agents) or L_B (for Arab agents) or remain at their current residence with probability $1 - L_J$ or $1 - L_B$. We assume a value of $L_J = 0.1$ per month and $L_B = 0.01$ per month, that is, L_B is ten times lower than L_J. The factor ten represents the ratio of areas available for resettlement of Jewish and Arab householders in Tel-Aviv, the latter having about ten times fewer options for resettlement than the former.

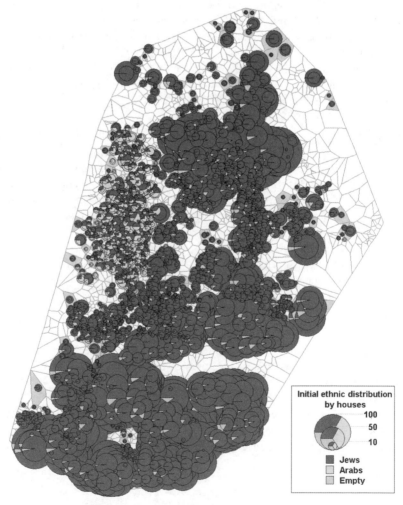

Figure 5.65 *Initial conditions of the Yaffo model: 3000 Arab citizens randomly located within the Adjami residential area; source Benenson et al. (2002)*

Based on partial data obtained by Itzhak Omer (Omer, 1996), the number of individuals who attempt to settle in Yaffo for the first time in the model is set at 300 householders (natural increase and in-migration combined), with the ratio of Arabs to Jews equal to 1:2. The actual number of new householders remains below 150 per year in all model scenarios studied. Agents that fail to enter Yaffo do not repeat the attempt.

Initial population distribution: According to Omer's data for 1955 (Omer, 1996), Yaffo's 3000 Arab residents were located in the Adjami neighborhood. In that year, the full capacity of Adjami was three times higher and Jewish householders populated the balance of the dwellings in Adjami as well as the remainder of Yaffo's territory (Figure 5.65).

Evaluation of the model: The Yaffo model is calibrated by changing coefficients α_h, α_n in formula (5.31) and by varying the attitude of agents from each ethnic group toward houses of an unfamiliar type or an unfamiliar neighborhood; the initial values are shown in Table 5.8. The model results are then compared with reality in Yaffo according to four characteristics.

- The fraction of Arab population, present in 1961, 1974, 1985, and 1995 (Figure 5.63).
- The level of segregation of Arab and Jewish agents is estimated by means of Moran's index I of spatial autocorrelation (Anselin, 1995) at the resolution of buildings.
- The noncorrespondence of the population with the architectural style of the buildings.
- The annual fraction of householders leaving a residence within Yaffo's boundaries.

The best correspondence is achieved with a scenario of low influence exerted by both factors explored ($\alpha_h - 0.05$ and $\alpha_n = 0.05$) and the tolerance of Arab agents adjusted two levels higher than initially suggested. In this "Arab Assimilation II" scenario, the dissonance between Arab agents and purely Jewish neighborhoods and "block" houses is set to "Low" instead of "High," (Table 5.8).

Table 5.9 and Figures 5.66 and 5.67 illustrate correspondence between real data and this scenario for 1995 and during the entire period of simulation. Confidence intervals of model characteristics are estimated based on one hundred model runs, run under identical conditions.

The correspondence between reality and the model in the case where Jews only experience dissonance within Arab neighborhoods coincides with the results of the theoretical models of Schelling type. Both in the theoretical and the Yaffo models, if members of one group either avoid only strangers, or only prefer members of their own

Table 5.9 *Characteristics of Yaffo's population distribution in 1995 versus the most likely scenario of "Arab Assimilation II" in model year 40, based on 100 runs; source Benenson et al. (2002)*

	Yaffo data	Model mean	Model 95% confidence interval
Overall percentage of Arab agents	32.2	34.8	(34.4, 35.2)
Moran index I of segregation for Arab agents	0.65	0.66	(0.63. 0.69)
Percentage of Arab agents in block houses	18.5	15.0	(12.8, 17.2)
Percentage of Jewish agents in oriental houses	28.1	8.0	(6.7, 9.3)
Actual annual percentage of immigrants	3.5	3.7	(3.5, 3.9)

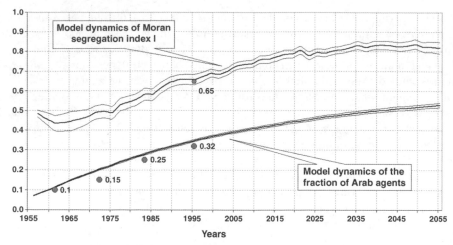

Figure 5.66 *Dynamics of global characteristics in the Yaffo model; source Benenson et al.* (2002)

group as neighbors, then residential segregation occurs. The "common sense" view that two-sided competition is necessary for segregation to take place is, thus, questionable.

Spatial determinism of Yaffo's residential dynamics: Table 5.9 and Figures 5.66 and 5.67 demonstrate local and global correspondence between the model and reality for the Yaffo case. The agent-based approach makes it possible to investigate the finer properties of real-world residential dynamics. The stochasticity of the model is extremely important in this analysis, as it represents the uncontrolled variation of local factors to which the agents react. To test the "inevitability" of the residential distribution observed in Yaffo, maps representing the probability that the fraction of Arab or Jewish agents in a house is above a given threshold $F = 0.9$ are constructed based on 100 runs of the "Arab Assimilation II" scenario (Figure 5.69).

These maps clearly indicate the areas where variation in local processes does not, or does only weakly, influence residential dynamics in Yaffo; these areas contain about 80% of Yaffo's populated houses. Variation in the fraction of Arab agents in each house in the "Arab" part of Yaffo area is higher, both relatively and absolutely, than that in the Jewish part. That is, the Arab area is more responsive to factors that the model does not account for. The ethnic structure within the houses over the remaining fifth of Yaffo's area (the boundaries between regions marked in Figure 5.68) is very sensitive to its agents' residential behavior; hence, it could be strongly influenced by other factors external to the model. The specific behavior of human agents in these boundary areas—for instance, exaggerated reactions to strangers and housing constructed for one specific population group—may significantly influence agents' residential choice and result in changing Yaffo's unique residential distribution.

5.9 MAS Models as Planning and Assessment Tools

Conceptually, MAS models are ideal tools for policy and planning; they enable formula-tion of possible changes in urban settings, couched in easily interpretable terms.

Figure 5.67 *Model (top) and real-world (bottom) population distribution in Yaffo in 1995 at the resolution of separate buildings (See also color section.) and statistical areas; source Benenson et al. (2002)*

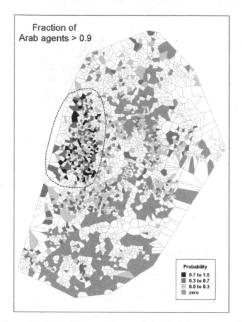
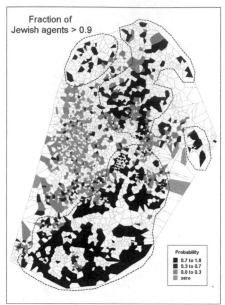

Figure 5.68   *Uncertainty of model forecast of Yaffo residential distribution; source Benenson (2004)*

Following this idea, Portugali and Benenson (1995) introduced the notion of "artificial planning experience" and proposed the use of MAS models as intuitively suitable planning tools, which can serve to translate the ideas in planners' minds into model rules, thereby facilitating the estimation of consequences of planning interventions at the level of the citizen. Since then, several, while still quite few, attempts at employing MAS for planning purposes have been performed (Ritsema van Eck and de Jong, 1999).

Any urban model can be used for planning purposes, and the more realistic the model is, the higher its potential use might be. The main problem of using any planning-support or decision-support system is in the balance struck between the rules and parameters that the system "holds internally" and those that are useful for user experimentation. The more parameters and rules the system demands to be defined by the user, the higher the risks concerning user-friendliness may be, making the system understandable and easily taught. We believe that MAS-based planning systems have potential for better balance between the flexibility necessary for realistic urban modeling and a level of automation that makes the endeavor worthwhile for planers.

Several attempts to find such a balance in MAS frameworks have been made recently. The CityDev-prototype MAS planning system, developed by Semboloni and colleagues (Semboloni *et al.*, 2003), is an interactive open-ended MAS that offers the ability to change the parameters of urban agents and the rules of their behavior. CityDev is not aimed at real-world modeling; rather for acquiring "artificial planning experience." In CityDev, multiple users manage agents, and actually serve as a special class of agents acting in the city themselves; user-agents interact with agents managed by the simulator. At the same time, users' intervention is not necessary, and a background model can run by itself.

CityDev dynamics follow a simple economic exchange. Agents of several types—householders, industrial firms, developers, etc.—locate themselves in a simulated city,

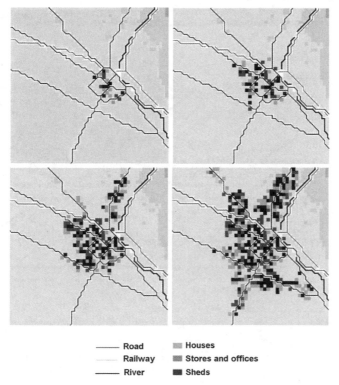

—— Road	▨ Houses
—— Railway	▨ Stores and offices
—— River	■ Sheds

Figure 5.69   *A typical City Dev output; source Semboloni et al. (2003)*

produce goods by using other goods, and exchange them at markets. Each good has a price, and when agents locate, produce goods, and interact in markets, the urban fabric is built and transformed as a by-product. The environment has 2D and 3D spatial output forms—standard and cubic cells with a superimposed road network (Figure 5.69).

CityDev is an Internet-based system and allows Web users to get involved in the functioning of the model. In addition, an administration board controls the development of the city trough a master urban plan, the building of new roads, and the location of public facilities. One can experiment with the model by browsing fs.urba.arch.unifi.it/eng/alto.html.

The state-of-the-art in the application of MAS models for real-world planning and assessment is characteristic for the situation in environmental simulation in general. The best that can be done is to formulate the system, allowing parts of this formalization to vary, and investigate the influence of those variations. The part of the system that is open to variation is usually too large to facilitate this investigation in regular form; instead, *scenarios* are investigated and compared. For example, Berger (2001) presents a spatial multiagent model that has been developed for assessing policy options in agricultural land-use in rural Chile. Unlike conventional simulation tools, the model considers farm-household agents and captures their social and spatial interactions explicitly. Individual agents choose among available production, consumption, investment, and marketing alternatives. Water is the main limiting factor in the system, and the integration of economic and hydrologic processes facilitates feedback effects in the use of water for

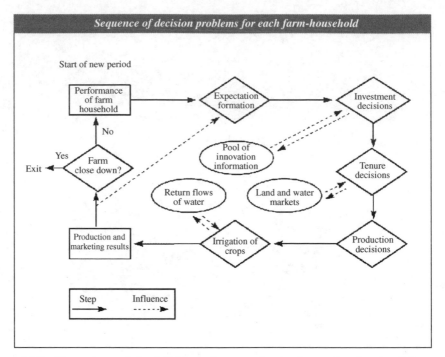

Figure 5.70 *Flow chart of the MAS-based system for land-use planning in Chile; source Berger (2001)*

irrigation (Figure 5.70). Simulation runs of the model are carried out with an empirical data set, which has been derived from various data sources for an agricultural region in Chile.

   Moving toward real-world applications, it is inevitable that less and less details may need to be presented, and this is also the case regarding several planning-support systems that involve agent-based methodology. As an example of this approach, let us point the reader to the UrbanSim environment (Waddell, 2000; Noth *et al.*, 2003). UrbanSim is modular, with modules that simulate land-use, transportation, and environmental impacts. It is oriented toward evaluating the long-term results of alternate plans for urbanization. UrbanSim is a large software system, which consists of a collection of models that represent different urban actors and processes. It is impossible to do the system justice with description in this book and the interested reader should consult a paper by Noth and colleagues (Noth *et al.*, 2003) for technical details; also, there is a wealth of documentation available online at www.urbansim.org.

5.10   Conclusions

There are some general conclusions that may be drawn from the (lengthy) discussion presented in this chapter. These conclusions relate to the interface between agent rules and design and our understanding of human behavior, agent-based models of systems and our knowledge about those systems, the use of agent rules as an exploratory tool in urban theory, and the relationship between agents, MAS, and complexity studies.

The concept of agents makes it possible for a model to directly reflect human behavior.

MAS geosimulations naturally reflect human capacities to perceive and react to information on different levels of urban spatial hierarchy, to assess opportunities before making a decision, to sort opportunities before exploring them on-site, and so on. All of these attributes can be implemented in a framework of weak agents, but may not be directly projected at aggregate levels; the concept of the agent thus allows us to avoid problems related to the scale at which we observe an urban system.

Agent-based models do not demand comprehensive knowledge of the phenomena being studied.

Changing the rules of agents' behavior does not demand changing relationships between MAS model components. This flexibility is a crucial asset for investigating consequences of different formulations of agents' behavior. We can begin with the simplest rules of agent behavior and increase their complexity while preserving the model's structure.

Formalization of behavioral rules reveals gaps in our knowledge of urban processes.

The ability to vary formal representation of agent behavior often reveals gaps in our knowledge, and MAS models reveal which behavioral parameters can make a more significant contribution to our understanding of urban residential dynamics, or the dynamics of any urban system.

Self-organizing consequences of human behavior can be investigated directly.

Many urban phenomena are outcomes of local disturbances, whose significance is recognized or acknowledged only long after the event. In the MAS framework, information regarding local changes is made available to agents and collective urban phenomena can be investigated directly through the study of the spatial patterns an agent-based model generates. MAS model's ability to reflect local indeterminacy opens the door to investigating urban self-organization and emergence at different spatial and temporal resolutions.

Needless to say, the field is in relatively early stages of exploration in geography, and much work remains to be completed on these issues.

Chapter 6

Finale: Epistemology of Geosimulation

The background to this book is somewhat a blizzard of ideas and methodologies—complex, as befitting complex systems. Urban systems are a world unto themselves, and there are a variety of ways of representing and modeling urban phenomena. Much of the background we have presented thus far reflects that. This book is about simulation, and we have concentrated on that topic—how to build and apply geosimulation to urban environments and systems. There is a subtext to any discussion of modeling of this nature, however, relating to the philosophy of science in general and urban science and modeling in particular.

6.1 Universal Questions

There are some basic epistemological questions that might be asked of urban modeling. Perhaps the most obvious is this: Why bother building models? We could also ask whether we need models at all. Considering a synoptic viewpoint, we might even ask questions about a scientific approach to research in urban geography: Is this science; are other alternatives perhaps better or more appropriate in uncovering "truth"; should we adopt a postmodern perspective, or remain with traditional views? We are going to abstract from much of that potential discussion, and we refer the reader to other areas of the literature where those ideas are explored and debated in much detail (Batty, 1979; Openshaw, 1979; Sayer, 1979; Batty, 1994; Harris, 1994; Klosterman, 1994; Smith, 1998; Batty and Torrens, 2001; Torrens and O'Sullivan, 2001). We are not going to let ourselves off completely, however; we would like to devote some attention to explaining our opinions on these issues.

Geosimulation: Automata-based Modeling of Urban Phenomena.   I. Benenson and P. Torrens
© 2004 John Wiley & Sons, Ltd   ISBN: 0-470-84349-7

So, here goes: we have a manifesto of sorts. We should preface this by saying that we are definitely traditional in our general views. Here is our manifesto: We say yes to (a) positivism; yes to (b) a systems approach; (c) we consider reductionism as an initial point of any research; (d) we concur with the proof that some phenomena cannot be reduced to lower-level synergetic action; (e) we believe that practice and experiment are criteria of a theory, what might seem logical can nonetheless be wrong. That is our manifesto; some readers might say it is just Popper's primitive view of natural science. We say yes.

Let us explore in some more detail what we think that means for the modeling of urban systems.

6.1.1   Social Phenomena are Repeatable

Continuing a thread that was started in the beginning of this book, we claim that urban systems are social systems. The laws that govern the functioning of an urban system are— at least partially and certainly in terms of human geography—social in nature. Almost tautologically, we also claim that formalization of *any* system presumes repeatability of the phenomena that we consider as representative of that system. Bringing that together, studying urban systems formally, we simply *must* assume that social phenomena are not unique; rather they are repeatable at the same degree as any other natural phenomena. Consequently, standard statistical views are applicable to the description and analysis of urban phenomena.

6.1.2   We are Interested in Urban Changes During Time Intervals Derived from Those of a Human Lifespan

There is another social dimension to urban systems, which comes from those who study such systems (us): we, ourselves, are also social creatures. This distinguishes urban modeling from similar research in other disciplines. We might use the case of agent-based modeling of fauna movement as an example. In that case, the researcher, as an observer of animal communities, can distinguish quite easily between herself and the system that she is investigating. The distinction is not as clear in the case of urban systems. The study of cities is beyond the frontier of clear separation, but, we claim, is still far from being considered as a complete mix of cognitive science or psychology, for example.

There is a rich literature on topics of involvement between researcher and the researched in human geography; the postmodern movement in geography is at the vanguard of this sort of inquiry, and the literature on cities in this context is voluminous; Soja and Dear have written some very interesting pieces (Soja, 1995, 1996; Dear, 2000; Soja, 2000). In terms of geosimulation, the notion of researcher as social creature factors significantly into the specification and interpretation of *time*—model time and time in urban systems. In studying an urban system, a researcher invariably applies *endogenous time scales* to her interpretation of urban phenomena, and, most broadly, the units of these scales vary from *an hour* to *a year*. The important consequence is that we accept an indeterminism of years, maybe one decade, but not more. We are unlikely to be interested in theory that says that something can happen, equally, right now, and in the next generation. And we can abstract details of, say, traffic accident dynamics, out of the domain of urban modeling and agree to account for the average rate of accidents per hour, day, month, or year. If we were ants, things might be different, we might think in minutes and hours. If we were rocks, our consideration might be of the order of millions of years.

6.1.3   Urban Systems are Unique because They are Driven by Social Forces

Urban geography—as a part of human geography—accounts for social properties of human actors in a city, directly or in a roundabout way, and interprets social forces as driving the development and evolution of an urban system. That is, the uniqueness of urban systems is measured in terms of the difference in social and physical elements, and in the action of social and physical laws. It is easy to accept that social entities and laws might induce qualitatively new phenomena, or at least it is difficult to argue that they may not: Physical entities do not have memory, do not know what happens far away, do not always foresee (correctly or wrongly, it does not matter) likely consequences, etc. This is not enough to prove that the action of social forces brings qualitatively new phenomena, but it is enough to suggest that and investigate the consequences of this suggestion. Simulation is a way to investigate the consequences of the uniqueness of "social" and "urban," as it provides a trial-and-error framework to compare dynamics of the system with and without human-specific mechanisms. And we can say in advance that, quite often, social forces can be adequately expressed in a purely physical frame.

6.1.4   The Uniqueness of Urban Systems is not Necessarily Exhibited

Of course, there are circumstances where the social uniqueness of urban systems can be ignored. We see two situations where this is likely the case. The first situation is as follows—when a physical view of the situation is sufficient (disregarding the social nature of the system's elements) to *approximate* urban phenomena, just as linear regression is often sufficient for approximating phenomena which are actually nonlinear.

The second situation is more genuine; the consequences of social forces on the system can be retained within an element of the system itself, and, metaphorically, we can point to a citizen of a totalitarian state as an example. The uniqueness is only of relevance to that particular entity, that is, until the "element" begins to make decisions—choices—about and between alternatives that the system provides. At this point, the "element" begins a relationship with the system, and its human abilities begin to matter for system dynamics. Members of a totalitarian system have no right to choose.

Conceptually, it is easy to connect between "physical" and "social" views of the urban system: The action of economic, cultural, and social factors can be easily interpreted in a nonsocial mode. Disregard of a social component is the necessary first step in studying urban phenomena, providing a comparative background. Outcomes of social and non-social models developed for the same situation should be compared; the differences will indicate when, where, and how social forces are important for the system.

6.1.5   Why do we Hope to Understand Urban Systems?

Consider a nontotalitarian urban system; in this system, system elements do encounter alternatives, and they are free to choose among them. In studying such a system, can we hope to understand the consequences of individual-level choice at a global level? This is, after all, one of the main goals of system studies—the search for mechanisms that link the macro to the micro, tracing an indelible path of cause and effect across scales. Is it enough to endow a model with independent actors; is that sufficient for imitating the behavior and decision-making we observe in the real world? On paper, it seems easy enough: specify some individual–based and social-like conditions (say, provide agents with memory, or let them share their knowledge with friends), map out links between microelements, build a model, wind it up, and let it go. It is considerably more complicated in practice. Consider a romantic example—we might specify the conditions for love, and all its subtle nuances, but that does not mean a computer model can help us in understanding how love makes the world go around!

Understanding the link between the individual and the system may not be simply about understanding based on the freedom of element's choice; it is as much to do with the opposite condition: the constraints an urban system imposes on element's behavior. The hidden assumption of geosimulation is that the spectrum of decisions that social elements are faced with, and that are important for urban systems *as we (urban geographers) see it*, is severely limited by the physical properties of the system.

Put in hard terms: Geosimulation assumes that the physical compounds of urban systems provide constraints—spatial, environmental, technological—that are so strong that the system—albeit social—can actually be considered to be much "simpler" than social systems of other types, be they political, ethnic, ethical, etc.

6.1.6   Tight-coupling between Urban Theory and Urban Data

If we think of an urban system as a "society of automata," the laws of behavior and interactions of which are essentially constrained by the system itself, isn't urban geography a part of physics then? Does urban reality bring a surplus of basic knowledge to the table that we, geographers, could use to take physicists down a peg? Isn't this book full of models, which are "physical" or can be so-called, and don't they work well?

We are still far from being able to answer this basic question. But we can approach it, and it is clear how to search for an answer with geosimulation. Indeed, thinking "positivistically," we have to supply examples of essentially urban phenomena, which cannot be explained until model objects start to exhibit really human features—bounded rationality, memory, ability to foresee. Geosimulation can provide the framework to test this non-reducibility of urban systems to physical ones: First, take experimental data on behavior of humans as participants of urban processes—data on individual residential choice, pedestrian way-finding, or building constructor bidding for parcels. Second, build the model, where objects behave according to experimentally revealed laws. Third, compare the model results to urban data, which are external regarding each specific human—data on urban residential pattern, pedestrian crowding, or urban sprawl.

As we mentioned several times in this book, we are overloaded with external data for comparison; they come from high-resolution GIS, remote sensing, and census databases. What we still lack are experimental data of human behavior in the city. The majority of the geosimulation models in this book is based on anthropomorphic interpretation of the "physical" behavior of objects, and not on experiments. Our limited success in explaining real-world urban phenomena is a good sign—interpretation of atoms as humans is insufficient; moreover some of the models yield a rationale for adapting an optimistic outlook.

6.1.7   Automata versus State Equations

System analysis of the mid-1990s began with complexity and uniqueness of the system, without always focusing on system elements. It was based primarily on analytical/ numerical studies of top-down *state equations*. The origin of state equations is in physics, where elementary system elements are simple, countless, react on either very local (nearest neighbors) or on very global (climate) characteristics, and do not make decisions. Think of dumb particles in a soup of interacting elements. Writing down equations, the modeler had to set up the aggregate—state—parameters of the system or "big" parts of it *before formulating the laws of dynamics*. In Chapter 3, we presented the results of this view of the world and the criticism it caused when applied to urban systems.

System science of the last two decades goes closer to concentrating on system elements, and focuses interest on the collective behavior of ensembles of objects. Many of the models we consider in Chapters 4 and 5 *do not* account for the social nature of urban systems and can be equally considered as general system models. And these models capture quite a lot of urban phenomena! But there is still some richness in the human approach that is missing. Geosimulation is a natural reaction to the "nonmolecular" abilities of humans as the driving entities of urban dynamics. The concept of geographic automata and Geographic Automata Systems more accurately reflects our *way of thinking* about reality, when an element has its own laws of behavior, and that behavior is as complex as we want to see it. Dynamics of a GAS, as a whole, are a derivative of behavior and interactions between unitary automata. The development of computer environments and programming literacy has made a routine task of translating the rules of automata behavior into a computer program.

6.2   The Future of Geosimulation

So, geosimulation deals with social systems, which are essentially constrained by "physical" laws. We are not talking about social engineering here; rather we mean to refer to the idea that there are certain constraints that can help us bound what would otherwise be considered a blizzard of potential factors in explaining cities. You can, however, think of some examples of *actual* laws: land-use zoning, developer impact fees, smart growth, etc. Other laws come from the laws of nature: You need communications to be close to a land parcel, in order to be able to construct a building there; you cannot build on slopes beyond a certain grade, even in the Hollywood Hills or the foothills of the Wasatch Mountains. Two motivations for a geosimulation modeling paradigm can be considered (Grimm, 1999): *Pragmatic*, which uses the automata approach as a tool without any reference to the theoretical issues that have emerged from urban, social, and general system theory, and *paradigmatic*, which aims at investigation of cities as sociospatial systems.

6.2.1   The Applied Power of Geosimulation

The geosimulation framework definitely responds to a pragmatic motivation, and some people would say that there are no limits. More and more researchers share an *optimistic* or *WYSIWYG* view of geosimulation modeling of urban systems. Real-world problems can be considered in terms of whatever kinds of automata seem appropriate: land parcels, buildings, cars, pedestrians, householders, developers, entrepreneurs, and many others. Behavior can be represented in rules, those rules can be written to code, experimental data can be utilized as-is, intuitive scenarios can be formulated, software can be built to investigate the output, and all that is followed by discussion of results with the external user in her own language.

But it should not be forgotten that the aim of geosimulation, as with any other type of geographic modeling, is not "intuitivism," but modeling!

6.2.2   The Theoretical Focus of Geosimulation

Geosimulation is a bottom-up paradigm, which starts with "entities" (i.e., humans and infrastructure elements) of an urban system and then tries to understand the system's properties that emerge from interaction among these entities, and what the properties of higher levels of the system are.

A *skeptical* or *systemic* view would consider the genuine capabilities of nonsocial systems, and profound results of general system theory as well as the enormous amount of computer experiments necessary to really understand the dynamics of liberally formulated automata systems. Skeptics always seek confirmation of the necessity of automata resolution and of the superiority of bottom-up geosimulation formulation over top-down description of certain phenomena based on state equations.

The state-of-the-art of urban modeling is essentially "optimistic," and for us this indicates that we are still in the beginning of establishing geosimulation as a theoretical framework. We easily make use of social properties in defining objects of GAS models—memory, knowledge, etc.—but seldom, if ever, investigate their necessity. A lot of research needed to establish a firm (skeptic) basis of geosimulation has yet to be performed. Until then, our convincing argument remains as follows: *it simply works*.

6.2.3   From Modeling of Urban Phenomena to Models of a City: Integration Based on a Hierarchy of Models

Reductionism exited fashionable consideration over a decade ago today the commonplace view is that the bottom-up approach alone will not lead to *theories*, i.e., the ability to pose relevant questions at a global level. But, a top-down view is necessary as well. The question of where to start from in relating top-down and bottom-up views is essentially a view based on taste.

Can geosimulation help in amalgamating these views?

As we see in this book, geosimulation begins with modeling particular phenomena, having their own spatial and temporal resolution. As Berec (2002) points out, finding an appropriate level of aggregation in a model is a decisive part of the modeling procedure. Adhering to system science, it is advisable to combine them in such a way that the variables of slower and less detailed (both temporally and spatially) models serve as control parameters for faster and more detailed models. In turn, detailed models forward averaged information to higher levels in the hierarchy, aggregated spatially and having slower time scales. And we present the first attempts of such constructions in this book (White and Engelen, 2000).

Ecological modeling, which also began with the "stocks and flows" models several decades ago, has definitely moved in this direction. Recent developments are oriented toward hierarchical model structures, a suite of models with increasing resolution that provides adequate model scale for a given purpose (Murdoch *et al.*, 1992; Eagleson *et al.*, 2002; Wu and David, 2002).

Geosimulation is based on urban objects and their behavior, and, thus, urban objects should be arranged based on levels of model hierarchy. It should be apparent that all of the models we deal with in this book consider objects acting at levels of very high resolution, both in time and space. Climbing up the model hierarchy, we encounter companies, institutions, municipalities, governments; all of these creatures have their own rules of behavior. If we want to model a city as a whole, we simply must include them in the models, some of them as autonomous agents, others as users that might actually play with the models in order to estimate the consequences of their planning, political, or economic decisions.

Many outer issues have to be addressed to make geosimulation analysis more reliable; they include methods of statistical description of urban spatial patterns, model calibration, verification, and investigation, etc. As the field develops from infancy to maturity, from an art to science, issues of simulation code sharing seem most important.

6.2.4   From Stand-alone Models to Sharing Code and Geosimulation Language

Any conceptual description of geosimulation model rules is always incomplete. A complete description of any GAS model demands both analytical presentation of behavioral rules *and* computer code (Berec, 2002). Publication of the latter in a printed journal is unfeasible. The experienced reader is always skeptical, and is never sure whether the analytical description, flow charts, and other abstract means satisfy the reported results; yet, she has no means to verify that. Even publishing code does not help. On the one hand, people use different computer languages, some of them even specific to some model environments; on the other, 90% of any model is not directly related to system dynamics.

Yet, the issue need not necessarily be irresolvable. A partial solution is to make the code of the models open, shareable. The latter demands quite an investment into programming and documentation, and it is hard to believe people will follow this line.

We are going to go out on a limb and argue for a *geosimulation metalanguage*, which can present the model's algorithms, as well as the characteristics important for model dynamics, as an exact sequence of model events, synchronous/asynchronous updating, criteria of self-organization involved, etc. A welcome by-product of such a metalanguage could be a library of GAS methods. Some similar method exists in global climate modeling; why not in urban modeling? That was a rhetorical question! The systems are very different, as are the models and applications of those models. And specification of a metalanguage, or languages, for urban modeling is not a simple endeavor. Nevertheless, the point is still salient. Currently, the cogs of the simulation infrastructure—if not the entire wheel—are being reinvented over and over by different groups, simultaneously. If converged to a common denominator, these attempts can give birth to such a language.

Bibliography

Adamatski, A. and O. Holland (1998). Phenomenology of excitation in 2-D cellular automata and swarm systems. *Chaos Solitons & Fractals* **9**(7): 1233–1266.

Adler, J. L. and V. J. Blue (2002). A cooperative multi-agent transportation management and route guidance system. *Transportation Research Part C* **10**: 433–454.

Adrianova, L. Y. (1995). *Introduction to Linear Systems of Differential Equations*. Providence, MI, American Mathematical Society.

Ajzen, I. (1988). *Attitudes, Personality, and Behaviour*. Milton Keynes, England, Open University Press.

Albin, P. S. (1975). *The Analysis of Complex Socioeconomic Systems*. Lexington, MS, Lexington Books.

Albrecht, R., G. Alefeld, and H. J. Stetter, Eds. (1993). *Validation Numerics: Theory and Applications*. Wien, Springer-Verlag.

Allen, E. (2001). INDEX: Software for community indicators. *Planning Support Systems: Integrating Geographic Information Systems, Models and Visualization Tools*. R. K. Brail and R. E. Klosterman. Redlands, California, ESRI: 229–261.

Allen, G. R. (1954). The 'courbe des populations'—a further analysis. *Oxford Bulletin of Statistics* **16**: 179–189.

Allen, P. M., G. Engelen, and M. Sanglier (1986). Self-organizing systems and the 'laws of socio-economic geography'. *European Journal of Operational Research* **37**: 42–57.

Allen, P. M. and M. Sanglier (1979). A dynamic model of growth in a central place system. *Geographical Analysis* **11**(3): 256–272.

Allen, P. M. and M. Sanglier (1981). Urban evolution, self-organization and decision making. *Environment and Planning A* **13**(1): 167–183.

Alligood, K. T., T. D. Sauer, and J. A. Yorke (1996). *Chaos—An Introduction to Dynamical Systems*. New York, Springer.

Alonso, W. (1964). *Location and Land Use*. Cambridge, Harward University Press.

Alves, S. G., N. M. Oliveira Neto, and M. L. Martins (2002). Electoral surveys' influence on the voting processes: a cellular automata model. *Physica A* **316**: 601–614.

Anas, A., R. Arnott, and K. A. Small (1998). Urban spatial structure. *Journal of Economic Literature* **36**(3): 1426–1464.

Andersson, C., K. Lindgren, S. Rasmussen, and R. White (2002). Urban growth simulation from first principles. *Physical Review E* **66**: 1–9.

Andow, D. A., P. M. Kareiva, S. A. Levin, and A. Okubo (1990). Spread of invading organisms. *Landscape Ecology* **4**(2/3): 177–188.

Anselin, L. (1995). Local Indicators of Spatial Association—LISA. *Geographical Analysis* **27**(2): 93–115.

Anselin, L. and M. Madden, Eds. (1990). *New Directions in Regional Analysis. Integrated and Multi-Regional Approach*. London, Belhaven Press.

Aracil, J., E. Ponce, and L. Pizarro (1997). Behavior patterns of logistic models with a delay. *Mathematics and Computers in Simulation* **44**: 123–141.

Arai, T. and T. Akiyama (2004). Empirical analysis of the land use transition potential in a CA based land use model: application to the Tokyo Metropolitan Region. *Computers, Environment and Urban Systems* **28**(1/2): 65–84.

Arnold, V. I. (1973). *Ordinary Differential Equations*. Boston, MIT Press.

Ashby, W. R. (1947). *Principles of the Self-Organizing Dynamic System*.

Ashby, W. R. (1956). *An Introduction to Cybernetics*. London, Chapman & Hall.

Avner, D., O. Biham, D. Lidar, and O. Malcai (1998). Is the geometry of nature fractal? *Science* **279**: 39, 40.

Back, T., H. Dornemann, U. Hammel, and P. Frankhouser (1996). *Modeling Urban Growth by Cellular Automata*. International Conference on Evolutionary Computation. 5th Conference on Parallel Problem Solving from Nature. Berlin, Germany, Springer-Verlag.

Bak, P. (1996). *How Nature Works*. Springer-Verlag.

Bak, P., C. Tang, and K. Wiesenfeld (1988). Self-organized criticality. *Physical Review A* **38**: 364–374.

Bandini, S., G. Mauri, and R. Serra (2001). Cellular automata: From a theoretical parallel computational model to its application to complex systems. *Parallel Computing* **27**(5): 539–553.

Bandura, A. (1986). *Social Foundations of Thought and Action: A Social Cognitive Theory*. Englewood Cliffs, NJ, Prentice Hall.

Banks, E. (1970). Cellular automata. *AI Memo. MIT Artificial Intelligence Laboratory*. Cambridge, MA.

Banzhaf, W. (1994). Self-organization in a system of binary strings. *Artificial Life IV*. R. A. Brooks and P. Maes. Cambridge, MIT Press: 109–118.

Barlovic, R., A. Schadschneider, and M. Schreckenberg (2001). Random walk theory of jamming in a cellular automatonmodel for traffic flow. *Physica A* **294**: 525–538.

Barredo, J. I., M. Kasanko, N. McCormick, and C. Lavalle (2003). Modelling dynamic spatial processes: simulation of urban future scenarios through cellular automata. *Landscape and Urban Planning* **64**: 145–160.

Barrett, C. L., R. J. Beckman, K. P. Berkbigler, K. R. Bisset, B. W. Bush, K. Campbell, S. Eubank, K. M. Henson, J. M. Hurford, D. A. Kubicek, M. V. Marathe, P. R. Romero, J. P. Smith, L. L. Smith, P. L. Speckman, P. E. Stretz, G. L. Thayer, E. V. Eeckhout, and M. D. Williams (2002). Chapter 2: Population Synthesizer. TRANSIMS: Transportation Analysis Simulation System. Version: TRANSIMS 3.0. Volume Three: Modules. Los Alamos, NM, Los Alamos National Laboratory.

Barrett, C. L., R. J. Beckman, K. P. Berkbigler, K. R. Bisset, B. W. Bush, K. Campbell, S. Eubank, K. M. Henson, J. M. Hurford, D. A. Kubicek, M. V. Marathe, P. R. Romero, J. P. Smith, L. L. Smith, P. E. Stretz, G. L. Thayer, E. Van Eeckhout, and M. D. Williams (2001). TRansportation ANalysis SIMulation System (TRANSIMS). Portland Study Reports. Los Alamos, Los Alamos National Laboratory.

Barrett, C. L., R. J. Beckman, K. P. Berkbigler, K. R. Bisset, B. W. Bush, S. Eubank, J. M. Hurford, G. Konjevod, D. A. Kubicek, M. V. Marathe, J. D. Morgeson, M. Rickert, P. R. Romero, L. L. Smith, M. P. Speckman, P. L. Speckman, P. E. Stretz, G. L. Thayer, and M. D. Williams (1999). TRANSIMS (TRansportation ANalysis SIMulation System). Volume 0: Overview. Los Alamos, Los Alamos National Laboratory.

Batten, D. F. (1982). On the dynamics of the industrial revolution. *Regional Science and Urban Economics* **12**: 449–462.

Batty, M. (1972). Recent developments in land use modeling: a review of British research. *Urban Studies* **9**: 151–177.

Batty, M. (1976). *Urban Modelling: Algorithms, Calibration, Predictions*. Cambridge, Cambridge University Press.

Batty, M. (1979). Progress, success, and failure in urban modeling. *Environment and Planning A* **11**: 863–878.

Batty, M. (1991). Generating urban forms from diffusive growth. *Environment and Planning A* **23**(4): 511–544.

Batty, M. (1994). A chronicle of scientific planning: the Anglo-American modeling experience. *Journal of the American Planning Association* **60**(1): 7–16.

Batty, M. (1998). Urban evolution on the desktop: simulation with the use of extended cellular automata. *Environment and Planning A* **30**(11): 1943–1967.

Batty, M. (2001a). Models in planning: technological imperatives and changing roles. *International Journal of Applied Earth Observation and Geoinformation* **3**(3): 252–266.

Batty, M. (2001b). Polynucleated urban landscapes. *Urban Studies* **38**(4): 635–655.

Batty, M., H. Couclelis, and M. Eichen (1997). Urban systems as cellular automata. *Environment and Planning B-Planning & Design* **24**(2): 159–164.

Batty, M. and P. Longley (1994). *Fractal Cities: A Geometry of Form and Function*. Academic Press.

Batty, M., P. Longley, and S. Fotheringham (1989). Urban-growth and form—scaling, fractal geometry, and diffusion-limited aggregation. *Environment and Planning A* **21**(11): 1447–1472.

Batty, M. and P. M. Torrens (2001). Modeling complexity: the limits to prediction. *CyberGeo* **201**.

Batty, M. and Y. Xie (1994). From Cells to Cities. *Environment and Planning B-Planning & Design* **21**: S31–S38.

Batty, M. and Y. Xie (1997). Possible urban automata. *Environment and Planning B-Planning & Design* **24**(2): 175–192.

Batty, M. and Y. C. Xie (1999). Self-organized criticality and urban development. *Discrete Dynamics in Nature and Society* **3**(2, 3): 109–124.

Baylor University (2002). GRASS GIS (Geographic Resources Analysis Support System). Waco, TX, Center for Applied Geographic and Spatial Research, Baylor University.

Bell, E. J. (1974). Markov analysis of land use change: an application of stochastic processes to remotely sensed data. *Socio-Economic Planning Science* **8**: 311–316.

Bell, M., C. Dean, and M. Blake (2000). Forecasting the pattern of urban growth with PUP: a web-based model interfaced with GIS and 3D animation. *Computers, Environment and Urban Systems* **24**: 559–581.

Benati, S. (1997). A cellular automaton for the simulation of competitive location. *Environment and Planning B-Planning & Design* **24**(2): 205–218.

Benenson, I. (1998). Multi-agent simulations of residential dynamics in the city. *Computers, Environment and Urban Systems* **22**: 25–42.

Benenson, I. (1999). Modeling population dynamics in the city: from a regional to a multi-agent approach. *Discrete Dynamics in Nature and Society* **3**(2, 3): 149–170.

Benenson, I. (2001). *OBEUS: Object-Based Environment for Urban Simulation*. 6th International Conference on GeoComputation, University of Queensland, Brisbane, Australia, GeoComputation CD-ROM, available at http://www.geocomputation.org/2001/papers/benenson.pdf

Benenson, I. (2004). Agent-based modeling: from individual residential choice to urban residential dynamics. *Spatially Integrated Social Science: Examples in Best Practice*. M. F. Goodchild and D. G. Janelle. Oxford, Oxford University Press: 67–95.

Benenson, I. and I. Omer (2003). High-resolution Census data: a simple way to make them useful. *CODATA Journal*: in press.

Benenson, I., I. Omer, and E. Hatna (2002). Entity-based modeling of urban residential dynamics— the case of Yaffo, Tel-Aviv. *Environment and Planning B* **29**: 491–512.

Benenson, I. and P. M. Torrens (2003). *Geographic Automata Systems: A New Paradigm for Integrating GIS and Geographic Simulation*. Association Geographic Information Laboratories Europe (AGILE), Lyons.

Benenson, I. and P. M. Torrens (2004). Geosimulation: object-based modeling of urban phenomena. *Computers, Environment and Urban Systems* **28**(1/2): 1–8.

Benenson, I., S. Aronovich, and S. Noam (2004). Let's talk objects: generic methodology for urban high-resolution simulation. *Computers, Environment and Urban Systems*, in press.

Benenson, I. and I. Omer (2003). High-resolution census data: a simple way to make them useful. *Data Science Journal (Spatial Data Usability Special Section)* **2**(26): 117–127.

Benenson, Y., T. Paz-Elizur, R. Adar, E. Keinan, Z. Livneh, and E. Shapiro (2001). Programmable and autonomous computing machine made of biomolecules. *Nature* **414**: 430–434.

Benguigi, L., D. Czamanski, and M. Marinov (2001). City growth as a leap-frogging process: an application to the Tel-Aviv metropolis. *Urban Studies* **38**(10): 1819–1839.

Benguigui, L. (1995). A new aggregation model: application to town growth. *Physica A* **219**: 13–26.

Benguigui, L. (1998). Aggregation models for town growth. *Philosophical Magazine B* **77**(4): 1269–1275.

Benguigui, L., D. Czamanski, and M. Marinov (2001). The dynamics of urban morphology: the case of Petah Tikvah. *Environment and Planning B: Planning and Design* **28**: 447–460.

Berec, L. (2002). Techniques of spatially explicit individual-based models: construction, simulation, and mean-field analysis. *Ecological Modelling* **150**: 55–81.

Berger, T. (2001). Agent-based spatial models applied to agriculture: a simulation tool for technology diffusion, resource use changes and policy analysis. *Agricultural Economics* **25**(2, 3): 245–260.

Berlekamp, E. R., J. H. Conway, and R. K. Guy (1982). *Winning Ways For Your Mathematical Plays*. London, Academic Press.

Berry, M. W., R. O. Flamm, B. C. Hazen, and R. L. MacIntyre (1996). The Land-Use Change and Analysis System (LUCAS) for Evaluating Landscape Management Decisions. *IEEE Computational Science & Engineering* **3**(1): 24–35.

Bertuglia, C. S., G. Leonardi, S. Occelli, G. A. Rabino, R. Tadei, and W. A. G., Eds. (1994). *Urban Systems: Contemporary approach to modeling*. London, Croom Helm.

Besussi, E., A. Cecchini, and E. Rinaldi (1998). The diffused city of the Italian North-East: Identification of urban dynamics using cellular automata urban models. *Computers, Environment and Urban Systems* **22**(5): 497–523.

Blank, A. and S. Solomon (2000). Power laws in cities population, financial markets and internet sites (scaling in systems with a variable number of components). *Physica A* **287**: 279–288.

Blodgett, J. (1998). Geographic Correspondence Engine. New York, Center for International Earth Science Information Network (CIESIN), Columbia University.

Blue, V. J. and J. L. Adler (2001). Cellular automata microsimulation for modeling bi-directional pedestrian walkways. *Transportation Research Part B—Methodological* **35**: 293–312.

Boerner, R. E. J., M. N. DeMers, J. W. Simpson, F. J. Artigas, A. Silva, and L. A. Berns (1996). Markov models of inertia and dynamism on two continuous Ohio landscapes. *Geographical Analysis* **28**(1): 56–66.

Booch, G. (1994). *Object-oriented Analysis and Design with Applications*. Menlo Park, CA, Addison-Wesley.

Booch, G., J. Rumbaugh, and I. Jacobson (1999). *The Unified Modeling Language User Guide*. Reading, MA, Addison-Wesley.

Bourne, L. S. (1971). Physical adjustment processes and land use succession: a conceptual review and central city example. *Economic Geography* **47**: 1–15.

Box, P. (2001). Spatial units as agents: making the landscape an equal player in agent-based simulations. *Integrating Geographic Information Systems and Agent-Based Modeling Techniques for Simulating Social and Ecological Processes.* H. R. Gimblett. Oxford, Oxford University Press: 59–83.

Braga, G., G. Cattaneo, P. Flocchini, and C. Q. Vogliotti (1995). Pattern growth in elementary cellular automata. *Theoretical Computer Science* **145**(1, 2): 1–26.

Britannica, E. (1982). *New Encyclopedia Britannica.* Chicago, Encyclopedia Britannica.

Brog, W. and E. Erl (1983). Application of a model of individual behaviour (situational approach) to explain household activity patterns in an urban area due to forecast behavioural changes. *Recent Advances in Travel Demand Analysis.* S. Carpenter and P. M. Jones. Gower, Aldershot.

Brookings Institution (2001). Ascape. Washington, D.C., Center on Social and Economic Dynamics.

Brown, A. L. and J. K. Affum (2002). A GIS-based environmental modelling system for transportation planners. *Computers, Environment and Urban Systems* **26**: 577–590.

Brown, D. G., B. C. Pijanowski, and J. D. Duh (2002). Modeling the relationships between land use and land cover on private lands in the Upper Midwest, USA. *Journal of Environmental Management* **59**: 247–263.

Brown, L. A. and E. A. Moore (1970). The intra-urban migration process: a perspective. *Geografiska Annaler* **52B**: 1–13.

Brown, P. J. (1993). *Measurement, Regression, and Calibration.* Oxford, Clarendon.

Bruckmann, G. (2001). Global modeling. *Futures* **33**: 13–20.

Burks, A. W., Ed. (1966). *Theory of Self-Reproducing Automata [by] John von Neumann.* Urbana, University of Illinois Press.

Burks, A. W., Ed. (1970). *Essays on Cellular Automata.* Champaign, IL, University of Illinois Press.

Burstedde, C., K. Klauck, A. Schadschneider, and J. Zittartz (2001). Simulation of pedestrian dynamics using a two-dimensional cellular automaton. *Physica A* **295**: 507–525.

Bush, B. W. (2001). Portland synthetic population. Los Alamos, Los Alamos National Laboratory.

Candau, J. T., S. Rasmussen, and K. C. Clarke (2000). *A Coupled Cellular Automata Model for Land Use/Land Cover Change Dynamics.* 4th International Conference on Integrating GIS and Environmental Modeling (GIS/EM4): Problems, Prospects and Research Needs, Banff, Alberta, Canada.

Capello, R. and A. Faggian (2002). An economic-ecological model of urban growth and urban externalities: empirical evidence from Italy. *Ecological Economics* **40**: 181–198.

Carley, K. M. (1996). Artificial intelligence within sociology. *Sociological Methods and Research* **25**(1): 3–25.

Cattaneo, G., P. Flocchini, G. Mauri, C. Quaranta Vogliotti, and N. Santoro (1997). Cellular automata in fuzzy backgrounds. *Physica D* **105**: 105–120.

Cattaneo, G., E. Formenti, L. Margara, and G. Mauri (1999). On the dynamical behavior of chaotic cellular automata. *Theoretical Computer Science* **217**: 31–51.

Cecchini, A. and F. Viola (1990). Eine Stadtbausimulation. *Wissenschaftliche Zeitschrift der Hochschule für Architektur und Bauwesen* **36**(4).

Centre for Computational Geography (2003). GeoTools. Leeds, University of Leeds, School of Geography.

Cetin, N., K. Nagel, B. Raney, and A. Voellmy (2001). Large-scale multi-agent transportation systems. *Computational Physics Communications* **147**(1, 2): 559–564.

Chapin, F. S. and S. F. Weiss (1962). Factors influencing land development. *An Urban Studies Research Monograph.* Chapel Hill, Institute for Research in Social Science, University of North Carolina: 68.

Chapin, F. S. and S. F. Weiss (1965). Some input refinements for a residential model. *An Urban Studies Research Monograph.* Chapel Hill, Institute for Research in Social Science, University of North Carolina: 68.

Chapin, F. S. and S. F. Weiss (1968). A probabilistic model for residential growth. *Transportation Research* **2**: 375–390.

Charles, C. Z. (2003). The Dynamics of Racial Residential Segregation. *Annual Revew of Sociology* **29**: 167–207.

Cheng, J. and I. Masser (2003). Urban growth pattern modeling: a case study of Wuhan city, People's Republic of China. *Landscape and Urban Planning* **62**: 199–217.

Chowdhury, D., L. Santen, and A. Schadschneider (2000). Statistical physics of vehicular traffic and some related systems. *Physics Reports* **329**: 199.

Chowdhury, D., D. E. Wolf, and M. Schreckenberg (1997). Particle hopping models for two-lane traffic with two kinds of vehicles: Effects of lane-changing rules. *Physica A* **235**: 417–439.

Christaller, W. (1933, 1966). *Central Places in Southern Germany, Translation of C.W. Baskin.* London, Prentice-Hall.

Chua, L. and L. Yang (1988). Cellular neural networks: theory and applications. *IEEE Transactions on Circuits and Systems*: 1257–1290.

Ciolek, M. T. (1978). Spatial behaviour in pedestrian areas. *Ekistics* **268**: 120, 121.

Claramunt, C., B. Jiang, and A. Bargiela (2002). A new framework for the integration, analysis and visualisation of urban traffic data within GIS. *Unknown*.

Clark, C. (1967). *Population Growth and Land Use*. London.

Clark, J. D., R. M. Itami, H. R. Gimblett, and G. L. Ball (1990). *A Framework for Modeling and Simulating Spatial Dynamics in Geographic Information Systems Using The Integrated Dynamic Ecological Analysis System.* Proceedings of American Association of Geographers Conference, Toronto, Ontario, Canada, http://www.srnr.arizona.edu/people/facultypage/gimblett_public_html/toronto.html

Clark Labs (2002). IDRISI. Worcester, MA, Clark Labs.

Clark, W. A. V. (1991). Residential preferences and neighborhood racial segregation: a test of the schelling segregation model. *Demography* **28**(1): 1–18.

Clark, W. A. V. and S. D. Withers (1999). Changing jobs and changing houses: Mobility outcomes of employment transitions. *Journal of Regional Science* **39**(4): 653–673.

Clarke, G. P., Ed. (1996). *Microsimulation for Urban and Regional Policy Analysis.* European Research in Regional Science 6. London, Pion.

Clarke, K. C. (1997). Land Transition Modeling With Deltatrons, http://www.geog.ucsb.edu/~kclarke/Papers/deltatron.html. 2002.

Clarke, K. C. and L. Gaydos (1998). Loose coupling a cellular automaton model and GIS: long-term growth prediction for San Francisco and Washington/Baltimore. *International Journal of Geographical Information Science* **12**(7): 699–714.

Clarke, K. C., S. Hoppen, and L. J. Gaydos (1997). A self-modifying cellular automata model of historical urbanization in the San Francisco Bay area. *Environment and Planning B: Planning and Design* **24**(2): 247–261.

Cochrane, M. A., A. Alencar, M. D. Schulze, C. M. Souza, Jr., D. C. Nepstad, P. Lefebvre, and E. A. Davidson (1999). Positive feedbacks in the fire dynamic of closed canopy tropical forests. *Nature* **284**: 1832–1835.

Codd, E. (1968). *Cellular Automata.* New York, NY, Academic Press.

Congalton, R. G. and R. A. Mead (1983). A quantitative method to test for consistency and correctness in photointerpretation. *Photogrammetric Engineering and Remote Sensing* **49**(1): 69–74.

Couclelis, H. (1985). Cellular worlds: a framework for modeling micro-macro dynamics. *Environment and Planning B* **17**: 585–596.

Couclelis, H. (1988). Of mice and men—what rodent populations can teach us about complex spatial dynamics. *Environment and Planning A* **20**(1): 99–109.

Creighton, R. L., J. D. Carroll, Jr., and G. S. Finney (1959). Data processing for city planning. *Journal of the American Institute of Planners* **25**(2): 96–103.

Culik, II, K., L. P. Hurd, and S. Yu (1990). Computation theoretic aspects of cellular automata. *Physica D* **45**(1–3): 357–378.

DaVanzo, J. (1981). Repeat migration, information cost, and location-specific capital. *Population and Environment* **4**(1): 45–73.

Davis, P. J. and P. Rabinowitz (1975). *MeThods Of Numerical Integration*. New York, Academic Press.

Day, R. H. (1982). Emergence of chaos from neoclassical growth. *Geographical Analysis* **13**(4): 315–327.

de Almeida, C. M., M. Batty, A. M. V. Monteiro, G. Camara, B. S. Soares-Filho, G. C. Cerqueira, and C. L. Pennachin (2003). Stochastic cellular automata modeling of urban land use dynamics: empirical development and estimation. *Computers, Environment and Urban Systems* **27**: 481–509.

De Angelis, D. and L. J. Gross (1992). *Individual-Based Models and Approaches in Ecology: Populations, Communities, and Ecosystems*. New York, Chapman and Hall.

De Angelis, D. L., W. M. Post, and C. C. Travis (1986). *PoSitive Feedbacks in Natural Systems*. Berlin, Springer.

de Koning, G. H. J., P. H. Verburg, A. Veldkamp, and L. O. Fresco (1999). Multi-scale modelling of land use change dynamics in Ecuador. *Agricultural Systems* **61**: 77–93.

de Salesa, J. A., M. L. Martins, and J. G. Moreira (1997). One-dimensional cellular automata characterization by the roughness exponent. *Physica A* **245**: 461–471.

Dear, M. (2000). The Postmodern Urban Condition. London, Blackwell.

Deitz, R. (1998). A joint model of residential and employment location in urban areas. *Journal of Urban Economics* **44**(2): 197–215.

Dendrinos, D. and H. Mullally (1982). Evolutionary patterns of urban populations. *Geographical Analysis* **13**: 328–344.

Dendrinos, D. and H. Mullally (1985). *Urban Evolution: Studies in the Mathematical Ecology of the Cities*. Oxford, Oxford University Press.

Dendrinos, D. and M. Sonis (1990). *Chaos and Socio-Spatial Dynamics*. New York, Springer.

Derrida, B. and V. Hakim (1996). Coarsening in the 1D Ising model evolving with Swendsen-Wang dynamics: an unusual scaling. *Jornal of Physics A: Mathematics and General* **29**: L589–L594.

Deutsch, A. (2000). A new mechanism of aggregation in a lattice-gas cellular automaton model. *Mathematical and Computer Modelling* **31**: 35–40.

Dietz, R. D. (2002). The estimation of neighborhood effects in the social sciences: an interdisciplinary approach. *Social Science Research* **31**: 539–575.

Dijkstra, J. and H. Timmermans (2002). Towards a multi-agent model for visualizing simulated user behavior to support the assessment of design performance. *Automation in Construction* **11**: 135–145.

Dijkstra, J., H. J. P. Timmermans, and A. J. Jessurun (2000). A multi-agent cellular automata system for visualising simulated pedestrian activity. *Theoretical and Practical Issues on Cellular Automata*. S. Bandini and T. Worsch. London, Springer-Verlag: 29–36.

Domain, C. and H. Gutowitz (1997). The topological skeleton of cellular automaton dynamics. *Physica D* **103**: 155–168.

Donnelly, T. G., F. S. Chapin, and S. F. Weiss (1964). A probabilistic model for residential growth. *An Urban Studies Research Monograph*. Chapel Hill, Institute for Research in Social Science, University of North Carolina: 65.

Dorigo, M., G. Di Caro and L. M. Gambardella (1999). Ant Algorithms for Discrete Optimization. *Artifical Life* **5**(3): 137–172.

Dupuis, A. and B. Chopard (1998). Parallel simulation of traffic in geneva using cellular automata. *Unknown*.

Durrett, R. (1999). Stochastic spatial models. *SIAM Review* **41**(4): 677–718.

Durrett, R. and S. A. Levin (1994). Stochastic spatial models: a user's guide to ecological applications. *Philosophical Transactions of the Royal Society of London, Series B: Biological Sciences* **343**: 329–350.

Eagleson, S., F. Escobar, and I. Williamson (2002). Hierarchical spatial reasoning theory and GIS technology applied to the automated delineation of administrative boundaries. *Computers, Environment and Urban Systems* **26**: 185–200.

Ellner, S. P., E. McCauley, B. E. Kendall, C. J. Briggs, P. R. Hosseinik, S. N. Wood, A. Janssen, M. W. Sabelis, P. Turchin, R. M. Nisbet, and W. W. Murdoch (2001). Habitat structure and population persistence in an experimental community. *Nature* **412**(2): 538–543.

Eloranta, K. (1997). Critical growth phenomena in cellular automata. *Physica D* **103**: 478–484.

Emmerich, H. and E. Rank (1995). Investigating traffic flow in the presence of hindrances by cellular automata. *Physica A* **216**: 435–444.

Engelen, G. (1988). The theory of self-organization and modeling complex urban system. *European Journal of Operational Research* **37**: 42–57.

Engelen, G., R. White, and I. Uljee (1997). Integrating constrained cellular automata models, GIS and decision support tools for urban planning and policy making. *Decision Support Systems in Urban Planning*. H. P. J. Timmermans. London, E&FN Spon: 125–155.

Engelen, G., R. White, and I. Uljee (2002). The MURBANDY and MOLAND models for Dublin. Submitted to European Comisssion Joint Research Center, Ispra, Italy. Dublin, ERA-Maptec: 172.

Engelen, G., R. White, I. Uljee, and P. Drazan (1995). Using cellular-automata for integrated modeling of socio-environmental systems. *Environmental Monitoring and Assessment* **34**(2): 203–214.

Epstein, J. M. (1999). Agent-based computational models and generative social science. *Complexity* **4**(5): 41–60.

Epstein, J. M. and R. Axtell (1996). *Growing Artificial Societies from the Bottom Up*. Washington, D.C., Brookings Institution.

Eradus, P., A. Schoemakers, and T. van der Hoorn (2002). Four applications of the TIGRIS model in the Netherlands. *Journal of Transport Geography* **10**: 111–121.

Erickson, B. and T. Lloyd-Jones (1997). Experiments with settlement aggregation models. *Environment and Planning B-Planning & Design* **24**(6): 903–928.

Ermentrout, G. B. and L. Edelstei-Keshet (1993). Cellular automata approaches to biological modeling. *Journal of Theoretical Biology* **160**: 97–133.

Evans, K. M. (2003). Larger than life: threshold-range scaling of Life's coherent structures. *Physica D*, in press.

Feigenbaum, M. J. (1980). The metric universal property of period doubling bifurcations and the spectrum for the route to turbulence. *Annals of the New York Academy of Sciences* **357**: 330–336.

Ferber, J. (1999). *Multi-agent systems: an introduction to distributed artificial intelligence*. Harlow, UK, Addison-Wesley.

Festinger, L. (1954). A theory of social comparision processes. *Human Relations* **7**: 117–140.

Flache, A. and R. Hegselmann (2001). Do irregular grids make a difference? relaxing the spatial regularity assumption in cellular models of social dynamics. *Journal of Artificial Societies and Social Simulation vol. 4, no. 4* **4**(4): ⟨http://www.soc.surrey.ac.uk/JASSS/4/4/6.html⟩.

Fokkema, T. and L. VanWissen (1997). Moving plans of the elderly: A test of the stress-threshold model. *Environment and Planning A* **29**(2): 249–268.

Forrester, J. W. (1961). *Industrial Dynamics*. Cambridge, Massachusetts and London, England, MIT Press.

Forrester, J. W. (1969). *Urban Dynamics*. Cambridge, MA, MIT Press.

Fotheringham, A. S., M. Batty, and P. A. Longley (1989). Diffusion-limited aggregation and the fractal nature of urban growth. *Papers of the Regional Science Association* **67**: 55–69.

Fotheringham, A. S. and M. E. O'Kelly (1989). *Spatial Interaction Models: Formulations and Applications*. Dordrecht, Kluwer Academic Publishers.

Fox, M. (1995). Transport planning and the human activity approach. *Journal of Trasport Geography* **3**(2): 105–116.

Frankhouser, P. (1991). Aspects fractals des structures structures urbaines. *L'Espace Geographique* **1**: 45–69.

Franklin, S. and A. Graesser (1996). *Is it an Agent, or Just a Program?: A Taxonomy for Autonomous Agents*. Intelligent Agents III: Proceedings of the Third International Workshop on Agent Theories, Architectures, and Languages, (ATAL'96), Lecture Notes in AI, Third International Workshop on Agent Theories, Architectures and Languages (ATAL'96), Springer-Verlag.

Fruin, J. J. and N. Y. York (1971). Pedestrian planning and design. New York, Metropolitan Association of Urban Designers and Environmental Planners.

Fukui, M., K. Nishinari, D. Takahashi, and Y. Ishibashi (2002). Metastable flows in a two-lane traffic model equivalent to extended Burgers cellular automaton. *Physica A* **303**: 226–238.

Gabaix, X. (1999). Zipf's law for cities: an explanation. *Quarterly Journal of Economics* **64**: 739–767.

Galam, S. (1997). Rational group decision making: A random field Ising model at T = 0. *Physica A* **238**: 66–80.

Galam, S. and J.-D. Zucker (2000). From individual choice to group decision-making. *Physica A* **287**: 644–659.

Gamma, E. R., H. R. Johnson, and J. Vlissides (1995). *Design Patterns: Elements of Reusable Object-Oriented Software*. Boston, Addison-Wesley.

Gardner, M. (1970). The fantastic combinations of john conway's new solitaire game "Life". *Scientific American* **223**: 120–123.

Gardner, M. (1971). Mathematical games: On cellular automata, self-reproduction, the Garden of Eden, and the game 'life'. *Scientific American* **224**(2): 112–117.

Garn, H. A. and R. H. Wilson (1972). A look at urban dynamics: The Forrester model and public policy. *Urban Dynamics: Extensions and Reflections*. K. Chen. San Francisco, CA, San Francisco Press, Inc.

Gerling, R. W. (1990a). Classification of three-dimensional cellular automata. *Physica A* **162**: 187–195.

Gerling, R. W. (1990b). Classification of triangular and honeycomb cellular automata. *Physica A* **162**: 196–209.

Gigerenzer, G. and D. G. Goldstein (1996). Reasoning the fast and frugal way: models of bounded rationality. *Psychological Review* **103**(4): 650–669.

Gilbert, G. N. and R. Conte, Eds. (1995). *Artificial Societies: The Computer Simulation of Social Life*. London, UCL Press.

Gillman, M. and R. Hails (1997). *An Introduction to Ecological Modelling: Putting Practice into Theory*. Oxford, Blackwell.

Gimblett, H. R. (2002). Integrating geographic information systems and agent-based technologies for modeling and simulating social and ecological phenomena. *Integrating Geographic Information Systems and Agent-Based Modeling Techniques for Simulating Social and Ecological Processes*. H. R. Gimblett. Oxford, Oxford University Press: 1–21.

Ginot, V., C. Le Page, and S. Souissi (2002). A multi-agents architecture to enhance end-user individual-based modeling. *Ecological Modelling* **157**(1): 23–41.

Gipps, P. G. (1981). A behavioural car-following model for computer simulation. *Transportation Research B* **15**: 105–111.

Gipps, P. G. (1986a). A model for the structure of lane-changing decisions. *Transportation Research B* **20**(5): 403–414.

Gipps, P. G. (1986b). MULTSIM: A model for simulating vehicular traffic on multi-lane arterial roads. *Mathematics and Computers in Simulation* **28**: 291–295.

Gipps, P. G. and B. Marksjo (1985). A micro-simulation model for pedestrian flows. *Mathematics and Computers in Simulation* **27**: 95–105.

Gleick, J. (2003). *Isaac Newton*. London, Fourth Estate.

Gobin, A., P. Campling, and J. Feyen (2002). Logistic modelling to derive agricultural land use determinants: a case study from southeastern Nigeria. *Agriculture, Ecosystems & Environment* **89**(3): 213–228.

Goldstein, N. C., J. T. Candau, and K. C. Clarke (2004). Approaches to simulating the "March of Bricks And Mortar". *Computers, Environment and Urban Systems* **28**(1/2): 125–147.

Golze, U. (1976). Differences between 1- and 2-dimensional cell spaces. *Automata, Languages, Development*. Amsterdam, The Netherlands, North-Holland Publishing Co.: 369–384.

Goodchild, M. F. (2001). Towards a location theory of distributed computing and e-commerce. *Worlds of E-Commerce: Economic, Geographical and Social Dimensions*. T. R. Leinbach and S. D. Brunn. New York, Wiley: 67–86.

Goodman, J. L. (1981). Information, uncertainty, and the microeconomic model of migration decision making. *Migration Decision Making: Multidisciplinary Approaches to Microlevel Studies in Developed and Developing Countries*. G. F. DeJong and R. W. Gardner. New York, Pergamon Press: 13–58.

Gora, P. and A. Boyarsky (1990). Higher-dimensional point transformations and asymptotic measures for cellular automata. *Computers & Mathematics with Applications* **19**(12): 13–31.

Gravner, J. and D. Griffeath (1998). Cellular automaton growth on Z2: theorems, examples, and problems. *Advances In Applied Mathematics* (21): 241–304.

Grimm, V. (1999). Ten years of individual-based modelling in ecology: what have we learned and what could we learn in the future? *Ecological Modelling* **115**: 129–148.

Gu, G. Q., K. H. Chung, and P. M. Hui (1995). Two-dimensional traffic flow problems in inhomogeneous lattices. *Physica A* **217**: 339–347.

Gulyas, L., T. Kozsik, and J. B. Corliss (1999). The multi-agent modelling language and the model design interface. *Journal of Artificial Societies and Social Simulation* **2**(3).

Guo, H. C., L. Liu., G. H. Huang, G. A. Fuller, R. Zou, and Y. Y. Yin (2001). A system dynamics approach for regional environmental planning and management: A study for the Lake Erhai basin. *Journal of Environmental Management* **61**: 93–111.

Guttorp, P. (1995). *Stochastic Modelling of Scientific Data*. London, Chapman and Hall.

Hägerstrand, T. (1952). The propagation of innovation waves. *Lund Studies in Geography B, Human Geography* **4**: 3–19.

Hägerstrand, T. (1967). *Innovation Diffusion as a Spatial Process, translated by Allan Pred from 'Innovationsforloppet ur Korologisk Synpunkt' Gleerup, Lund, Sweeden, 1953*. Chicago, University of Chicago Press.

Haggett, P., A. D. Cliff, and A. Frey (1977). *Locational models*. London, Edward Arnold.

Haken, H. (1983). *Synergetics. An Introduction*. Berlin, Springer.

Haken, H. (1993). Synergetics as a strategy to cope with complex systems. *Interdisciplinary Approaches to Non-Linear Complex Systems*. H. Haken and A. Mikhailov. Berlin, Springer.

Haken, H. and J. Portugali (1995). A synergetic approach to the self-organization of cities and settlements. *Environment and Planning B* **22**: 35–46.

Haken, H. and J. Portugali (1996). Synergetics, inter-representation networks and cognitive maps. *The Construction of Cognitive Maps*. J. Portugali. Dordrecht, Kluwer Academic Publishers: 45–67.

Haklay, M., D. O'Sullivan, M. Thurstain-Goodwin, and T. Schelhorn (2001). "So go downtown": simulating pedestrian movement in town centres. *Environment and Planning B: Planning and Design* **28**: 343–359.

Halls, P. J., M. Bulling, P. C. L. White, L. Garland, and S. Harris (2001). Dirichlet neighbors: revisiting Dirichlet tessellation for neighborhood analysis. *Computers, Environment and Urban Systems* **25**: 105–117.

Hansen, W. G. (1959). How accessability shapes land use. *Journal of the American Institute of Planners* **25**: 73–76.

Harary, F. and G. Gupta (1997). Dynamic graph models. *Mathematical and Computer Modelling* **25**(7): 79–87.

Harris, B. (1994). The real issues concerning lees requiem. *Journal of the American Planning Association* **60**(1): 31–34.

Harvey, D. (1969). *Explanation in Geography.* London, Edward Arnold Ltd.

Hathout, S. (2002). The use of GIS for monitoring and predicting urban growth in east and west St Paul, Winnipeg, Manitoba, Canada. *Journal of Environmental Management* **66**: 229–238.

Hayes-Roth, B. (1995). An architecture for adaptive intelligent systems. *Artificial Intelligence: Special Issue on Agents and Interactivity* **72**: 329–365.

Hegselmann, R. and A. Flache (1999). Understanding complex social dynamics: a plea for cellular automata based modeling. *Journal of Artificial Societies and Social Simulation* **1**(3): ⟨http://www.soc.surrey.ac.uk/JASSS/1/3/1.html⟩.

Helbing, D. (1991). A mathematical modeling for the behavior of pedestrians. *Behavioral Science* **36**: 298–310.

Helbing, D. (1992). A fluid-dynamic for the movement of pedestrians. *Complex Systems* **6**: 391–415.

Helbing, D., I. Farkas, and T. Vicsek (2000). Simulating dynamical features of escape panic. *Nature* **47**(28): 487–490.

Helbing, D. and P. Mulnar (1995). Social force model for pedestrian dynamics. *Physical Review E* **51**: 4282–4286.

Helbing, D., P. Mulnar, and P. Schweitzer (1988). Computer simulations of pedestrian dynamics and trail formation, arXiv: cond-mat/ 9805074 v2 7 May 1998. 2002.

Helly, W. (1969). A time-dependent intervening opportunity land use model. *Socio-Economic Planning Science* **3**: 65–73.

Helly, W. (1975). *Urban Systems Models.* New York, Academic Press.

Hemmingsson, J. (1995). Consistent results on "Life". *Physica D* **80**: 151–153.

Herold, M. (2002). *Remote Sensing Based Analysis of Urban Dynamics in the Santa Barbara Region Using the SLEUTH Urban Growth Model and Spatial Metrics.* Third Symposium on Remote Sensing of Urban Areas, Istanbul, Turkey.

Herrin, W. E. and C. R. Kern (1992). Testing the standard urban model of residential choice—an implicit markets approach. *Journal of Urban Economics* **31**(2): 145–163.

Hester, D. J. (1998). Modeling Albuquerque's Urban Growth. Case Study: Isleta, New Mexico, 1:24,000-scale quadrangle, US Department of the Interior, US Geological Survey, Rocky Mountain Mapping Center. 2000.

Hidas, P. (1998). A car following model for urban traffic simulation. *Traffic Engineering & Control* **39**(5): 300–309.

Hidas, P. (2002). Modelling lane changing and merging in microscopic traffic simulation. *Transportation Research Part C* **10**: 351–371.

Hogeweg, P. (1988). Cellular automata as a paradigm for ecological modeling. *Applied Mathematics And Computation* **27**: 81–100.

Holland, J. H. (1995). *Hidden Order: How Adaptation Builds Complexity.* Reading, MA, Addison-Wesley.

Holland, J. H. (1998). *Emergence: From Chaos to Order.* Reading, MA, Perseus Books.

Hortsmann, C. S. and G. Cornell (2001). *Core Java 2, Volume 1: Fundamentals.* Saddle River, NJ, Sun Microsystems Press & Prentice Hall.

Huberman, B. and N. Glance (1993). Evolutionary games and computer simulations. *Proceedings of the National Academy of Sciences* **90**: 7716–7718.

Huffaker, C. B. (1958). Experimental studies on predation: dispersion factors and predator-prey oscillations. *Hilgardia* **27**(14): 343–383.

ICBS (2000). *Socio-Economic Characteristics of Population and Households in Localities and Statistical Areas.* Jerusalem, State of Israel, Central Bureau of Statistics Publications.

Ingerson, T. E. and R. L. Buvel (1984). Structure in asynchronous cellular automata. *Physica D* **10**: 59–68.

Ioannides, Y. M. and H. G. Overman (2003). Zipf's law for cities: an empirical examination. *Regional Science and Urban Economics* **33**: 127–137.

Iooss, G. and D. D. Joseph (1980). *Elementary Stability and Bifurcation Theory.* New York, Springer.

Irwin, E. G. and N. E. Bockstael (2002). Interacting agents, spatial externalities and the evolution of residential land-use patters. *Journal of Economic Geography* **2**: 31–54.

Irwin, E. G. and J. Geoghegan (2001). Theory, data, methods: developing spatially explicit economic models of land use change. *Agriculture, Ecosystems & Environment* **85**(1-3): 7–23.

Itami, R. (1988). Cellular worlds: models for dynamic conception of landscapes. *Lanscape Architecture* **78**(5): 52–57.

Jager, W., M. A. Janssen, H. J. N. de Vries, J. de Greef, and C. A. J. Vlek (2000). Behavior in commons dilemmas: *Homo economicus and Homo psychologicus* in an ecological-economic model. *Ecological Economics* **35**: 357–379.

Jahan, S. (1986). The determination of stability and similarity of markovian land use change processes: a theoretical and empirical analysis. *Socio-enonomic Planning Sciences* **20**(4): 243–251.

Janis, I. L. and L. Mann (1977). Decision making; a psychological study of conflict, choice and commitment. New York, The Free Press.

Janssen, M. and W. Jager (1999). An integrated approach to simulating behavioral processes: A case study of the lock-in of consumption patterns. *Journal of Artificial Societies and Social Simulation* **2**(2): ⟨http://www.soc.surrey.ac.uk/JASSS/2/2/2.html⟩.

Jarvis, R. A. (1983). Growing polyhedral obstacles for planninq collision-free paths. *The Australian Computer Journal* **15**: 103–111.

Jensen, H. J. (1998). *Self-Organized Criticality*. Cambridge, Cambridge University Press.

Johnson, S. (2002). *Emergence: The Connected Lives of Ants, Brains, Cities, and Software*, Touchstone/Simon and Schuster.

Jones, R. W. (1973). *Principles of Biological Regulation: An Introduction to Feedback Systems*. New York, Academic Press.

Kanenko, K. (1995). Chaos as a source of complexity and diversity in evolution. *Artificial Life. An Overview*. C. G. Langton. Oxford, England, MIT Press: 163–177.

Kauffman, S. A. (1995). *At Home in the Universe*, Oxford University Press.

Kayama, Y., M. Tabuse, H. Nishimura, and T. Horiguchi (1993). Characteristic parameters and classification of one-dimensional cellular automata. Solitons & Fractals **3**(6): 651–665.

Kerridge, J., J. Hine, and M. Wigan (2001). Agent-based modelling of pedestrian movements: the questions that need to be asked and answered. *Environment and Planning B: Planning and Design* **28**: 327–341.

Kirchner, A. and A. Schadschneider (2002). Simulation of evacuation processes using a bionics-inspired cellular automaton model for pedestrian dynamics. *Physica A* **312**: 260–276.

Kitamura, R., S. Nakayama, and T. Yamamoto (1999). Self-reinforcing motorization: can travel demand management take us out of the social trap? *Transport Policy* **6**: 135–145.

Klosterman, R. E. (1994). Symposium: Large-scale urban models: twenty years later. *Journal of the American Planning Association* **60**(1): 3–44.

Klosterman, R. E. (2001). The What If? Planning support system. *Planning Support Systems: Integrating Geographic Information Systems, Models and Visualization Tools*. R. K. Brail and R. E. Klosterman. Redlands, California, ESRI: 264–284.

Kohler, T. A. (2000). Putting social sciences together again: an introduction to the volume. *Dynamics in Human and Primate Societies*. T. A. Kohler and G. Gumerman. New York, Oxford University Press: 1–18.

Kohler, T. A. and G. Gumerman (2001). *Dynamics in Human and Primate Societies*. New York, Oxford University Press.

Krumpus, M. A. (2001). Overview of the Evo Artificial Life Framework, Omicron Group, http://omicrongroup.org/evo/overview/html/overview.html.

Kuhn, W. (2001). Ontologies in support of activities in geographic space. *International Journal of Geographic Information Science* **15**(7): 613–631.

Kwartler, M. and R. N. Bernard (2001). CommunityViz: an integrated planning support system. *Planning Support Systems: Integrating Geographic Information Systems, Models and Visualization Tools*. R. K. Brail and R. E. Klosterman. Redlands, California, ESRI: 285–308.

LaGro, J. A. and S. D. DeGloria (1992). Land use dynamics within an urbanizing non-metropolitan county in New York State (USA). *Landscape Ecology* 7(4): 275–289.

Lambin, E. F., M. D. A. Rounsevell, and H. J. Geist (2000). Are agricultural land-use models able to predict changes in land-use intensity? *Agriculture Ecosystems & Environment* 82(1–3): 321–331.

Landis, J. (2001). CUF, CUF II, and CURBA: A family of spatially explicit urban growth and land-use policy simulation models. *Planning Support Systems: Integrating Geographic Information Systems, Models and Visualization Tools*. R. K. Brail and R. E. Klosterman. Redlands, California, ESRI: 157–200.

Landis, J. and M. Zhang (1998a). The second generation of the California urban futures model: Part 1: model logic and theory. *Environment and Planning B: Planning & Design* 25(5): 657–666.

Landis, J. and M. Zhang (1998b). The second generation of the california urban futures model. Part 2: specification and calibration results of the land-use change submodel. *Environment and Planning B: Planning & Design* 25(6): 795–824.

Langlois, A. and M. Phipps. (1997). Automates Cellulaires: *Application a la Simulation Urbaine*. Paris, Editions Hermes.

Langton, C. G. (1986). Studying artificial life with cellular automata. *Physica D* 22.

Langton, C. G. (1992). Life at the edge of chaos. *Artificial Life* II. C. G. Langton, C. Taylor, J. D. Farmer, and S. Rasmussen. Redwood City, CA, Addison-Wesley: 41–93.

La Salle, J. and S. Lefschetz (1961). *Stability by Liapunov's Direct Method with Applications*. New York, Academic Press.

Latrop, G. T. and J. R. Hamburg (1965). An opportunity-accessability model for allocating regional growth. *Journal of the American Institute of Planners* 31: 95–103.

Laurie, A. J. and K. J. Narendra (2003). Role of vision in neighbourhood racial segregation: a variant of the schelling segregation model. *Urban Studies* 40(13): 2687–2704.

Leao, S., I. Bishop, and D. Evans (2001). Assessing the demand of solid waste disposal in urban region by urban dynamics modelling in a GIS environment. *Resources, Conservation and Recycling* 33: 289–313.

Lee, D. B. (1973). A requiem for large scale modeling. *Journal Of The American Institute Of Planners* 39(3): 163–178.

Lee, D. B. (1994). Retrospective on large-scale urban models. *Journal of the American Planning Association* 60(1): 35–40.

Leonard, A. (1997). *Bots: The Origin of a New Species*. San Francisco, Hardwired.

Levy, S. (1992). *Artificial Life. The Quest for a New Creation*. Penguin.

Lewin, K. (1951). *Field Theory in Social Science*. Harper & Brothers.

Li, S. M. and Y. M. Siu (2001). Residential mobility and urban restructuring under market transition: A study of Guangzhou, China. *Professional Geographer* 53(2): 219–229.

Li, X. and A. G.-O. Yeh (2000). Modelling sustainable urban development by the integration of constrained cellular automata and GIS. *International Journal of Geographical Information Science* 14(2): 131–152.

Lieberson, S. (1981). An asymmetrical approach to segregation. *Ethnic Segregation in the Cities*. C. Peach, V. Robinson, and S. Smith. London, Croom Helm: 61–82.

Ligtenberg, A., A. K. Bregt, and R. Van Lammeren (2001). Multi-actor-based land use modelling: Spatial planning using agents. *Landscape and Urban Planning* 56(1, 2): 21–33.

Lindenmayer, A. (1968). Mathematical models for cellular interaction in development, Parts I and II. *Journal of Theoretical Biology* 18: 280–315.

Liu, X.-H. and C. Andersson (2004). Assessing the impact of temporal dynamics on land-use change modeling. *Computers, Environment and Urban Systems* 28(1/2): 107–124.

Liu, Y. and S. R. Phinn (2001). *Developing a Cellular Automaton Model of Urban Growth Incorporating Fuzzy Set Approaches*. 6th International Conference on GeoComputation,

University of Queensland, Brisbane, Australia, GeoComputation CD-ROM, available also at http://www.geocomputation.org/2001/papers/liu.pdf.

Liu, Y. and S. R. Phinn (2003). Modelling urban development with cellular automata incorporating fuzzy-set approaches. *Computers, Environment and Urban Systems* **27**: 637–658.

Logi, F. and S. G. Ritchie (2002). A multi-agent architecture for cooperative inter-jurisdictional traffic congestion management. *Transportation Research Part C* **10**: 507–527.

Logsdon, M. G., E. J. Bell, and F. V. Westerlund (1996). Probability mapping of land use change: A GIS interface for visualizing transition probabilities. *Computers, Environment and Urban Systems* **20**(6): 381–398.

Longley, P. A., M. Batty, and J. Shepherd (1991). The size, shape and dimension of urban settlements. *Transactions of the Institute of British Geographers* **16**(1): 75–94.

Longley, P. A. and T. V. Mesev (2001). Measuring urban morphology using remotely-sensed imagery. *Remote Sensing and Urban Analysis*. J.-P. Donnay, M. Barnsley, and P. A. Longley. London, Taylor and Francis.

Lopez, E., G. Bocco, M. Mendoza, and E. Duhau (2001). Predicting land-cover and land-use change in the urban fringe A case in Morelia city, Mexico. *Landscape and Urban Planning* **55**: 271–285.

Losch, A. (1940, 1954). *The Economics of Location, Translation of W.H. Woglam and W. F. Stolper.* New Haven, Yael University Press.

Lotka, A. J. (1925/1956). *Elements of Physical Biology.* Baltimore, Williams & Wilkins Co./New York, Dover.

Louviere, J. and H. Timmermans (1990). A review of recent advances in decompositional preference and choice models. *Tijdschrift Voor Economische En Sociale Geografie* **81**(3): 214–224.

Louviere, J. J., D. A. Hensher, and J. D. Swatt (2000). *Stated Choice Methods: Analysis and Application.* Cambridge, Cambridge University Press.

Lowry, I. S. (1964). A model of metropolis. Santa Monica, CA, The RAND Corporation: 136.

Lozano-Perez, L. and M. A. Wesley (1979). An algorithm for planning collision-free paths among polyhedral obstacles. *Communications of the ACM* **22**(10): 560–570.

Luna, F. and B. Stefansson, Eds. (2000). *Economic Simulation in Swarm: Agent-based Modelling and Object Oriented Programming.* Dordrecht, Kluwer.

Machi, A. and F. Mignosi (1993). Garden of Eden configurations for cellular automata on cayley graphs of groups. *SIAM Journal on Discrete Mathematics* **6**(1): 44–56.

Maes, P. (1995a). Artificial life meets entertainment: life like autonomous agents. *Communications of the ACM* **38**(11): 108–114.

Maes, P. (1995b). Modeling adaptive autonomous agents. *Artificial Life, An Overview.* C. G. Langton. Oxford, MIT Press: 135–162.

Magnier, M., C. Lattaud, and J. Heudin (1997). Complexity classes in the two-dimensional life cellular automata subspace. *Complex Systems* **11**(6): 419–436.

Makse, H. A., J. S. Andrade, M. Batty, S. Havlin, and H. E. Stanley (1998). Modeling urban growth patterns with correlated percolation. *Physical Review E* **58**(6): 7054–7062.

Makse, H. A., S. Havlin, and H. E. Stanley (1995). Modeling urban growth patterns. *Nature* **377**(6): 608–612.

May, R. M. (1976). Simple mathematical models with very complicated dynamics. *Nature* **261**: 459.

May, R. M. and A. Oster (1976). Bifurcation and dynamic complexity in simple ecological models. *The American Naturalist* **110**: 573–599.

McCulloch, W. S. and W. H. Pitts (1943). A logical calculus of the ideas immanent in nervous activity. *Bulletin of Mathematical Biophysics* **5**: 115–133.

McMillen, D. P. (1989). An empirical model of urban fringe land use. *Land Economics* **65**(2): 138–145.

Meadows, D. H., D. L. Meadows, J. Randers, and W. W. Behrens, III (1972). *The Limits to Growth: A Report for the Club of Rome's Project on the Predicament of Mankind.* New York, Universe Books.

Meyer, C. F., A. B. Corbeau, and H. L. Mack (1975). A computer-oriented land use forecasting model with mapping capability. *Computers and Urban Society* **1**: 31–48.

Meyer, W. B. (2000). The other Burgess model. *Urban Geography* **21**(3): 261–270.

Microsoft Corporation and Digital Equipment Corporation (1995). The Component Object Model Specification. Redmond, WA, Microsoft Corporation and Digital Equipment Corporation.

Mills, E. S. and B. W. Hamilton (1989). *Urban Economics*. Glenview, IL, Scott, Foresman.

Milsum, J., Ed. (1968). *Positive Feedback: A General Systems Approach to Positive/Negative Feedback and Mutual Causality*. Oxford, Pergamon Press.

Minar, N., R. Burkhart, C. Langton, and M. Askenazi (1996). The Swarm simulation system: A toolkit for building multi-agent simulations. Santa Fe, Santa Fe Institute.

Molin, E., H. Oppewal, and H. Timmermans (1999). Group-based versus individual-based conjoint preference models of residential preferences: a comparative test. *Environment and Planning A* **31**(11): 1935–1947.

Moore, E. G., Ed. (1964). *Sequential Machines. Selected Papers*. Redwood City, CA, Addison-Wesley Publishing.

Morale, D. (2001). Modeling and simulating animal grouping. Individual-based models. *Future Generation Computer Systems* **17**: 883–891.

Murdoch, W. W., E. McCauley, R. M. Nisbet, W. S. C. Gurney, and A. M. de Roos (1992). Individual-based models: combining testability and generality. *Individual-Based Models and Approaches in Ecology—Populations, Communities and Ecosystems*. D. L. De Angelis and L. J. Gross. New York, Chapman & Hall: 18–35.

Murray, J. D. (1989). *Mathematical Biology*. Berlin, Springer-Verlag.

Nagatani, T. (1993). Jamming transition induced by a stagnant street in a traffic-flow model. *Physica A* **198**: 108–116.

Nagatani, T. (1996). Effect of car acceleration on traffic flow in 1D stochastic CA model. *Physica A* **223**: 137–148.

Nagel, K., R. J. Beckman, and C. L. Barrett (1999). TRANSIMS for urban planning. Los Alamos, NM, Los Alamos National Laboratory.

Nagel, K., M. Rickert, P. M. Simon, and M. Pieck (2000). The dynamics of iterated transportation simulations, http://www.santafe.edu/sfi/publications/Working-Papers/00-02-012.pdf. 2003.

Nagel, K. and M. Schreckenberg (1992). Cellular automaton model for freeway traffic. *J. Physique I (Paris)* **2**: 2221–2229.

Nagel, K. and M. Schrekenberg (1995). Traffic jams in stochastic cellular automata. Los Alamos, NM, Los Alamos National Laboratory.

Nagel, K., M. Shubik, M. Paczuski, and P. Bak (2000). Spatial competition and price formation. *Physica A* **287**: 546–562.

Nakajima, T. (1977). Application de la théorie de l'automate à la simulation de l'évolution de l'espace urbain. Congrès Sur La Méthodologie De L'Aménagement Et Du Dévelopment. Montreal, Association Canadienne-Française Pour L'Avancement Des Sciences et Comité De Coordination Des Centres De Recherches En Aménagement, Développement Et Planification (CRADEP): 154–160.

Negroponte, N. (1995). *Being Digital*. London, Coronet.

Nicolis, G. and I. Prigogine (1977). *Self-Organization in Nonequilibrium Systems*. New York, Wiley.

Nijkamp, P. and A. Reggiani (1995). Non-linear evolution of dynamic spatial systems. The relevance of chaos and ecologically-based models. *Regional Science and Urban Economics* **25**(2): 183–210.

Ninagawa, S., M. Yoneda, and S. Hirose (1998). 1/f fluctuation in the "Game of Life". *Physica D* **118**: 49–52.

Noble, J. (2000). Chapter 6. Basic relationship patterns. *Pattern Languages of Program Design* 4. N. Harrison, B. Foote, and H. Rohnert, Addison-Wesley: 73–89.

Noth, M., A. Borning, and P. Waddell (2003). An extensible, modular architecture for simulating urban development, transportation, and environmental impacts. *Computers, Environment and Urban Systems* **27**: 181–203.

Oliveira, G. M. B., P. P. B. de Oliveira, and N. Omar (2001). Definition and application of a five-parameter characterization of one-dimensional cellular automata rule space. *Artificial Life* **7**: 277–301.

Omer, I. (1996). Ethnic residential segregation as a structuration process. *Geography*. Tel-Aviv, Tel-Aviv University.

Omer, I. and I. Benenson (2002). GIS as a tool for studying urban fine-scale segregation. *Geography Research Forum*.

O'Neill, W. (1985). Estimation of a logistic growth and diffusion model describing neighborhood change. *Geographical Analysis*: 389–397.

Openshaw, S. (1979). A methodology for using models for planning purposes. *Environment and Planning A* **11**: 879–896.

Openshaw, S. (1983). *The Modifiable Areal Unit Problem*. Norwich, GeoBooks.

Orishimo, I. (1987). An approach to urban dynamics. *Geographical Analysis* **19**(3): 200–210.

O'Sullivan, D. (2001). Exploring spatial process dynamics using irregular cellular automaton models. *Geographical Analysis* **33**(1): 1–18.

O'Sullivan, D. and P. M. Torrens (2000). Cellular models of urban systems. *Theoretical and Practical Issues on Cellular Automata*. S. Bandini and T. Worsch. London, Springer-Verlag: 108–117.

Otter, H. S., A. van der Veen, and H. J. de Vriend (2001). ABLOoM: Location behaviour, spatial patterns, and agent-based modelling. *Journal of Artificial Societies and Social Simulation* **4**(4): ⟨http://www.soc.surrey.ac.uk/JASSS/4/4/2.html⟩.

Packard, N. (1985). Two-dimensional cellular automata. Journal of Statistical Physics **30**: 901–942.

Page, S. E. (1999). On the emergence of cities. *Journal of Urban Economics* **45**: 184–208.

Pancs, R. and N. J. Vriend (2003). Schelling's spatial proximity Model of Segregation Revisited. London, Working Paper 487, Queen Mary Colledge University of London: 1–56.

Peckham, J., B. MacKellar, and M. Doherty (1995). Data models for extensible support of explicit relationships in design databases. *VLDB Journal* **4**: 157–191.

Pedersen, P. O. (1967). Multivariate models of urban development. *Socio-Economic Planning Science* **1**: 101–116.

Phipps, A. G. and J. E. Carter (1984). An individual-level analysis of the stress-resistance model of household mobility. *Geographical Analysis* **16**(1): 176–189.

Phipps, M. (1989). Dynamic behavior of cellular automata under the constraint of neighborhood coherence. *Geographical Analysis* **21**: 197–215.

Phipps, M. and A. Langlois (1997). Spatial dynamics, cellular automata, and parallel processing computers. *Environment and Planning B: Planning and Design* **24**(2): 193–204.

Pijanowski, B. C., D. G. Brown, B. A. Shellito, and G. A. Manik (2002). Using neural networks and GIS to forecast land use changes: a Land Transformation Model. *Computers, Environment and Urban Systems*. In Press.

Pijanowski, B. C., D. T. Long, S. H. Gage, and W. E. Cooper (1997). *A Land Transformation Model: Conceptual Elements, Spatial Object Class Hierarchies, GIS Command Syntax and an Application for Michigan's Saginaw Bay Watershed*. Land Use Modeling Workshop, EROS Data Center, Sioux Falls, South Dakota, http://www.ncgia.ucsb.edu/conf/landuse97/main.html.

Pitowsky, I. (1996). Laplace's demon consults an oracle: The computational complexity of prediction. *Studies in History and Philosophy of Science Part B: Studies in History and Philosophy of Modern Physics*. **27B**(2): 161–180.

Polhill, J. G., N. M. Gotts, and A. N. R. Law (2001). Imitative versus non-imitative strategies in a land use simulation. *Cybernetics and Systems* **32**(1, 2): 285–307.

Portugali, J. (1991). An arab segregated neighborhood in Tel-Aviv: The case of Adjami. *Geography Research Forum* **11**: 37–50.

Portugali, J. (1996). Inter-representation networks and cognitive maps. *The Construction of Cognitive Maps*. J. Portugali. Dordrecht, Kluwer Academic Publishers: 11–43.

Portugali, J. (2000). *Self-Organization and the City*. Berlin, Springer.

Portugali, J. and I. Benenson (1994). Competing order parameters in a self-organizing city. *Managing and Marketing of Urban Development and Urban Life*. G. O. Braun. Berlin, Dietrich Reimer: 669–681.

Portugali, J. and I. Benenson (1995). Artificial planning experience by means of a heuristic sell-space model: simulating international migration in the urban process. *Environment and Planning B* **27**: 1647–1665.

Portugali, J. and I. Benenson (1997). Human agents between local and global forces in a self-organizing city. *Self-Organization of Complex Structures: From Individual to Collective Dynamics*. F. Schweitzer. London, Gordon and Breach: 537–546.

Portugali, J., I. Benenson and I. Omer (1994). Socio-spatial residential dynamics: stability and instability within a self-organized city. *Geographical Analysis* **26**(4): 321–340.

Portugali, J., I. Benenson, and I. Omer (1997). Spatial cognitive dissonance and sociospatial emergence in a self-organizing city. *Environment and Planning B-Planning & Design* **24**(2): 263–285.

Prigogine, I. (1967). Introduction to Thermodynamics of Irreversible Processes. New York, Inter-science.

Prigogine, I. (1980). *From Being to Becoming*. San Francisco, Freeman & Co.

Principia Cybernetica (2003). Principia Cybernetica, http://pespmc1.vub.ac.be.

Putman, S. (1970). Developing and testing an intraregional model. *Regional Studies* **4**: 473–490.

Putman, S. (1990). Equilibrium solutions and dynamics of integrated urban models. *New Directions in Regional Analysis. Integrated and Multi-regional Approach*. L. Anselin and M. Madden. London, Belhaven Press: 48–65.

Rabenstein, A. L. (1975). *Elementary Differential Equations with Linear Algebra*. New York, Academic Press.

Rajewsky, N. and M. Schreckenberg (1997). Exact results for one-dimensional cellular automata with different types of updates. *Physica A* **245**: 139–144.

RePast (2003). University of Chicago, RePast 2.0. (Software), Chicago: Social Science Research Computing Program.

Resnick, M. (1994). Changing the centralized mind. *Technology Review*(July): 32–40.

Reynolds, C. (1987). Flocks, birds, and schools: a distributed behavioral model. *Computer Graphics* **21**: 25–34.

Reynolds, C. (1999). *Steering Behaviors for Autonomous Characters*. Game Developers Conference, San Jose, CA.

Rickert, M. and K. Nagel (2001). Dynamic traffic assignment on parallel computers in TRANSIMS. *Future Generation Computer Systems* **17**: 637–648.

Rickert, M., K. Nagel, M. Schreckenberg, and A. Latour (1996). Two lane traffic simulations using cellular automata. *Physica A* **231**: 534–550.

Riggs, D. S. (1970). *Control Theory and Physiological Feedback Mechanisms*. Baltimore, Williams and Wilkins.

Ritsema van Eck, J. R. and T. de Jong (1999). Accessibility analysis and spatial competition efects in the context of GIS-supported service location planning. *Computers, Environment and Urban Systems* **23**: 75–89.

Roache, P. J. (1998). *Verification and Validation in Computational Science and Engineering*. Albuquerque, NM, Hermosa.

Roka, Z. (1994). One-way cellular automata on Cayley graphs. *Theoretical Computer Science* **132**(1, 2): 259–290.

Roka, Z. (1999). Simulations between cellular automata on Cayley graphs. *Theoretical Computer Science* **225**: 81–111.

Rosa, R., A. Pugliese, A. Villani, and A. Rizzol (2003). Individual-based vs. deterministic models for macroparasites: host cycles and extinction. *Theoretical Population Biology* **63**: 295–307.

Rosenfield, H. G. and K. Fitzpatrick-Lins (1986). A coefficient of agreement as a measure of thematic classification accuracy. *Photogrammetric Engineering and Remote Sensing* **52**(2): 223–227.

Rosser, J. B. J. (1994). Dynamics of emergent urban hierarchy. *Chaos Solitons & Fractals* **4**(4): 553–561.

Rossetti, R. J. F., R. H. Bordini, A. L. C. Bazzan, S. Bampi, R. Liu, and D. Van Vliet (2002). Using BDI agents to improve driver modelling in a commuter scenario. *Transportation Research Part C* **10**: 373–398.

Royama, T. (1992). *Analytical Population Dynamics*. London, Chapman & Hall.

Rucker, R. (1999). *Seek! Selected Nonfiction by Rudy Rucker*. New York, Four Walls Eight Windows.

Russell, S. J. and P. Norvig (1995). *Artificial Intelligence: A Modern Approach*. Englewood Cliffs, NJ, Prentice Hall.

Sakoda, J. M. (1971). The checkerboard model of social interaction. *Journal of Mathematical Sociology* **1**: 119–132.

Salomaa, A., D. Wood, and S. Yu, Eds. (2001). *A Half-Century of Automata Theory: Celebration and Inspiration*. New Jersey, World Scientific.

Sampson, R. J., J. D. Morenoff, and T. Gannon-Rowley (2002). Assessing Neighborhood Effects: Social processes and new directions in research. *Annual Revew of Sociology* **28**: 443–478.

Sanders, L., D. Pumain, H. Mathian, F. GuerinPace, and S. Bura (1997). SIMPOP: A multiagent system for the study of urbanism. *Environment and Planning B—Planning & Design* **24**(2): 287–305.

Sayer, R. A. (1979). Understanding urban models versus understanding cities. *Environment and Planning A* **11**: 853–862.

Saysel, A. K., Y. Barlas, and O. Yenigun (2002). Environmental sustainability in an agricultural development project: a system dynamics approach. *Journal of Environmental Management* **64**: 247–260.

Schadschneider, A. (2000). Statistical physics of traffic flow. *Physica A* **285**: 101–120.

Schellekens, M. P. G. and H. J. P. Timmermans (1997). A conjoint-based simulation model of housing-market clearing processes: theory and illustration. *Environment and Planning A* **29**(10): 1831–1846.

Schelling, T. C. (1969). Models of segregation. *American Economic Review* **59**(2): 488–493.

Schelling, T. C. (1971). Dynamic models of segregation. *Journal of Mathematical Sociology* **1**: 143–186.

Schelling, T. C. (1974). On the ecology of micro-motives. *The Corporate Society*. R. Marris. London, Macmillan: 19–55.

Schelling, T. C. (1978). *Micromotives and Macrobehavior*. New York, WW Norton and Company.

Schneider, M. (1959). Gravity models and trip distribution theory. *Papers and Proceedings of the Regional Science Association* **5**: 51–56.

Schonfisch, B. (1997). Anisotropy in cellular automata. *BioSystems* **41**: 29–41.

Schonfisch, B. and A. M. de Roos (1999). Synchronous and asynchronous updating in cellular automata. *Biosystems* **51**: 123–143.

Schonfisch, B. and K. P. Hadeler (1996). Dimer automata and cellular automata. *Physica D* **94**: 188–204.

Schumacher, M. (2001). *Objective Coordination in Multi-Agent System Engineering*. Berlin, Springer.

Schweitzer, F. and L. Schimansky-Geier (1994). Clastering of 'active' walkers in a two-component system. *Physica A: Statistical Mechanics and its Applications* **206**: 359–379.

Schweitzer, F., J. Zimmermann, and H. Muhlenbein (2002). Coordination of decisions in a spatial agent model. *Physica A: Statistical Mechanics and its Applications* **303**: 189–216.

Semboloni, F. (1997). An urban and regional model based on cellular automata. *Environment and Planning B-Planning & Design* **24**(4): 589–612.

Semboloni, F. (2000a). *The Dynamic of an Urban Cellular Automata Model in a 3-D Spatial Pattern*. XXI National Conference Aisre: Regional and Urban Growth in a Global Market, Palermo.

Semboloni, F. (2000b). The growth of an urban cluster into a dynamic self-modifying spatial pattern. *Environment and Planning B-Planning & Design* **27**(4): 549–564.

Semboloni, F., J. Assfalg, S. Armeni, R. Gianassi, and F. Marsoni (2003). CityDev, an interactive multi-agents urban model on the web. *Computers, Environment and Urban Systems* **28**(1/2): 45–64.

Sermons, M. W. (2000). Influence of race on household residential utility. *Geographical Analysis* **32**(3): 225–246.

Shabazian, D. and R. Johnston (2000). UPLAN—Urban Growth Model (UC Davis, Information Center for the Environment), http://snepmaps.des.ucdavis.edu/uplan/. access 2002.

Shannon, C. E. and W. Weaver (1949). *The Mathematical Theory of Communication*. Urbana, University of Illinois Press.

Shaw, S.-L. and X. Xin (2003). Integrated land use and transportation interaction: a temporal GIS exploratory data analysis approach. *Journal of Transport Geography* **11**: 103–115.

Sheynin, O. B. (1988). A Markov's work on probability. *Archive for History of Exact Science* **39**: 337–377.

Shi, W. and M. Y. C. Pang (2000). Development of Voronoi-based cellular automata—an integrated dynamic model for geographical information systems. *International Journal of Geographical Information Science* **14**(5): 455–474.

Shnerb, N. M., Y. Louzoun, E. Bettelheim, and S. Solomon (2003). The importance of being discrete—life always wins on the surface. *arXiv: adap- org/ 9912005 v1 29 Dec 1999*.

Silva, E. A. and K. C. Clarke (2002). Calibration of the SLEUTH urban growth model for Lisbon and Porto, Portugal. *Computers, Environment and Urban Systems* **26**: 525–552.

Silva, H. S. and M. L. Martins (2003). A cellular automata model for cell differentiation. *Physica A* **322**: 555–566.

Silver, S. (2003). Life Lexicon Home Page, http://www.argentum.freeserve.co.uk/lexhome.htm. 2003.

Silverberg, G. and B. Verspagen (1994). Collective learning, innovation and growth in a boundedly rational, evolutionary world. *Journal of Evolutionary Economics* **4**: 207–226.

Simon, H. A. (1956). Rational choice and the structure of the environment. *Psychological Review* **63**: 129–138.

Simon, H. A. (1976). *Administrative Behavior: A Study of Decision Making Processes in Administrative Organizations*. New York, Harper.

Simon, H. A. (1982). *Models of Bounded Rationality*. Cambridge, MIT Press.

Sipper, M. (1997). *Evolution of Parallel Cellular Machines: The Cellular Programming Approach*. Berlin, Springer.

Skellam, J. G. (1951). Random dispersal in the theoretical populations. *Biometrika* **38**.

Smith, III, A. (1976). Introduction to and survey of polyautomata theory. *Automata, Languages, Development*. Amsterdam, The Netherlands, North-Holland Publishing Co.

Smith, M. (1998). Painting by numbers—mathematical models of urban systems. *Environment and Planning B* **25**: 483–493.

Smith, S. and A. Kandel (1993). *Verification and Validation of Rule-based Expert Systems*. Boca Raton, FL, CRC Press.

Soares-Filho, B. S., G. C. Cerqueira, and C. L. Pennachin (2002). DINAMICA—a stochastic cellular automata model designed to simulate the landscape dynamics in an Amazonian colonization frontier. *Ecological Modelling* **154**: 217–235.

Soja, E. (1995). Postmodern urbanization: the six restructurings of Los Angeles. *Postmodern Cities and Spaces*. S. Watson and K. Gibson. Oxford, Blackwell.

Soja, E. (1996). *Thirdspace: Journeys to Los Angeles and Other Real and Imagined Places*. London, Blackwell.

Soja, E. (2000). *Postmetropolis. Critical Studies of Cities and Regions*. London, Blackwell.

Sonis, M. (1991). Innovation diffusion, schumpeterian competition and dynamic choice: a new synthesis. *Journal of Scientific and Industrial Research* **51**: 172–186.

Sonis, M. (1992). Innovation diffusion, Schumpeterian competition and dynamic choice: a new synthesis. *Journal of Scientific & Industrial Research.* J. Nehru University, New Delhi **51**.

Speare, A. (1974). Residential satisfaction as an intervening variable in residential mobility. *Demography* **11**: 173–188.

Speare, A., S. Goldstein, and W. H. Frey (1975). *Residential Mobility, Migration and Metropolitan Change.* Cambridge.

Stanley, H. E., J. S. J. Andrade, S. Havlin, H. A. Makse, and B. Suki (1999). Percolation phenomena: a broad-brush introduction with some recent applications to porous media, liquid water, and city growth. *Physica A* **266**: 5–16.

Stauffer, A. and M. Sipper (1998). On the relationship between cellular automata and L-systems: The self-replication case. *Physica D* **116**: 71–80.

Stauffer, D. (2001). Monte Carlo simulations of Sznajd models. *Journal of Artificial Societies and Social Simulation* **5**(1): ⟨http://www.soc.surrey.ac.uk/JASSS/5/1/4.html⟩.

Steinitz, C. and P. Rogers (1970). *A System Analysis Model of Urbanization and Change: An Experiment in Interdisciplinary Education.* Cambridge, MA, MIT Press.

Stern, N. (1980). John von Neumann's influence on electronic digital computing, 1944–1946. *Annals of the History of Computing* **2**(4): 349–362.

Stonebraker, M. (1972). A simplification of Forrester's model of an urban area. *IEEE Transactions on Systems, Man and Cybernetics* **4**: 468–472.

Straatman, B., R. White, and G. Engelen (2004). Towards an automatic calibration procedure for constrained cellular automata. *Computers, Environment and Urban Systems* **28**(1/2): 149–170.

Sui, D. Z. and H. Zeng (2001). Modeling the dynamics of landscape structure in Asia's emerging desakota regions: a case study in Shenzhen. *Landscape and Urban Planning* **53**(1–4): 37–52.

Summers, J. (2000). Game of Life status page, http://home.mieweb.com/jason/life/status.html.

Sun Microsystems (2002). Java Development Kit. Mountainview, CA, Sun Microsystems.

Taub, A. H., Ed. (1961). *John von Neumann: Collected Works. Volume V: Design of Computers, Theory of Automata and Numerical Analysis.* Oxford, Pergamon Press.

Terna, P. (1998). Simulation tools for social scientists: Building agent based models with SWARM. *Journal of Artificial Societies and Social Simulation* **1**(2).

Tesfatsion, L. (1997). How economists can get alife. *The Economy as an Evolving Complex System II.* W. B. Arthur, S. Durlaf, and D. Lane. Reading, MA, Addison-Wesley: 533–564.

The Executive Committee of the Club of Rome (1973). The Club of Rome: the new threshold. *Technological Forecasting and Social Change* **5**: 335–348.

Timmermans, H. and R. G. Golledge (1990). Applications of behavioral-research on spatial problems 2. Preference and Choice. *Progress in Human Geography* **14**(3): 311–354.

Timmermans, H. and L. van Noortwijk (1995). Context dependencies in housing choice behavior. *Environment and Planning A* **27**(2): 181–192.

Tobler, W. (1959). Automation and cartography. *Geographical Review* **49**(4): 526–534.

Tobler, W. (1970). A computer movie simulating urban growth in the Detroit region. *Geographical Analysis* **46**(2): 234–240.

Tobler, W. (1979). Cellular geography. *Philosophy in Geography.* S. Gale and G. Ollson. Dordrecht, Kluwer: 379–386.

Torrens, P. M. (2001). New tools for simulating housing choices. Program on Housing and Urban Policy Conference Paper Series C01-006. Berkeley, CA: University of California Institute of Business and Economic Research and Fisher Center for Real Estate and Urban Economics.

Torrens, P. M. (2002). Cellular automata and multi-agent systems as planning support tools. *Planning Support Systems in Practice.* S. Geertman and J. Stillwell. London, Springer-Verlag: 205–222.

Torrens, P. M. (2003). Automata-based models of urban systems. *Advanced Spatial Analysis.* P. A. Longley and M. Batty. Redlands, CA, ESRI Press.

Torrens, P. M. (2004a). Geosimulation approaches to traffic modeling. *Transport Geography and Spatial Systems*. P. Stopher, K. Button, K. Haynes, and D. Hensher. London, Pergamon. In Press.

Torrens, P. M. (2004b). Looking forward: remote sensing as dataware for human settlement simulation. *Remote Sensing of Human Settlements*. M. Ridd. New York, John Wiley and Sons. In Press.

Torrens, P. M. and D. O'Sullivan (2001). Cellular automata and urban simulation: where do we go from here? *Environment and Planning B-Planning & Design* **28**(2): 163–168.

Transportation Research Board (1985). Highway Capacity Manual, Chapter 13, Special Report 209. Washington, D.C., Transportation Research Board.

Tu, Y. and J. Goldfinch (1996). A two-stage housing choice forecasting model. *Urban Studies* **33**(3): 517–537.

Turing, A. M. (1936). On computable numbers with an application to the Entscheidungsproblem. *Proceedings of the London Mathamatical Society* **42**(2): 230–265.

Turing, A. M. (1950). Computing machinery and intelligence. *Mind* **59**(236): 433–460.

Turing, A. M. (1952). The chemical basis for morphogenesis. *Philosophical Transactions of the Royal Society of London* **37–72**: 37–72.

Turner, A. and A. Penn (2002). Encoding natural movement as an agent-based system: an investigation into human pedestrian behaviour in the built environment. *Environment and Planning B: Planning and Design.* 473–490.

Turner, M. G. (1988). A spatial simulation model of land use changes in a piedmont county in Georgia. *Applied Mathematics and Computation* **27**: 39–51.

Ulam, S. (1952). *Random Processes and Transformations*. International Congress of Mathematics.

Ungerer, M. J. and M. F. Goodchild (2002). Integrating spatial data analysis and GIS: a new implementation using the Component Object Model (COM). *International Journal of Geographical Information Science* **16**(1): 41–53.

University of Chicago (2003). RePast. Chicago, Social Science Research Computing Program.

van de Vyvere, Y. (1994). Stated preference decompositional modeling and residential choice. *Geoforum* **25**(2): 189–202.

van de Vyvere, Y., H. Oppewal, and H. Timmermans (1998). The validity of hierarchical information integration choice experiments to model residential preference and choice. *Geographical Analysis* **30**(3): 254–272.

van Dyke Parunak, H., R. Savit, and R. L. Riolo (1998). Agent-based modeling vs equation-based modeling: a case study and user's guide. *Multi-Agent systems and Agent-Based Simulation*. J. S. Sichman, R. Conte, and G. N. Gilbert. Berlin, Springer. **1534**: 10–26.

van Ommeren, J., P. Rietveld, and P. Nijkamp (1996). Residence and workplace relocation: A bivariate duration model approach. *Geographical Analysis* **28**(4): 315–329.

Veldhuisen, J. and H. Timmermans (1984). Specification of individual residential utility function: a comparative analysis of three measurement procedures. *Environment and Planning A* **16**: 1573–1582.

Verburg, P. H., G. H. J. de Koning, K. Kok, A. Veldkamp, and J. Bouma (1999). A spatial explicit allocation procedure for modelling the pattern of land use change based upon actual land use. *Ecological Modelling* **116**: 45–61.

Vernadsky, V. I. (1926/1997). *The Biosphere: complete annotated edition*. New York, Springer-Verlag.

Volterra, V. (1926). Variazioni e fluttuazioni del numero d'individui in specie animali conviventi. *Memorie Roma Accademia Nazionale dei Lincei, Ser. VI* **2**.

von Bertalanffy, L. (1968). *General System Theory. Foundations, Development, Applications*. New York, George Braziller.

von Foerster, H. (1960). On self-organizing systems and their environments. *Self-Organizing Systems*. M. C. Yovits and S. Cameron. Oxford, Pergamon Press: 31–50.

von Neumann, J. (1951). The general and logical theory of automata. *Cerebral Mechanisms in Behavior—The Hixon Symposium, 1948.* L. A. Jeffress. Pasadena, CA, New York: Wiley: 1–41.

von Neumann, J. (1961). Chapter 9: The general and logical theory of automata. *John von Neumann: Collected Works. Volume V: Design of Computers, Theory of Automata and Numerical Analysis.* A. H. Taub. Oxford, Pergamon Press: 288–328.

von Neumann, J. (1966). The theory of automata: Construction, reproduction, homogeneity. *Theory of Self-Reproducing Automata [by] John von Neumann.* A. W. Burks. Urbana, University of Illinois Press: 89–250.

von Thunen, J. H. (1826). *Der Isolierte Staat in Beziehung auf Landwirtschaft und Nationalokonomie.* Hamburg, F. Perthes.

Waddell, P. (2000). A behavioral simulation model for metropolitan policy analysis and planning: residential location and housing market components of UrbanSim. *Environment and Planning B-Planning & Design* **27**(2): 247–263.

Waddell, P. A. (2002). UrbanSim: modeling urban development for land use, transportation and environmental planning. *Journal of the American Planning Association* **68**(3): 297–314.

Wagner, D. F. (1997). Cellular automaton and geographic information systems. *Environment and Planning B: Planning and Design* **24**(1): 219–234.

Wagner, P., K. Nagel, and D. E. Wolf (1997). Realistic multi-lane traffic rules for cellular automata. *Physica A* **234**: 687–698.

Wahle, J., A. L. C. Bazzan, F. Klugl, and M. Schreckenberg (2000). Decision dynamics in a traffic scenario. *Physica A* **287**: 669–681.

Wahle, J., A. L. C. Bazzan, F. Klugl, and M. Schreckenberg (2002). The impact of real-time information in a two-route scenario using agent-based simulation. *Transportation Research Part C* **10**: 399–417.

Wahle, J., L. Neubert, J. Esser, and M. Schreckenberg (2001). A cellular automaton traffic flow model for online simulation of traffic. *Parallel Computing* **27**(5): 719–735.

Wang, R. and H. J. Ruskin (2002). Modeling traffic flow at a single-lane urban roundabout. *Computer Physics Communications* **147**: 570–576.

Wang, Y. and X. Zhang (2001). A dynamic modeling approach to simulating socioeconomic effects on landscape changes. *Ecological Modelling* **140**: 141–162.

Ward, D. P., A. T. Murray, and S. R. Phinn (2000). A stochastically constrained cellular model of urban growth. *Computers, Environment and Urban Systems* **24**: 539–558.

Weber, A. (1909). *Uber den Standort der Industrien.* Tubingen.

Webster, C. J. (1995). Urban morphological fingerprints. *Environment and Planning B* **22**: 279–297.

Wegener, M. (1994). Operational Urban Models—State-of-the-Art. *Journal of the American Planning Association* **60**(1): 17–29.

Wegener, M. (2001). New spatial planning models. *International Journal of Applied Earth Observation and Geoinformation* **3**(3): 224–237.

Weidlich, W. (2000). *Sociodynamics: A Systematic Approach to Mathematical Modelling in the Social Sciences.* Amsterdam, Harwood Academic Publishers.

Weifeng, F., Y. Lizhong, and F. Weicheng (2002). Simulation of bi-direction pedestrian movement using a cellular automata model. *Physica A.* In press.

Weng, Q. (2002). Land use change analysis in the Zhujiang Delta of China using satellite remote sensing, GIS and stochastic modelling. *Journal of Environmental Management* **64**.

White, R. (1977). Dynamic central place theory: results of a simulation approach. *Geographical Analysis* **9**: 226–243.

White, R. (1998). Cities and cellular automata. *Discrete Dynamics in Nature and Society* **2**: 111–125.

White, R. and G. Engelen (1993). Cellular-Automata and Fractal Urban Form—a Cellular Modeling Approach to the Evolution of Urban Land-Use Patterns. *Environment and Planning A* **25**(8): 1175–1199.

White, R. and G. Engelen (1994). Urban Systems Dynamics and Cellular-Automata—Fractal Structures between Order and Chaos. *Chaos Solitons & Fractals* **4**(4): 563–583.

White, R. and G. Engelen (1997). Cellular automata as the basis of integrated dynamic regional modelling. *Environment and Planning B-Planning & Design* **24**(2): 235–246.

White, R. and G. Engelen (2000). High-resolution integrated modelling of the spatial dynamics of urban and regional systems. *Computers, Environment and Urban Systems* **24**(5): 383–400.

White, R., G. Engelen, and I. Uljee (1993). *Cellular Automata Model of Fractal Urban Land-use Pattern: Forecasting Change for Planning Applications.* Eighth European Colloquium on Theoretical and Quantitative Geography, Stockholm.

White, R., G. Engelen, and I. Uljee (1997). The use of constrained cellular automata for high-resolution modelling of urban land-use dynamics. *Environment and Planning B-Planning & Design* **24**(3): 323–343.

Wiener, N. (1948/1961). *Cybernetics, or Control and Communication in the Animal and the Machine.* Cambridge, MA, MIT Press.

Wiener, N. and A. Rosenblueth (1946). The mathematical formulation of the problem of conduction of impulses in a network of connected excitable elements, specifically in cardiac muscle. *Archivos Del Insituts De Cardiologia De Mexico* **16**: 205–265.

Willson, S. (1978). On convergence of configurations. *Discrete Mathematics* **23**: 279–300.

Willson, S. (1981). Growth patterns of ordered cellular automata. *Journal of Computer and System Sciences* **22**: 29–41.

Wilson, A. G. (1968). Models in urban planning: a synoptic review of recent literature. *Urban Studies* **5**: 249–276.

Wilson, A. G. (1969). Developments of some elementary residential locations models. *Journal of Regional Science* **9**: 377–385.

Wilson, A. G. (1979). From comparative statics to dynamics in urban-systems theory. *Geoforum* **10**: 283–296.

Wilson, A. G. (2000). *Complex Spatial Systems: The Modelling Foundations of Urban and Regional Analysis.* Harlow, Prentice Hall.

Wilson, S. W. (1985). *Knowledge Growth in an Artificial Animal.* Proceedings of the International Conference on Genetic Algorithms and their Application, Pittsburgh PA, Lawrence Erlbaum.

Wilson, W. G. (1998). Resolving discrepancies between deterministic population models and individual-based simulations. *American Naturalist* **151**(2): 116–134.

Wissen, L. and A. van Rima (1986). *Modeling Urban Housing Market Dynamics. Evolutionary Pattern of Households and Housing in Amsterdam.* Elsevier.

Witten, T. A. J. and L. M. Sander (1981). Diffusion-limited aggregation, a kinetic critical phenomenon. *Physical Review Letters* **47**(19): 1400–1403.

Wolf, D. E. (1999). Cellular automata for traffic simulations. *Physica A* **263**: 438–451.

Wolfram, S. (1983). Statistical mechanics of cellular automata. *Reviews of Modern Physics* **55**: 601–644.

Wolfram, S. (1984a). Computation theory of cellular automata. *Communications in Mathematical Physics* **96**: 15–57.

Wolfram, S. (1984b). Universality and complexity in cellular automata. *Physica D* **10**: 1–35.

Wolfram, S. (1994). *Cellular Automata and Complexity.* Reading, MA, Addison-Wesley.

Wolfram, S. (2002). A new kind of science, Wolfram Media, Incorporated.

Wolpert, J. (1965). Behavioral aspects of the decision to migrate. *Papers and Proceedings of the Regional Science Association* **15**: 159–169.

Wooldridge, M. J. and N. R. Jennings (1995). Intelligent agents: theory and practice. *Knowledge Engineering Review* **10**(2): 115–152.

Wrigley, N., T. Holt, D. Steel, and M. Tranmer (1996). Analyzing, modeling, and resolving the ecological fallacy. *Spatial Analysis: Modeling in a GIS Environment.* P. Longley and M. Batty. Cambridge, UK, Geoinformation International: 25–40.

Wu, F. (1997). A linguistic cellular automata simulation approach for sustainable land development in a fast growing region. *Computers, Environment and Urban Systems* **20**(6): 367–387.

Wu, F. (1998a). An experiment on the generic polycentricity of urban growth in a cellular automatic city. *Environment and Planning B-Planning & Design* **25**(5): 731–752.

Wu, F. (1998b). SimLand: a prototype to simulate land conversion through the integrated GIS and CA with AHP-derived transition rules. *International Journal of Geographical Information Science* **12**(1): 63–82.

Wu, F. (1998c). Simulating urban encroachment on rural land with fuzzy-logic controlled cellular automata in a geographical information system. *Journal of Environmental Management* **53**(4): 293–308.

Wu, F. and D. Martin (2002). Urban expansion simulation of Southeast England using population surface modelling and cellular automata. *Environment and Planning A* **34**: 1855–1876.

Wu, F. and C. J. Webster (1998). Simulation of land development through the integration of cellular automata and multicriteria evaluation. *Environment and Planning B-Planning & Design* **25**(1): 103–126.

Wu, F. L. and A. G. O. Yeh (1997). Changing spatial distribution and determinants of land development in Chinese cities in the transition from a centrally planned economy to a socialist market economy: a case study of Guangzhou. *Urban Studies* **34**(11): 1851–1879.

Wu, J. and J. L. David (2002). A spatially explicit hierarchical approach to modeling complex ecological systems: theory and applications. *Ecological Modelling* **153**: 7–26.

Wuensche, A. (1998). Classifying Cellular Automata automatically, Santa Fe Working Paper 98-02-018.

Wylie, B., G. Cameron, W. Matthew, and D. McArthur (1993). *PARAMICS: Parallel Microscopic Traffic Simulator*. Second European Connection Machine Users Meeting, Meudon, Paris.

Xie, Y. C. (1996). A generalized model for cellular urban dynamics. *Geographical Analysis* **28**(4): 350–373.

Yates, F. E., Ed. (1987). *Self-Organizing Systems*, Plenum.

Yeh, A. G. O. and X. Li (1998). Sustainable land development model for rapid growth areas using GIS. *International Journal of Geographical Information Science* **12**(2): 169–189.

Yeh, A. G. O. and X. Li (2002). A cellular automata model to simulate development density for urban planning. *Environment and Planning B-Planning & Design* **29**(3): 431–450.

Yeh, A. G.-O. and X. Li (2001). A constrained CA model for the simulation and planning of sustainable urban forms by using GIS. *Environment and Planning B: Planning & Design* **28**(5): 733–753.

Zhabotinsky, A. M. (1964). Periodic liquid phase reactions. *Proceedings of the Academy of Sciences of the USSR* **157**: 392.

Zhang, W. B. (1989). Coexistence and separation of the two residential groups—an interactional spatial dynamic approach. *Geographical Analysis* **21**: 91–102.

Zhang, W. B. (1993). Location choice and land-use in an isolated state—endogenous capital and knowledge accumulation. *Annals of Regional Science* **27**(1): 23–39.

Zhang, W. B. (1994). Dynamics of interacting spatial economies. *Chaos Solitons & Fractals* **4**(4): 595–604.

Index

Geosimulation: *Automata-based Modeling of Urban Phenomena*.   I. Benenson and P. Torrens
© 2004 John Wiley & Sons, Ltd   ISBN: 0-470-84349-7